GEOMETRY, TOPOLOGY AND DYNAMICS OF CHARACTER VARIETIES

LECTURE NOTES SERIES
Institute for Mathematical Sciences, National University of Singapore

Series Editors: Louis H. Y. Chen and Wing Keung To
Institute for Mathematical Sciences
National University of Singapore

ISSN: 1793-0758

Published

Vol. 13 Econometric Forecasting and High-Frequency Data Analysis
edited by Roberto S Mariano & Yiu-Kuen Tse

Vol. 14 Computational Prospects of Infinity — Part I: Tutorials
edited by Chitat Chong, Qi Feng, Theodore A Slaman, W Hugh Woodin & Yue Yang

Vol. 15 Computational Prospects of Infinity — Part II: Presented Talks
edited by Chitat Chong, Qi Feng, Theodore A Slaman, W Hugh Woodin & Yue Yang

Vol. 16 Mathematical Understanding of Infectious Disease Dynamics
edited by Stefan Ma & Yingcun Xia

Vol. 17 Interface Problems and Methods in Biological and Physical Flows
edited by Boo Cheong Khoo, Zhilin Li & Ping Lin

Vol. 18 Random Matrix Theory and Its Applications: Multivariate Statistics and Wireless Communications
edited by Zhidong Bai, Yang Chen & Ying-Chang Liang

Vol. 19 Braids: Introductory Lectures on Braids, Configurations and Their Applications
edited by A Jon Berrick, Frederick R Cohen, Elizabeth Hanbury, Yan-Loi Wong & Jie Wu

Vol. 20 Mathematical Horizons for Quantum Physics
edited by H Araki, B-G Englert, L-C Kwek & J Suzuki

Vol. 21 Environmental Hazards: The Fluid Dynamics and Geophysics of Extreme Events
edited by H K Moffatt & E Shuckburgh

Vol. 22 Multiscale Modeling and Analysis for Materials Simulation
edited by Weizhu Bao & Qiang Du

Vol. 23 Geometry, Topology and Dynamics of Character Varieties
edited by W Goldman, C Series & S P Tan

*For the complete list of titles in this series, please go to
http://www.worldscibooks.com/series/LNIMSNUS

Lecture Notes Series, Institute for Mathematical Sciences,
National University of Singapore

Vol. 23

GEOMETRY, TOPOLOGY AND DYNAMICS OF CHARACTER VARIETIES

Editors

William Goldman
University of Maryland, USA

Caroline Series
University of Warwick, UK

Ser Peow Tan
National University of Singapore, Singapore

World Scientific

NEW JERSEY • LONDON • SINGAPORE • BEIJING • SHANGHAI • HONG KONG • TAIPEI • CHENNAI

Published by

World Scientific Publishing Co. Pte. Ltd.

5 Toh Tuck Link, Singapore 596224

USA office: 27 Warren Street, Suite 401-402, Hackensack, NJ 07601

UK office: 57 Shelton Street, Covent Garden, London WC2H 9HE

British Library Cataloguing-in-Publication Data
A catalogue record for this book is available from the British Library.

Cover image:
This represents the Maskit slice of the moduli space of once punctured torus groups in complex probability coordinates. The coloured tessellation corresponds to the combinatorial structure of the Ford regions.
It was created by Masaaki Wada of Osaka University using his OPTi program, available from http://delta.math.sci.osaka-u.ac.jp/OPTi/
Reproduced with kind permission.

Lecture Notes Series, Institute for Mathematical Sciences, National University of Singapore — Vol. 23
GEOMETRY, TOPOLOGY AND DYNAMICS OF CHARACTER VARIETIES

ISBN-13 978-981-4401-35-7
ISBN-10 981-4401-35-8

Printed in Singapore by Mainland Press Pte Ltd.

CONTENTS

FOREWORD

The Institute for Mathematical Sciences at the National University of Singapore was established on 1 July 2000. Its mission is to foster mathematical research, both fundamental and multidisciplinary, particularly research that links mathematics to other disciplines, to nurture the growth of mathematical expertise among research scientists, to train talent for research in the mathematical sciences, and to serve as a platform for research interaction between the science community in Singapore and the wider international community.

The Institute organizes thematic programs which last from one month to six months. The theme or themes of a program will be chosen from areas at the forefront of current research in the mathematical sciences and their applications.

Generally, for each program there will be tutorial lectures followed by workshops at the research level. Notes on these lectures are usually made available to the participants for their immediate benefit during the program. The main objective of the Institute's Lecture Notes Series is to bring these lectures to a wider audience. Occasionally, the Series may also include the proceedings of workshops and expository lectures organized by the Institute.

The World Scientific Publishing Company has kindly agreed to publish the Lecture Notes Series. This Volume, "Geometry, Topology and Dynamics of Character Varieties", is the twenty-third of this Series. We hope that through the regular publication of these lecture notes the Institute will achieve, in part, its objective of promoting research in the mathematical sciences and their applications.

February 2012

Louis H.Y. Chen
Wing Keung To
Series Editors

PREFACE

This volume is based on a series of expository lectures presented at the highly successful graduate student summer school which kicked off the program *Geometry, Topology and Dynamics of Character Varieties* held at the National University of Singapore's Institute for Mathematical Sciences in July and August 2010. The theme was the character varieties of representations in a Lie group G of a discrete group Γ, the primary example being the case in which Γ is the fundamental group of a surface and G is $SL(2,\mathbb{R})$ or $SL(2,\mathbb{C})$. Character varieties lie at the confluence of many important areas of mathematics including algebraic geometry, hyperbolic geometry and Teichmüller theory, Kleinian groups and three-dimensional topology, dynamical systems and gauge theory. They have rich geometry and are related to interesting topological objects such as locally homogeneous geometric structures on manifolds and moduli spaces arising in gauge theory. Their study reveals many deep connections between these fields.

The summer school, which took place over three weeks with about 40 graduate students from over 12 countries participating, featured nine series of four lectures each, chosen so as to introduce important aspects of the background to the program. Following highly positive feedback, we felt it was important to preserve all this great material and enable it to be shared more widely. All but one of the lecturers were able to contribute, to which we have added two excellent expository articles based on lectures by students who attended the program (Koberda and Palesi). We hope that this resulting volume of edited and refereed articles will serve as a portal to a vibrant and many faceted area of mathematics.

Aimed at graduate students, much of the material in the volume is otherwise available only in specialized texts. Following Zhang's introduction to basic hyperbolic geometry in the synthetic axiomatic style, Aramayona and Leininger give an easily accessible introduction to the central topics of hyperbolic structures on surfaces and their degenerations via measured foliations and the Thurston compactification of Teichmüller space. Koberda explains the important technique of ping-pong lemmas, with applications in the context of hyperbolic geometry and mapping class groups of surfaces.

An important and unique feature of the summer school was Yamashita's specially designed lecture course which described how to write a program to draw limit sets and fundamental domains for Kleinian groups. Programming is done using the software python, and everything is explained from scratch without assuming any previous computing knowledge. Many people are interested in how to make the resulting beautiful pictures, and as far as we know, his is the only published article along these lines.

We then turn from the primary example of hyperbolic geometry to a more general context. Articles by Parker and Drumm provide introductions to complex hyperbolic and Lorentzian geometry, respectively. Palesi's article introduces a more general discussion about the representation space of surfaces groups into $SL(2, \mathbb{R})$ and its connected components, a topic taken further by Kim in the context of general Lie groups. The final article by Xia is an introduction to the topic of abelian and non-abelian cohomology which provides powerful analytical tools for the study of structures on the representation and character varieties.

The IMS program would not have been possible without contributions from many people. It was generously funded not only by NUS (from the IMS program funding as well as ARF grant R-146-000-133-112), but also aided by a grant from the NSF which enabled a substantial participation by US-based students and participants, and the Global Center of Excellence (Compview) of the Tokyo Institute of Technology which provided support for the Japanese contingent. We are grateful to the director of the IMS, Professor Louis Chen for his unwavering support and tremendous enthusiasm for the program, and all the staff at IMS, in particular, Emily, Claire, Stephen, Jolyn, Nurleen and Agnes for their help in running the program.

We would also like to thank the other scientific organizers of the program for their work and input: Javier Aramayona, Craig Hodgson, Sadayoshi Kojima, Yair Minsky, Makoto Sakuma, Jean-Marc Schlenker, Yan Loi Wong, Yasushi Yamashita and Ying Zhang. The contributions, of

Aramayona and Zhang in particular, to the smooth running of the program were invaluable.

December 2011

William Goldman
University of Maryland, USA

Caroline Series
University of Warwick, UK

Ser Peow Tan
National University of Singapore, Singapore
Editors

AN INVITATION TO ELEMENTARY HYPERBOLIC GEOMETRY

Ying Zhang
School of Mathematical Sciences, Soochow University
Suzhou, 215006, China
yzhang@suda.edu.cn

We offer a short invitation to elementary hyperbolic plane geometry. We first examine the contents of Book I of Euclid's Elements and obtain a hyperbolic plane from the Euclidean one by negating Euclid's parallel postulate and keeping all of his other axioms. Then we explore the fundamentals of hyperbolic plane geometry, and study the structure of its isometries. Finally, we obtain certain identities involving the isometries and evaluate them in the upper half-plane model to derive some trigonometric laws for hyperbolic triangles.

Keywords: Non-Euclidean, hyperbolic plane, isometry, trigonometry

Mathematics Subject Classification 2000: 51M10

Introduction

In this short invitation to elementary hyperbolic geometry we choose to follow the approach of the discoverers of non-Euclidean geometry. Thus without giving any account of the history of the discovery of non-Euclidean geometry, we start by examining the contents of Book I of Euclid's Elements and obtain a hyperbolic plane from the Euclidean one by negating Euclid's parallel postulate and keeping all of his other axioms. We then explore the fundamentals of the hyperbolic plane geometry, study the structures of the isometries of a hyperbolic plane, and finally apply certain identities of isometries of the hyperbolic plane to derive trigonometric laws for triangles.

We choose this synthetic approach and as far as possible use no analytic models, because we believe this will provide the reader with more feeling for the geometry and thus enable him or her to explore the subject him or herself. In the more easily accepted analytic approaches, as this author has experienced, the reader has to rely on the chosen models to obtain any results and thus loses his or her precious geometric motivation. We hope that by studying the material presented in these notes, the reader will be able to develop the geometric ideas he/she has in mind analytically with no essential difficulties in any preferred model of a hyperbolic plane.

We have to warn the reader that we do not cover the trigonometry in detail and do not even touch the rich solid hyperbolic geometry. To master this omitted material, the reader is referred to some advanced textbooks as briefly discussed at the very end of the notes (in §4.6).

1. Euclid's Elements, Book I and Neutral Plane Geometry

1.1. *A brief review of contents of Elements, Book I*

In Book I of Elements, Euclid treated the fundamentals of the Euclidean plane geometry, including theories of triangles, parallels, and area. Precisely, Book I consists of 23 definitions, 5 postulates, 5 common notions, and 48 propositions. The definitions describe certain basic terms, of which we list only a few, such as point, line (curve), straight line, and surface, and then define some others based on them. The postulates are fundamental assumptions on the plane geometry while the common notions are commonly accepted assumptions on algebra or scientific reasoning. And, after that, the propositions (some are construction problems), including the famous Pythagorean Theorem (I.47), are presented in logical order. In what follows our phrasing of the contents of Book I of Elements is taken from Heath [13] or as appeared in Joyce's website http://aleph0.clarku.edu/∼djoyce/java/elements/bookI/bookI.html".

Definitions (listed below are D1–D4 and D23):

D1. A point is that which has no part.

D2. A line is breadthless length.

D3. The ends of a line are points.

D4. A straight line is a line which lies evenly with the points on itself.

D23. (A pair of) parallel straight lines are straight lines which, being in the same plane and being produced indefinitely in both directions, do not meet one another in either direction.

Postulates (P1–P5): Let the following be postulated.

P1. To draw a (finite) straight line from any point to any point.

P2. To produce a finite straight line continuously in a straight line.

P3. To describe a circle with any center and radius.

P4. That all right angles equal one another.

P5. That, if a straight line falling on two straight lines makes the interior angles on the same side less than two right angles, then the two straight lines, if produced indefinitely, meet on that side of the straight line on which are the two interior angles less than the two right angles.

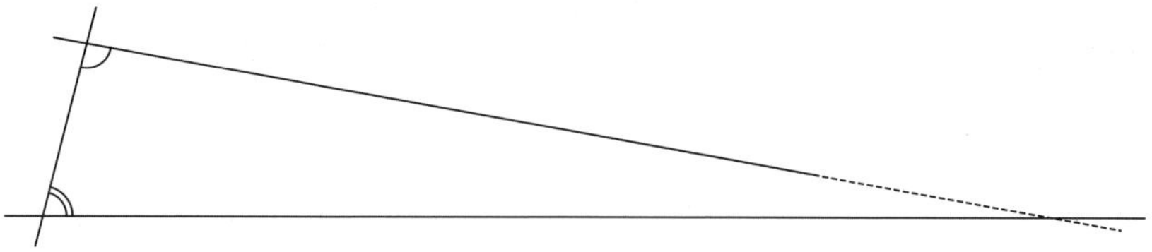

Fig. 1. Euclid's Postulate 5

Common Notions (CN1–CN5):

CN1. Things which equal the same thing also equal one another.

CN2. If equals are added to equals, then the wholes are equal.

CN3. If equals are subtracted from equals, then the remainders are equal.

CN4. Things which coincide with one another equal one another.

CN5. The whole is greater than the part.

Propositions (listed below are I.1–I.32 and I.47–I.48):

I.1. To construct an equilateral triangle on a given finite straight line.

I.2. To place a straight line equal to a given straight line with one end at a given point.

I.3. To cut off from the greater of two given unequal straight lines a straight line equal to the less.

I.4. If two triangles have the two sides equal to two sides respectively, and have the angles contained by the equal sides equal, then they also have the base equal to the base, the triangle equal to the triangle, and the remaining angles equal to the remaining angles respectively, namely those opposite the equal sides.

I.5. In isosceles triangles the angles at the base equal one another.

I.6. If in a triangle two angles equal one another, then the sides opposite the equal angles also equal one another.

I.7. Given two straight lines constructed from the ends of a straight line and meeting in a point, there cannot be constructed from the ends of the same straight line, and on the same side of it, two other straight lines meeting in another point and equal to the former two, namely each equal to that from the same end.

I.8. If two triangles have the two sides equal to two sides respectively, and also have the base equal to the base, then they also have the angles equal which are contained by the equal straight lines.

I.9. To bisect a given rectilinear angle.

I.10. To bisect a given finite straight line.

I.11. To draw a straight line at right angles to a given straight line from a given point on it.

I.12. To draw a straight line perpendicular to a given infinite straight line from a given point not on it.

I.13. If a straight line stands on a straight line, then it makes either two right angles or angles whose sum equals two right angles.

I.14. If with any straight line, and at a point on it, two straight lines not lying on the same side make the sum of the adjacent angles equal to two right angles, then the two straight lines are in a straight line with one another.

I.15. If two straight lines cut one another, then they make the vertical angles equal to one another.

I.16. In any triangle the exterior angle is greater than either of the interior and opposite angles.

I.17. In any triangle the sum of any two angles is less than two right angles.

I.18. In any triangle the angle opposite the greater side is greater.

I.19. In any triangle the side opposite the greater angle is greater.

I.20. In any triangle the sum of any two sides is greater than the remaining one.

I.21. If from the ends of one of the sides of a triangle two straight lines are constructed meeting within the triangle, then the sum of the straight lines so constructed is less than the sum of the remaining two sides of the triangle, but the constructed straight lines contain a greater angle than the angle contained by the remaining two sides.

I.22. To construct a triangle out of three straight lines which equal three given straight lines: thus it is necessary that the sum of any two of the straight lines should be greater than the remaining one.

I.23. To construct a rectilinear angle equal to a given rectilinear angle on a given straight line and at a point on it.

I.24. If two triangles have two sides equal to two sides respectively, but have one of the angles contained by the equal straight lines greater than the other, then they also have the base greater than the base.

I.25. If two triangles have two sides equal to two sides respectively, but have the base greater than the base, then they also have the one of the angles contained by the equal straight lines greater than the other.

I.26. If two triangles have two angles equal to two angles respectively, and one side equal to one side, namely, either the side adjoining the equal angles, or that opposite one of the equal angles, then the remaining sides equal the remaining sides and the remaining angle equals the remaining angle.

I.27. If a straight line falling on two straight lines makes the alternate angles equal to one another, then the two straight lines are parallel to one another.

I.28. If a straight line falling on two straight lines makes the exterior angle equal to the interior and opposite angle on the same side, or, equivalently, the sum of the interior angles on the same side equal to two right angles, then the two straight lines are parallel to one another.

I.29. A straight line falling on two parallel straight lines makes the alternate angles equal to one another, the exterior angle equal to the interior and opposite angle, and the sum of the interior angles on the same side equal to two right angles.

I.30. Straight lines parallel to the same straight line are also parallel to one another.

I.31. To draw a straight line through a given point parallel to a given straight line.

I.32. In any triangle, if one of the sides is produced, then the exterior angle equals the sum of the two interior and opposite angles, and the sum of the three interior angles of the triangle equals two right angles.

I.47. In right-angled triangles the square on the side opposite the right angle equals the sum of the squares on the sides containing the right angle.

I.48. If in a triangle the square on one of the sides equals the sum of the squares on the remaining two sides, then the angle contained by the remaining two sides is right.

1.2. *A useful lemma*

The following useful lemma, called the Bow Tie Lemma by some authors, is an easy consequence of the SAS congruence criterion.

Lemma 1.1. *Given $\triangle ABC$, let M be the midpoint of side BC. Produce*

AM to D so that $|AM| = |MD|$. Then $\triangle ACM \cong \triangle DBM$.

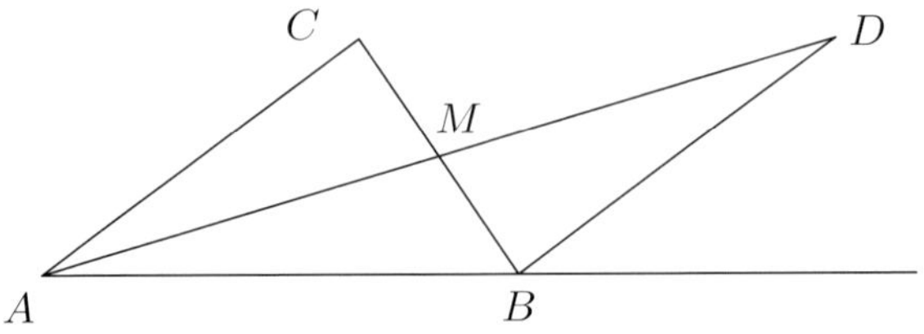

Fig. 2. Figure for Lemma 1.1

Euclid used Lemma 1.1 in Elements to give a clean proof of Proposition I.16, an important proposition in Book I. (We leave it to the reader to try to find a not-so-clean but easier proof for Proposition I.16.)

1.3. *A figure-free proof of Proposition I.7*

Proposition I.7 is a stronger version of I.8, the SSS congruence criterion. The proof that Euclid gave relies, however, on how the figure is drawn, and omits a case. Here we present a complete, figure-free proof.

Let $\triangle ABC$ and $\triangle ABD$ be two triangles such that C and D are distinct points on the same side of the straight line AB, and such that $|AC| = |AD|$ and $|BC| = |BD|$. We proceed to deriving a contradiction. Let M be the midpoint of CD. Then M lies on the same side of AB as C and D. By Propositions I.5 and I.4, we conclude that straight lines AM and BM are both perpendicular to CD and therefore coincide. Hence M lies on straight line AB, which is absurd. This proves Proposition I.7.

1.4. *More notes on Elements, Book I*

In our discussion of plane geometry, we do not pursue a pure axiomatic way and assume that everything under discussion occurs in the same plane which is, topologically, the usual plane.

Taken as granted, Euclid assumed that each finite straight line has its definite magnitude, *length*, which is additive when two finite straight lines lying in the same straight line are juxtaposed. He also assumed that an angle has its definite measure which is also additive when two angles with the same vertex are juxtaposed.

For the angles, Euclid further required in Postulate 4 that the full round angles (four right angles) at all points in the plane equal one another.

The definition, D4, of straight line is a vague description since one cannot conclude a line (curve) be a straight line merely by this definition.

Postulate 1, according to the way Euclid used it, should be interpreted as: Given two distinct points in the plane, there exists in the plane one and *only one* straight line which connects them. Therefore, if two distinct straight lines in the plane intersect, they intersect in only one point. Consequently, we conclude that each interior angle of a triangle is less than two right angles. Furthermore, as a consequence of Postulate 1 and Proposition I.4, we conclude that the perpendicular straight line drawn from a given point to a given straight line is unique (note that the existence of the perpendicular is guaranteed by Propositions I.14 and I.15).

We notice that Euclid's system of axioms for the plane geometry is incomplete. For example, Proposition I.4, the SAS congruence criterion, should be regarded as an axiom instead of a theorem; in fact, Euclid's proof for I.4 is not satisfactory. In so doing, Proposition I.5 follows as an easy consequence of (axiom) I.4. One also needs to include a continuity axiom to confirm that if a straight line enters the interior of a triangle then it will leave the region when the straight line is indefinitely produced in that direction. It is well known that a complete system of axioms for the Euclidean plane geometry was given by Hilbert in [14].

Proposition I.20 is the so-called triangle inequality.

Proposition I.21 can be rewritten as: If point D lies within triangle ABC, then $|AD| + |DB| < |AC| + |CB|$ and $\measuredangle ADB > \measuredangle ACB$.

Proposition I.26 establishes the ASA and AAS congruence criteria.

Proposition I.27 gives, without making use of Postulate 5 in its proof, parallel straight lines: If one straight line falling on two straight lines makes the alternate interior angles equal to one another, then the two straight lines are parallel. This can be easily proved using the SAS congruence criterion and the uniqueness of intersection points of two straight lines. Hence I.27 can be put just after I.4 if one wishes.

A careful reader will notice that Euclid did not use his Postulate 5 until in the proof of Proposition I.29, the Euclidean Parallel Theorem, which asserts that whenever one straight line falls on two parallel straight lines, the alternate angles equal one another. In other words, each of the first 28 propositions in Elements, Book I can be proved using only the axioms (including the hidden ones to be added in) of the plane geometry other than Euclid's Postulate 5. This leads to the term "neutral plane geometry" which we shall discuss shortly.

Of the very last two propositions in Book I, I.47 is the famous

Pythagorean Theorem, and I.48 is the converse of I.47. The proofs of them, of course, essentially make use of Postulate 5, at least in the way Euclid had proved them.

1.5. *Playfair's axiom*

One would wonder why Euclid made his Postulate 5 so complicated a statement compared with the other four, and, in particular, Euclid even did not mention parallels in it. This is partly because the ancient Greek philosophy avoids any unnecessary use of infinity which lies in the nature of parallels. For the same reason Euclid never used the term "infinite straight line".

On the other hand, with the presence of all axioms of the Euclidean plane geometry other than Euclid's Postulate 5, it is easy to show that Postulate 5 is equivalent to the so-called Playfair's Axiom below.

Playfair's axiom. Through a given point P not on a given straight line l, there passes at most one straight line which is parallel to l.

Since, by I.27, there exists at least one parallel, the phrase "at most one" in Playfair's Axiom can be replaced, if one wishes, by "exactly one".

1.6. *Neutral plane geometry*

By neutral geometry, for which J. Bolyai used the term "absolute geometry", we mean the geometry obtained from the Euclidean geometry by dropping just Euclid's Postulate 5.

In particular, all the first 28 propositions in Euclid's Elements, Book I are indeed theorems (possibly with I.4 chosen as an axiom) of neutral plane geometry, since their proofs make no use of Postulate 5.

1.7. *Angle-sums of triangles and Legendre's Theorems*

We have seen that, with the presence of all other axioms for the plane geometry, Euclid's Postulate 5 is equivalent to Playfair's Axiom. It is not hard to see that they are also equivalent to the Euclidean Angle-Sum Axiom (Proposition I.32).

Euclidean angle-sum axiom. The sum of the interior angles of every triangle equals two right angles.

Regarding angle-sums of triangles in a neutral plane, we have the well-known theorems of Saccheri and Legendre.

Theorem 1.2 (Legendre's First Theorem). *In a neutral plane the sum of the angles of a triangle is less than or equal to two right angles.*

Proof. Suppose on the contrary that there exists a triangle, $\triangle ABC$, with angle-sum greater than two right angles. With no loss of generality, we may assume that the angle at A is the smallest angle of $\triangle ABC$. Then by Lemma 1.1, we obtain a triangle $\triangle ABD$ with the same angle-sum as $\triangle ABC$, such that the smallest angle of $\triangle ABD$ is at most half the smallest angle of $\triangle ABC$. By Archimedes's Axiom, after a finite number of steps one arrives at a triangle with the same angle-sum as $\triangle ABC$ and with smallest angle less than the excess of its angle-sum over two right angles. Then this triangle has sum of certain two angles greater than two right angles, contradicting Proposition I.17. This proves Theorem 1.7. □

By virtue of Legendre's First Theorem, we may define the defect of a triangle in a neutral plane to be the deficiency of its angle-sum to two right angles. Similarly, we define the defect of a (simple) quadrilateral in a neutral plane to be the deficiency of its angle-sum to four right angles.

Definition 1.3. The *defect* $\delta(\triangle ABC)$ of triangle ABC equals two right angles minus the sum of the angles of $\triangle ABC$ and is therefore nonnegative. Similarly, the *defect* $\delta(\square ABCD)$ of (simple) quadrilateral $ABCD$ equals four right angles minus the sum of the interior angles of $\square ABCD$ and is also nonnegative.

The defect so defined is additive under subdivision of a triangle or quadrilateral into smaller triangles and quadrilaterals. Below we list two simple cases.

Proposition 1.4. *In triangle ABC let D be a point within side BC. Then*

$$\delta(\triangle ABC) = \delta(\triangle ABD) + \delta(\triangle ADC).$$

Proposition 1.5. *If quadrilateral $ABCD$ is obtained as the union of triangles ABC and ACD which share no common interior points, then*

$$\delta(\square ABCD) = \delta(\triangle ABC) + \delta(\triangle ACD).$$

The following theorem establishes the universality of the Euclidean Angle-Sum Axiom, namely, if the axiom is satisfied by one triangle in the plane then it is satisfied by every triangle.

Theorem 1.6 (Legendre's Second Theorem). *In a neutral plane if one triangle has angle-sum equal to two right angles, then so does every triangle.*

Proof. Suppose one triangle has angle-sum equal to two right angles and hence vanishing defect. Then by dissecting this triangle into two right triangles by one of its altitudes, we obtain one right triangle which has vanishing defect. Then by suitably juxtaposing copies of this right triangle we may obtain right triangles with arbitrarily large legs and with vanishing defect. We prove that every right triangle has vanishing defect. Given an arbitrary right triangle, we may place a congruent copy of it inside some right triangle with legs large enough and with vanishing defect, so that its right-angle coincides with that of the large triangle. Then by the additivity of defect, we conclude that the given right triangle has vanishing defect. It follows that every triangle has vanishing defect. This proves Theorem 1.6. □

As a corollary of Theorems 1.7 and 1.6 combined, we have

Theorem 1.7. *In a neutral plane if one triangle has angle-sum less two right angles, then so does every triangle.*

We thus call a neutral plane *Euclidean* or *Non-Euclidean* according as whether triangles in it have angle-sum equal to or less than two right angles.

1.8. *Quadrilaterals with two consecutive right angles*

For a quadrilateral with two consecutive right angles specified we call the common side of the right angles its *base*, the two sides perpendicular to the base its *legs*, the remaining side its *summit*, and the angles contained by the summit and the legs its *summit angles*.

Theorem 1.8. *Let $ABCD$ be a quadrilateral with consecutive right angles at A and B. Then $|AD| = |BC|$ if and only if $\measuredangle D = \measuredangle C$.*

Hint: To prove the "if" part, erect perpendicular bisector of AB to cut CD. Apply first the SAS congruence criterion and then AAS criterion to conclude that $|AD| = |BC|$.

Theorem 1.9. *Let $ABCD$ be a quadrilateral with consecutive right angles at A and B. Then $|AD| < |BC|$ if and only if $\measuredangle D > \measuredangle C$.*

Hint: To prove the "only if" part, take E within the longer leg BC so that $|BE| = |AD|$. Then use Theorem 1.8 and Proposition I.17 to conclude that $\measuredangle D > \measuredangle C$.

1.9. *Saccheri and Lambert quadrilaterals*

A *Saccheri quadrilateral* is a quadrilateral with two consecutive tight angles and with equal legs (see Figure 3(a)), while a *Lambert quadrilateral* is a quadrilateral with three right angles (See Figure 3(b)).

In a Saccheri quadrilateral by the *base-to-summit midline* we mean the straight line joining the midpoints of the base and the summit. The following properties of Saccheri quadrilaterals are ready to observe.

Theorem 1.10. *In a Saccheri quadrilateral the summit angles equal one another and the base-to-summit midline is perpendicular to both the base and the summit.*

It follows that the base-to-summit midline of a Saccheri quadrilateral bisects it into two congruent Lambert quadrilaterals. See Figure 3(a).

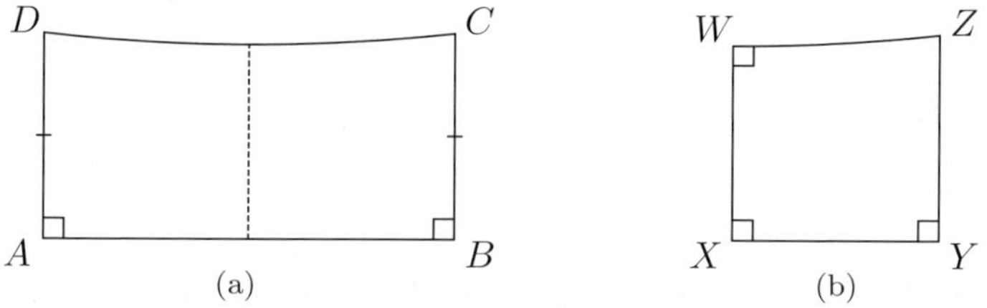

Fig. 3. Saccheri and Lambert quadrilaterals

As an immediate corollary of Theorem 1.7, we obtain

Proposition 1.11. *The summit angles of a Saccheri quadrilateral are acute or right, and so is the remaining angle in a Lambert quadrilateral.*

1.10. *Variation of triangles in a neutral plane*

By the SAS congruence theorem, (measures of) the sides and angles of a triangle are determined by any two sides and the angle contained by them. Euclid addressed in Proposition I.24 the variation problem for triangles when two of the sides remain fixed and the angle contained by them varies. Further in Proposition I.25 he considered the problem when two of the sides remain fixed and the remaining side varies.

For convenience of reference let us rewrite I.24 and I.25 as follows.

Proposition I.24. *If in* $\triangle ABC$ *and* $\triangle A'B'C'$ *there hold* $|AB| = |A'B'|$, $|BC| = |B'C'|$ *and* $\measuredangle B < \measuredangle B'$, *then* $|CA| < |C'A'|$.

Proposition I.25. *If in $\triangle ABC$ and $\triangle A'B'C'$ there hold $|AB| = |A'B'|$, $|BC| = |B'C'|$ and $|CA| < |C'A'|$, then $\measuredangle B < \measuredangle B'$.*

We remark that Theorem 1.34 in §1.12 serves an analog of the SAS criterion for congruence of triangles in the setting of quadrilaterals with two consecutive right angles, and Theorem 1.35 comes as an analog of I.24 and I.25 combined.

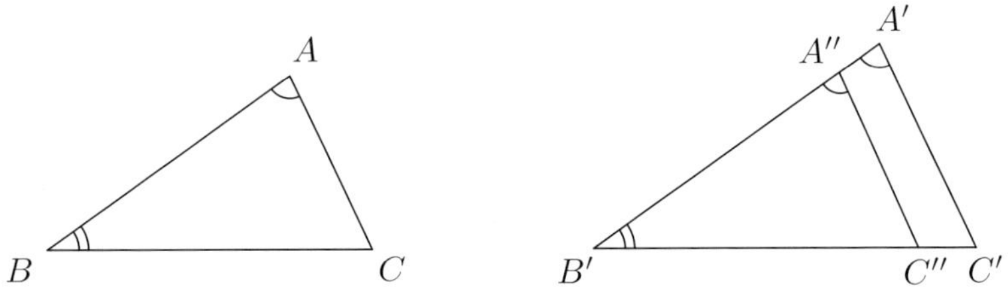

Fig. 4. Figure for the proof of Theorem 1.12

Now let us consider other variation problems for triangles in a neutral plane, noticing that a triangle is determined by its side-angle data SSS, SAS, ASA or AAS. First we let the side in data ASA vary.

Theorem 1.12. *If in $\triangle ABC$ and $\triangle A'B'C'$, $\measuredangle A = \measuredangle A'$, $\measuredangle B = \measuredangle B'$ and $|AB| < |A'B'|$, then $\measuredangle C' \leqslant \measuredangle C$, $|AC| < |A'C'|$ and $|BC| < |B'C'|$.*

Proof. Choose point A'' within side $A'B'$ of $\triangle A'B'C'$ so that $|A''B'| = |AB|$. Draw a straight line, $A''C''$ through A'', where C'' is a point on the same side of $A'B'$ as C', so that $\measuredangle C''A''B' = \measuredangle C'A'B'$. Then $A''C''$ is parallel to $A'C'$ and hence must cut side $B'C'$ at a point, which we may denote by C''. Then $|B'C''| = |BC|$ since $\triangle A''B'C'' \cong \triangle ABC$. Hence $|BC| = |B'C''| < |B'C'|$. A similar argument gives $|AC| < |A'C'|$. The remaining inequality, $\measuredangle C \leqslant \measuredangle C'$, follows from the inequality $\delta(A'B'C') \geqslant \delta(ABC)$ on defects of triangles. This proves Theorem 1.12. □

The following comes as an immediate corollary of Theorem 1.12.

Theorem 1.13. *If in $\triangle ABC$ and $\triangle A'B'C'$, $\measuredangle A = \measuredangle A'$, $\measuredangle B = \measuredangle B'$ and $|BC| < |B'C'|$, then $\measuredangle C' \leqslant \measuredangle C$, $|AB| < |A'B'|$ and $|AC| < |A'C'|$.*

Similarly, we obtain the following theorems on variation of triangles.

Theorem 1.14. *If in $\triangle ABC$ and $\triangle A'B'C'$, $\measuredangle A = \measuredangle A'$, $\measuredangle B < \measuredangle B'$ and $|BC| = |B'C'|$, then $\measuredangle C' < \measuredangle C$.*

Theorem 1.15. *If in $\triangle ABC$ and $\triangle A'B'C'$, $\measuredangle A < \measuredangle A'$, $\measuredangle B = \measuredangle B'$ and $|BC| = |B'C'|$, then $\measuredangle C' < \measuredangle C$ and $|A'B'| < |AB|$.*

Theorem 1.16. *If in $\triangle ABC$ and $\triangle A'B'C'$, $|BC| = |B'C'|$, $\measuredangle C = \measuredangle C'$ and $|CA| < |C'A'|$, then $\measuredangle A' < \measuredangle A$ and $\measuredangle B' > \measuredangle B$.*

Theorem 1.17. *If in $\triangle ABC$ and $\triangle A'B'C'$, $\measuredangle A = \measuredangle A'$, $|AB| = |A'B'|$ and $\measuredangle B < \measuredangle B'$, then $\measuredangle C' < \measuredangle C$ and $|C'A'| > |CA|$.*

We may short-hand abbreviate the above theorems, for example, Theorem 1.12 as $(A_1^0 S_1^+ A_2^0 \Rightarrow A_3^{0-} S_2^+ S_3^+)$, where we denote the area, the angles (vertices) and the sides of a given triangle by $\triangle$, by A_1, A_2, A_3 and by S_1, S_2, S_3, respectively, so that S_1 is opposite to A_1, and so on; we use superscript $()^+$, $()^-$ and $()^0$ to mean that the quantity in consideration gets increased, decreased and remains fixed, respectively; and we use $()^{0-}$ to mean "either $()^0$ or $()^-$". Furthermore, let us use notation $()^\times$ to mean that the tendency of the quantity in consideration cannot be determined to be independent of the initial data in the variation problem.

We have the following summarizing theorem on variation of triangles.

Theorem 1.18 (Variation of Triangles). *For variation of a single quantity in its side-angle data SSS, SAS, ASA, or AAS of a triangle in a neutral plane:*

(1) $S_1^0 S_2^0 S_3^+ \Rightarrow A_1^\times A_2^\times A_3^+ \triangle^\times$ (Proposition I.25);
(2) $S_1^0 A_3^+ S_2^0 \Rightarrow A_1^\times A_2^\times S_3^+ \triangle^\times$ (Proposition I.24);
(3) $S_1^0 A_3^0 S_2^+ \Rightarrow A_1^- A_2^+ S_3^\times \triangle^+$ (Theorem 1.16);
(4) $A_1^0 S_3^0 A_2^+ \Rightarrow A_3^- S_1^\times S_2^+ \triangle^+$ (Theorem 1.17);
(5) $A_1^0 S_3^+ A_2^0 \Rightarrow A_3^{0-} S_1^+ S_2^+ \triangle^+$ (Theorem 1.12);
(6) $A_1^0 A_2^0 S_1^+ \Rightarrow A_3^{0-} S_2^+ S_3^+ \triangle^+$ (Theorem 1.13);
(7) $A_1^0 A_2^+ S_1^0 \Rightarrow A_3^- S_2^\times S_3^\times \triangle^\times$ (Theorem 1.14);
(8) $A_1^+ A_2^0 S_1^0 \Rightarrow A_3^- S_2^\times S_3^- \triangle^-$ (Theorem 1.15).

Remark. It is easy to see that A_3^{0-} in (4) and (5) is A_3^0 in Euclidean case, and A_3^- in non-Euclidean case since triangles have positive defects.

Let us look at a case of Theorem 1.13 for right triangles to be used later.

Theorem 1.19. *Suppose C and D are points on different sides of straight line AB such that $BC \perp AB$ and $AD \perp AB$. Suppose CD and AB intersect in O. If $|AD| < |BC|$ then $|AO| < |BO|$, $|DO| < |CO|$ and $\measuredangle D \geqslant \measuredangle C$.*

Exercise. Try to obtain more theorems on variation of quadrilaterals with two consecutive right angles, similar to those of triangles, in a neutral plane.

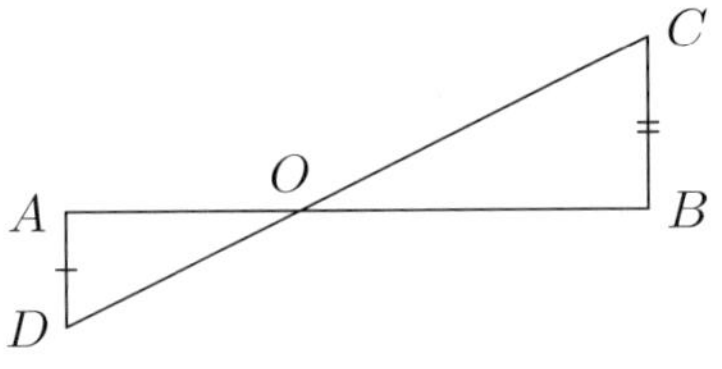

Fig. 5. Figure for Theorem 1.19

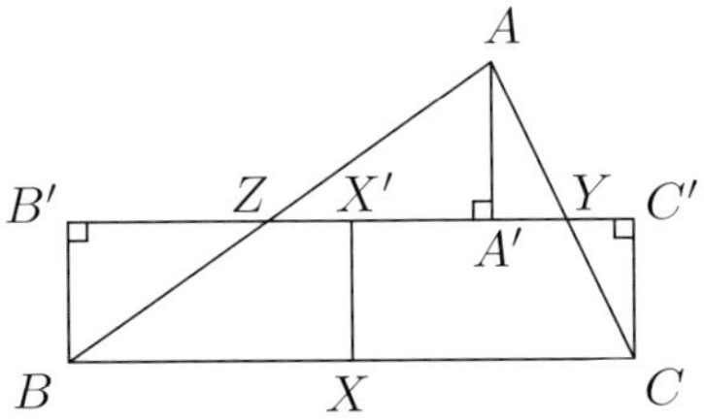

Fig. 6. A midline configuration for triangles

1.11. *A midline configuration for triangles*

A *midline* of a triangle is, in our terminology, a straight line joining the midpoints of two sides of the triangle. Thus a triangle has three midlines. We obtain an important configuration by dropping perpendiculars from the vertices of a triangle to one of its midlines.

Given $\triangle ABC$, let X, Y and Z be the midpoints of sides BC, CA and AB, respectively. Drop perpendicular straight lines AA', BB' and CC' to midline YZ, with A', B' and C' the feet. By twice applications of the AAS criterion we obtain $\triangle AA'Z \cong \triangle BB'Z$ and $\triangle AA'Y \cong \triangle CC'Y$. It follows that $\square B'C'CB$ is a Saccheri quadrilateral with base $B'C'$ and with the sum of its summit angles equal to the sum of the angles of $\triangle ABC$.

In particular, straight line XX', with X' the midpoint of $B'C'$, is a common perpendicular to BC and YZ. See Figure 6.

1.12. *More theorems of neutral plane geometry*

In what follows we include some more theorems of neutral plane geometry and leave their proofs as exercises. Note that we continue to use Euclid's definition of parallels.

Theorem 1.20. *If in $\triangle ABC$ and $\triangle A'B'C'$ there hold equality $\measuredangle C = \measuredangle C'$, with the angle right or obtuse, and two inequalities, $|BC| < |B'C'|$ and $|CA| < |C'A'|$, at least one of which is strict, then $|AB| < |A'B|$.*

Hint: We may assume $\triangle ABC$ sits inside $\triangle A'B'C'$ so that the equal angles agree and suppose $|CA| < |CA'|$. Then $|AB| < |A'B| \leqslant |A'B'|$.

Theorem 1.21. *If one straight line falling on two others makes the alternate angles equal to one another, then the two straight lines have a common perpendicular.*

Hint: Through the midpoint of the two intersection points draw perpendiculars to the two lines and prove that the two perpendiculars agree.

Theorem 1.22. *Two straight lines are parallel if they have a common perpendicular. Hence two intersecting straight lines have no common perpendiculars.*

Hint: Proposition I.27 gives existence of parallel straight lines.

Theorem 1.23. *Given a triangle, there exists a Saccheri quadrilateral such that the sum of its summit angles equals the sum of the angles of the triangle.*

Hint: A consequence of the important configuration in §1.11.

Theorem 1.24. *In a triangle a midline and the remaining side have a common perpendicular, the perpendicular bisector of the remaining side.*

Hint: Another consequence of the important configuration in §1.11.

Theorem 1.25. *Two right triangles are congruent if they have hypotenuse equal to hypotenuse and one other side equal to one other side.*

Hint: Prove it independently or as a special case of Theorem 1.26 below.

Theorem 1.26. *Two non-obtuse triangles are congruent if they have two sides equal to two sides and one angle equal to one angle, the equal angles being opposite to equal sides.*

Hint: Otherwise, we would be able to construct a congruent copy of the triangle with shorter remaining side, which sits suitably inside the other triangle, and this implies that the former triangle would be obtuse.

Exercise. Following the hint above, give an example to ensure that in the statement of Theorem 1.26 one cannot drop the word "non-obtuse".

Theorem 1.27. *In $\triangle ABC$ let M be the midpoint of side BC. If $AM \perp BC$ then $|AB| = |AC|$.*

Hint: Prove $\triangle AMB \cong \triangle AMC$ by the SAS criterion.

Theorem 1.28. *Suppose that $|AB| > |AC|$ in $\triangle ABC$ and let M be the midpoint of side BC. Then $\measuredangle BAM < \measuredangle CAM$.*

Hint: Try to obtain two proofs as follows. For the first proof, produce AM to D so that $|AM| = |MD|$. Then in $\triangle ABD$ there holds $|BD| = |AC| < |AB|$. It follows that $\measuredangle BAM < \measuredangle BDM = \measuredangle CAM$. For the second proof, drop perpendiculars from B and C to straight line AM with feet B' and C', respectively. Then apply Theorem 1.16 on variation to $\triangle ABB'$ and $\triangle ACC'$ to give $\measuredangle BAM < \measuredangle CAM$.

Corollary 1.29. *In $\triangle ABC$ let M be the midpoint of side BC. If AM bisects the angle at A then $|AB| = |AC|$.*

Theorem 1.30. *Suppose that $|AB| > |AC|$ in $\triangle ABC$ and the internal bisector of the angle at A cuts BC in X. Then $|BX| > |CX|$.*

Hint: Let D be in AB so that $|AD| = |AC|$. Prove that $|BX| > |DX|$.

Theorem 1.31. *In $\triangle ABC$ let X be an arbitrary point within side BC. Then $|AX| < \max\{|AB|, |AC|\}$.*

Hint: First prove the conclusion for the case where $|AB| = |AC|$.

Theorem 1.32. *Suppose quadrilateral $ABCD$ has consecutive right angles at A and B, with $|AD| \neq |BC|$. Then $|AB| < |CD|$.*

Hint: Suppose $|AD| < |BC|$ and drop perpendicular from D to BC.

Theorem 1.33. *Suppose quadrilateral $ABCD$ has consecutive right angles at A and B, with $|AD| < |BC|$. Draw perpendicular from any point F within side CD to AB with foot E. Then $|EF| < |BC|$.*

Hint: First deal with the case where $|AD| = |BC|$.

Theorem 1.34. *Suppose $ABCD$ and $A'B'C'D'$ are quadrilaterals with consecutive right angles at A, B and at A', B'. If $|AD| = |A'D'|$, $|BC| = |B'C'|$ and $|AB| = |A'B'|$, then $|CD| = |C'D'|$.*

Hint: First prove $\triangle DAB \cong \triangle D'A'B'$ by SAS congruence criterion and then $\triangle DBC \cong \triangle D'B'C'$ again by SAS criterion.

Theorem 1.35. *Suppose $ABCD$ and $A'B'C'D'$ are quadrilaterals with consecutive right angles at A, B and at A', B'. If $|AD| = |A'D'|$, $|BC| = |B'C'|$ and $|AB| < |A'B'|$, then $|CD| < |C'D'|$.*

Hint: We may assume that A' and D' agree with A and D, respectively, and that B is within side AB'. Then prove $|CD| < |C'D|$ in $\triangle C'CD$.

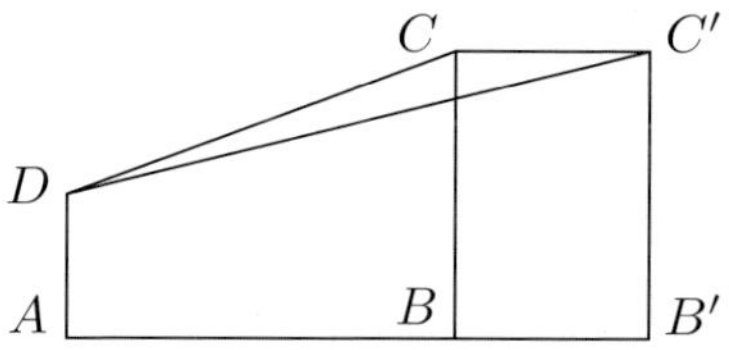

Fig. 7. Figure for the proof of Theorem 1.35

Theorem 1.36. *Suppose $ABCD$ and $A'B'C'D'$ are quadrilaterals with consecutive right angles at A, B and at A', B'. If $|AD| = |A'D'|$, $|BC| = |B'C'|$ and $|CD| = |C'D'|$, then $|AB| = |A'B'|$.*

Hint: Corollary of Theorem 1.35.

Theorem 1.37. *Suppose $ABCD$ and $A'B'C'D'$ are quadrilaterals with consecutive right angles at A, B and at A', B'. If $|AD| = |A'D'|$, $|BC| = |B'C'|$ and $|CD| < |C'D'|$, then $|AB| < |A'B'|$.*

Hint: Corollary of Theorems 1.34 and 1.35 combined.

Theorem 1.38. *Suppose $ABCD$ is a quadrilateral with consecutive right angles at A and B. Let F be the midpoint of side CD and drop perpendicular from F to AB with foot E. If $FE \perp CD$ then $|AD| = |BC|$.*

Hint: First prove $\triangle EFC \cong \triangle EFD$ by SAS congruence criterion and then $\triangle BEC \cong \triangle AED$ by AAS criterion.

Exercise. Try to give a Euclidean example such that, in the setting of Theorem 1.38, foot E is the midpoint of side AB but $|AD| \neq |BC|$.

Exercise. Corresponding to the SAS, SSS, ASA and AAS criteria for congruence of triangles, try to write down similar criteria for quadrilaterals $ABCD$ with consecutive right angles at A and B, and justify them.

The following theorems establish a few special points for triangles.

Theorem 1.39. *In a triangle the internal bisectors of the three angles meet in a point within the triangle, called the incenter of the triangle.*

Hint: The proof is the same as that in the Euclidean case.

Theorem 1.40. *In a triangle if two of the perpendicular bisectors of the sides meet, then all three meet in the same point, called the circumcenter of the triangle.*

Hint: The proof is the same as that in the Euclidean case.

Theorem 1.41. *In a triangle if two of the altitude-lines meet, then all three altitude-lines meet in the same point, called the orthocenter of the triangle.*

Hint: The proof is the same as that in the Euclidean case.

Theorem 1.42. *In a triangle the three medians meet in a point within the triangle, called the centroid of the triangle.*

Hint: The proof is quite challenging at this stage. Usually, to conclude the concurrency of cevians (i.e., straight lines through the vertices) of a triangle one needs (a metric form of) the Theorem of Ceva.

1.13. *Small angles*

By Proposition I.9, an arbitrary angle can be bisected. We can repeat the process and obtain angles as small as one wishes, by Archimedes's Axiom. On the other hand, Lobachevsky established the existence of arbitrarily small angles with a given arm line and a given point on the other arm line.

Theorem 1.43. *(cf. §21 in Lobachevsky* [17]*) Through a given point, a straight line can be drawn to make with a given straight line an angle arbitrary small.*

Proof. We shall apply Archimedes's Axiom. First, draw perpendicular from the given point A to the given straight line l, with foot B on l, and take point C on l so that $|AB| = |BC|$. Then $\measuredangle ACB$ is less than half the right angle. Next, produce BC to D so that $|CD| = |CA|$. Then $\measuredangle ADB < \frac{1}{2}\measuredangle ACB$, and so on. □

2. Hyperbolic Plane Geometry

In this section we turn to the non-Euclidean case of plane geometry, namely, the plane geometry in which a triangle has angle-sum less than two right angles. Klein named the non-Euclidean geometry in this sense the *hyperbolic geometry.* We remark that, in the non-strict sense, non-Euclidean geometries also include the elliptic geometry introduced by Riemann, although topologically an elliptic plane is no longer the usual plane.

2.1. *Hyperbolic plane*

A *hyperbolic plane* is a neutral plane for which the following Hyperbolic Parallel Postulate holds.

Hyperbolic parallel postulate. Given a point P and a straight line l in the plane, such that P is not on l, there exist in the plane more than one straight lines which pass through P and are parallel to l.

From now on we shall assume the hyperbolic parallel postulate above and thus work in hyperbolic geometry, unless otherwise stated.

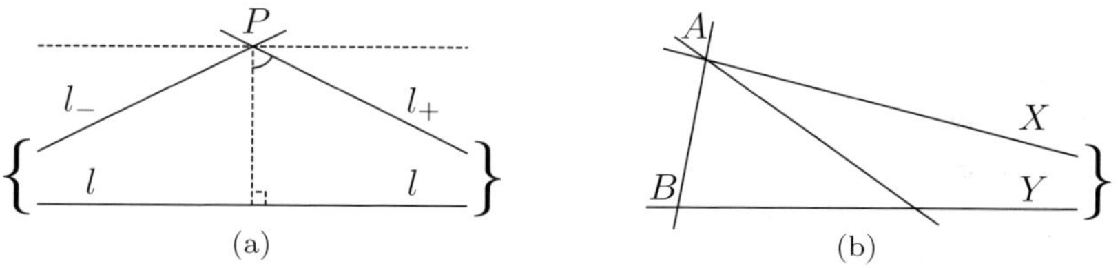

Fig. 8. Asymptotic parallels

2.2. *Asymptotic Parallelism*

Given a straight line l and a point P not on l, among all the straight lines through P and parallel to l, there are two boundary ones (l_- and l_+ in Figure 8(a)) which we shall call the *asymptotic parallels* to l. Historically, formal definitions of these special parallels were given by Gauss, J. Bolyai, and Lobachevsky independently and in essentially the same way.

To state the definition we shall need the notion of a directed straight line, and we use notation $\overline{AB}$ for directed straight line AB with chosen direction from point A to point B.

Definition 2.1 (Asymptotic Parallelism). A directed straight line $\overline{AX}$ is said to be *asymptotically parallel to* directed straight line $\overline{BY}$, denoted $\overline{AX} \parallel^{\mathrm{a}} \overline{BY}$, if they do not intersect and there exist a point A on straight line AX and a point B on straight line BY so that the points X and Y lie on the same side of AB and so that every straight line through A falling between rays AX and AB cuts BY. See Figure 8(b).

It is easy to check that the definition does not depend on the choices of A and B involved. Furthermore, one obtains the reciprocity: if $\overline{AX} \parallel^{\mathrm{a}} \overline{BY}$

then $\overline{BY} \parallel^{\mathrm{a}} \overline{AX}$, and also the transitivity: if $\overline{AX} \parallel^{\mathrm{a}} \overline{BY}$ and $\overline{BY} \parallel^{\mathrm{a}} \overline{CZ}$ then $\overline{AX} \parallel^{\mathrm{a}} \overline{CZ}$.

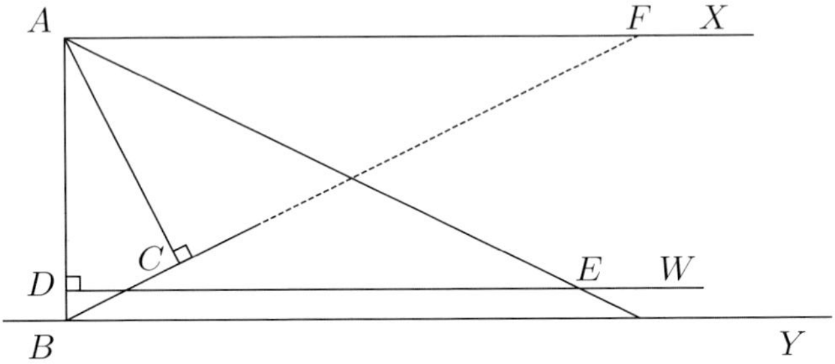

Fig. 9. Lobachevsky's proof of reciprocity of asymptotic parallels

Here we present Lobachevsky's proof of the reciprocity of asymptotic parallelism. Suppose $\overline{AX} \parallel^{\mathrm{a}} \overline{BY}$. To show $\overline{BY} \parallel^{\mathrm{a}} \overline{AX}$, we may assume $AB \perp BY$ and consider an arbitrary ray BC between rays BA and BY. Drop the perpendicular ray from A to BC and let C be the foot. Draw ray AE between AB and AX so that $\measuredangle BAE = \measuredangle CAX$. Take point D within AB so that $|AD| = |AC| < |AB|$ and draw ray DW perpendicular to AB towards the side of X and Y. Since $\overline{AX} \parallel^{\mathrm{a}} \overline{BY}$, we conclude that AE will cut BY and hence will cut DW (in E, say). Now take point F on ray AX so that $|AF| = |AE|$. Then we have $\triangle ACF \cong \triangle ADE$, $\measuredangle ACF$ is right angle and hence point F lies on ray BC. This gives $\overline{BY} \parallel^{\mathrm{a}} \overline{AX}$.

We remark that Gauss's proof of the reciprocity of asymptotic parallelism is quite similar to Lobachevsky's whereas J. Bolyai proved it by finding (corresponding) points A and B so that $\measuredangle BAX = \measuredangle ABY$ and then proceeding by symmetry.

Proposition 2.2. *If two straight lines have a common perpendicular then they are not asymptotically parallel in either direction.*

Proof. Suppose $l \perp l_1$ and $l \perp l_2$, with feet P_1 and P_2, respectively. Considering the asymptotic parallels drawn through P_1 to l_2 in two directions, one concludes by definition that l_1 is not asymptotically parallel to l_2 in either direction. □

Definition 2.3 (Ultra-Parallelism). A straight line AX is said to be *ultra-parallel to* a (different) straight line BY, denoted $AX \parallel^{\mathrm{u}} BY$, if they do not intersect and are, with all choices of directions, not asymptotically parallel.

The reciprocity of ultra-parallelism follows easily from that of asymptotic parallelism. However, there is no transitivity for ultra-parallelism. We shall establish in Theorem 2.4 that the converse of Proposition 2.2 is also true, namely, if two straight lines are ultra-parallel to each other then they have a common perpendicular.

2.3. *Angle of parallelism*

Given a straight line l and a point P not on l, we have seen that there exist two straight lines, l_+ and l_-, through P which are asymptotically parallel to l directed appropriately. Note that l_+ and l_- form with the perpendicular, $l_\perp$, from P to l equal acute angles; see Figure 8(a). It is clear that any parallel l' through P other than l_+ and l_- cannot be asymptotically parallel to l since there are no other parallels within l' and l. We shall call the acute angle formed by l_+ and $l_\perp$ (or l_-) the *angle of parallelism* of the pair (l, P).

It is not hard to show that the angle of parallelism of (l, P) depends only on the distance d of P from l. We can thus speak of the angle of parallelism of a finite distance d, or a straight line segment, and denote it by $\Pi(d)$.

In this way the angle of parallelism defines a function of one positive variable taking values the radian measures of acute angles

$$\Pi : (0, \infty) \to (0, \pi/2). \tag{2.1}$$

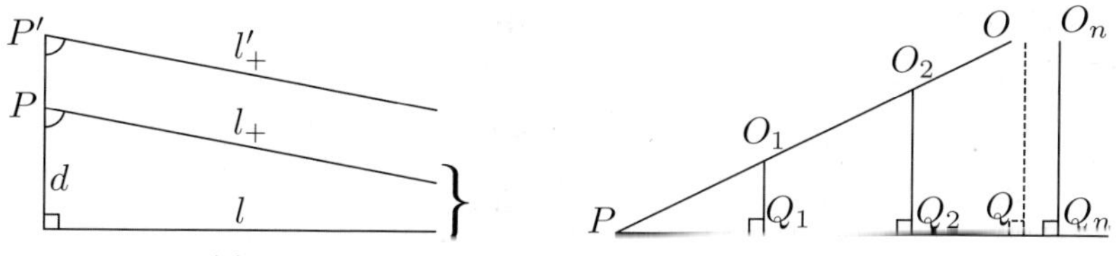

Fig. 10. Continuity and ontoness of function Π

Theorem 2.4 (Properties of Function Π). *The angle-of-parallelism function* $\Pi : (0, \infty) \to (0, \pi/2)$ *is strictly decreasing and onto (hence continuous).*

Proof. To prove Π is strictly decreasing, suppose points P and P' are distance d and d' (where $d < d'$) away from straight line l, respectively, such that PP' is perpendicular to l. See Figure 10(a). Let l_+ be the (right) asymptotic parallel through P to L, and let l'_+ be the straight line through

P' which forms angle $\Pi(d)$ with PP'. Then, by Proposition 2.2, l'_+ is not asymptotically parallel to l_+, hence not asymptotically parallel to l. It follows that $\Pi(d') < \Pi(d)$.

Now we prove that Π is onto, namely, any acute angle, $\angle OPQ$, is the angle of parallelism of some finite distance. For this, take any point O_1 on PO and draw perpendicular $O_1Q_1 \perp PQ$ with foot Q_1. See Figure 10(b). Thus we have right triangle $\triangle O_1PQ_1$. Next produce PQ_1 to $Q_2, Q_3, \cdots$ so that

$$|PQ_1| = |Q_1Q_2| = |Q_2Q_3| = \cdots .$$

Draw $Q_nO_n \perp PQ$ for $n = 1, 2, \cdots$. If Q_nO_n cuts PO, denote the cutting point by O_n. Then it is not hard to show that there holds the estimate on defect:

$$\delta(\triangle O_nPQ_n) > n\,\delta(\triangle O_1PQ_1).$$

It follows that when n is large enough the straight line Q_nO_n will eventually not cut PO. By a continuity argument, there exists point Q on PQ so that the perpendicular to PQ at Q is asymptotically parallel to PO. Therefore the function Π is onto. □

2.4. *The variation in the distance between two straight lines*

Besides the angle-sums of triangles, another important feature in which hyperbolic geometry differs significantly from Euclidean geometry is the divergence and convergence of two parallel straight lines. It is well known that in Euclidean geometry two parallel straight lines are always asymptotically parallel and everywhere equidistant. In hyperbolic geometry, on the contrary, two ultra-parallel straight lines diverge in both directions, while two asymptotically parallel straight lines converge in the direction of their asymptotic parallelism and diverge in the opposite direction.

Theorem 2.5. *In a Lambert quadrilateral the sides adjacent to the acute angle are greater than their respective opposite sides.*

Hint: Apply Theorem 1.9 to this case, since an acute angle is less than a right angle.

Corollary 2.6. *In a Saccheri quadrilateral the base-to-summit midline segment is less than each of the two equal legs.*

Hint: The midline is perpendicular to both the base and the summit.

Theorem 2.7. *In a right triangle ABC with right angle at B, let E be the midpoint of the side CA. Draw $EF \perp AB$ with foot F. Then $|AF| > |FB|$ and $|BC| > 2|EF|$.*

Hint: This is Theorem 2.8 below in the case where D and A agree.

Theorem 2.8. *Suppose a quadrilateral $ABCD$ has consecutive right angles at A and B, with $|AD| < |BC|$. Let E be the midpoint of side CA and draw $EF \perp AB$ with foot F. Then $|AF| > |FB|$ and $|AD| + |BC| > 2|EF|$.*

Proof. Draw $CC' \perp EF$ and $DD' \perp EF$ with feet C' and D' on straight line EF as in Figure 11. Then we have $\triangle C'EC \cong \triangle D'ED$ by the AAS criterion. Thus in Lambert quadrilaterals $CC'FB$ and $DD'FA$ there hold $|BC| > |FC'|$ and $|AD| > |FD'|$. It follows that $|AD| + |BC| > |FC'| + |FD'| = 2|EF|$.

Now we can apply Theorem 2.9 to Lambert quadrilaterals $FD'DA$ and $FC'CB$ since $|FD'| < |FC'|$ and $|DD'| = |CC'|$. It follows that $|FA| > |FB|$. □

Theorem 2.9 (A Variation of Lambert Quadrilaterals). *Suppose two Lambert quadrilaterals $ABCD$ and $A'B'C'D'$ have acute angles at C and C'. If $|AB| < |A'B'|$ and $|BC| = |B'C'|$, then $|A'D'| < |AD|$.*

Hint: Otherwise, try to derive a contradiction in defects of quadrilaterals.

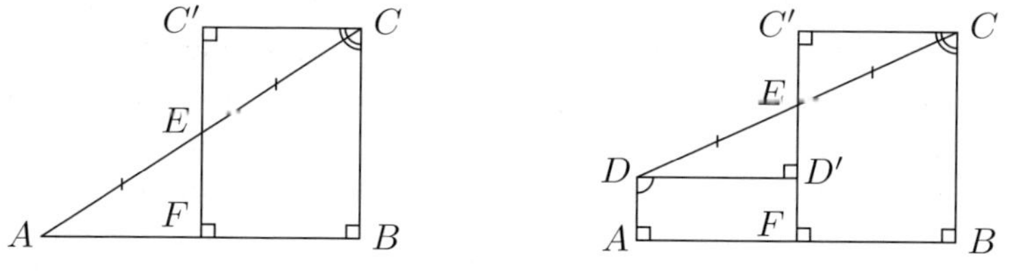

Fig. 11. Figures for Theorems 2.7 and 2.8

Theorem 2.10 (Continuity of Distance from One Straight Line to Another). *The distance function that assigns to each point in a given straight line the distance from it to another given straight line is continuous.*

Hint: Though not very difficult to obtain, the proof is a bit subtle.

Theorem 2.11 (Divergence of Two Intersecting Straight Lines). *If two straight lines intersect, then the distance from a point in one of the straight lines to the other increases without upper bounds as the point moves away from the intersection point of the two straight lines towards either end of the first straight line.*

Proof. If the two straight lines are perpendicular, then the conclusion is obviously true. Otherwise, suppose the two straight lines form an acute angle $\measuredangle POQ$ with $PQ \perp OQ$. By Theorem 1.9, as P moves in ray OP away from O, the distance $|PQ|$ increases. Now take a point P_1 on ray OP and produce OP_1 to a sequence of points $P_2, \cdots, P_n, \cdots$ on ray OP so that

$$|OP_1| = |P_1P_2| = |P_2P_3| = \cdots = |P_{n-1}P_n| = \cdots.$$

For each $n = 1, 2, \cdots$, draw $P_nQ_n \perp OQ$ with foot Q_n on OQ. By Theorem 2.7, we have $|P_2Q_2| > 2|P_1Q_1|$, and inductively, $|P_{2^n}Q_{2^n}| > 2^n|P_1Q_1|$ for $n = 1, 2, \cdots$. Therefore $|PQ|$ tends to infinity as P moves along the ray OP to infinity. □

Theorem 2.12 (Divergence and Convergence between Two Asymptotic Parallels). *If two straight lines are asymptotically parallel, then the distance from a point in one of the straight lines to the other diminishes and tends to zero as the point moves towards the end of their asymptotic parallelism, and the distance increases without upper bounds as the point moves towards the other end of the first line.*

Proof. Let $\overline{PP'}$ and $\overline{QQ'}$ be directed straight lines such that $\overline{PP'} \parallel^{\mathrm{a}} \overline{QQ'}$ with common ideal endpoint Ω and with $PQ \perp QQ'$ and $P'Q' \perp QQ'$. It is easy to see that $\measuredangle \Omega PQ$ and $\measuredangle \Omega P'Q'$ are acute. It follows that $|PQ| > |P'Q'|$ and hence $|PQ|$ decreases as P moves along PP' towards Ω, and, by the same argument as in the proof of Theorem 2.11, that $|PQ|$ increases without upper bound as P moves along PP' towards the opposite direction. It remains to prove that $|PQ|$ tends to zero as P moves along PP' towards Ω. For this, set $P_0 = P$, $P_1 = P'$, $Q_0 = Q$ and $Q_1 = Q'$. Produce Q_0Q_1 successively to Q_n, $n = 2, 3, \cdots$ so that

$$|Q_0Q_1| = |Q_1Q_2| = |Q_2Q_3| = \cdots.$$

Draw $Q_nP_n \perp Q\Omega$, $n = 1, 2, \cdots$; then Q_nP_n cuts ray $P\Omega$ in P_n, say. Suppose $|Q_nP_n| > d$, $n = 0, 1, 2, \cdots$, for some $d > 0$. Then, for each $n = 1, 2, \cdots$, the defect $\delta(Q_{n-1}Q_nP_nP_{n-1})$ of quadrilateral $Q_{n-1}Q_nP_nP_{n-1}$ is greater than δ, the defect of some quadrilateral. It follows that $\delta(Q_0Q_nP_nP_0) > n\delta$, contradicting that $\delta(Q_0Q_nP_nP_0)$ is less that four right angles. This proves Theorem 2.4. □

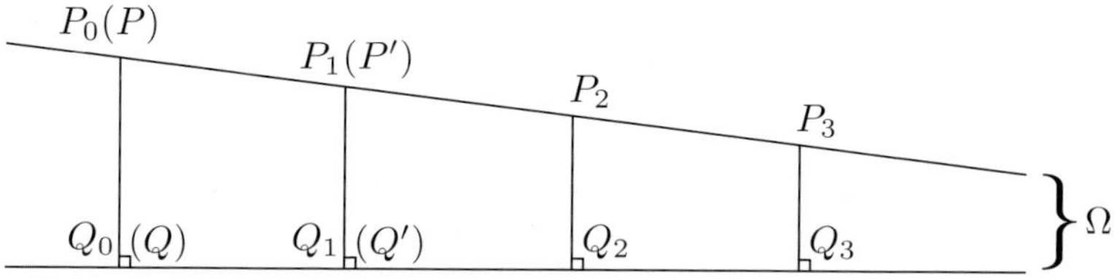

Fig. 12. Figures for Theorem 2.4

Theorem 2.13 (Divergence in either Direction between Two Ultra-Parallels). *If two straight lines are ultra-parallel to each other, then they have a common perpendicular, and the distance from a point in one of the two straight lines to the other increases without upper bounds as the point moves away from their common perpendicular towards either end of the first line.*

Proof. Suppose straight lines l^* and l are ultra-parallel. Choose a point P^* on l^* and let l_- and l_+ be the straight lines through P^* and asymptotically parallel to l, ending in ideal endpoints Ω_- and Ω_+, respectively, of l. Then, by definition, l^* falls between l_- and l_+. Let Q^* be the perpendicular foot on l of P^*. Now let P be a moving point in l^* and let Q be its perpendicular foot on l.

If P is located on the same side of straight line P^*Q^* as Ω_-, the $|PQ|$ is greater than the distance from P to l_-. Since the latter distance tends to infinity as P moves along L^* towards Ω_-, we conclude that $|PQ|$ tends to infinity as P moves towards Ω_-. Similarly, $|PQ|$ tends to infinity as P moves along L^* towards Ω_+. It follows that l^* and l have a common perpendicular, $l_\perp$. It is then easy to prove that $|PQ|$ increases as P moves along l^* away from $l_\perp$ in either direction. □

2.5. *Some more theorems in hyperbolic plane geometry*

Let us add some more theorems of hyperbolic plane geometry (again with proofs left as exercises).

Theorem 2.14. *Two straight lines have at most one common perpendicular.*

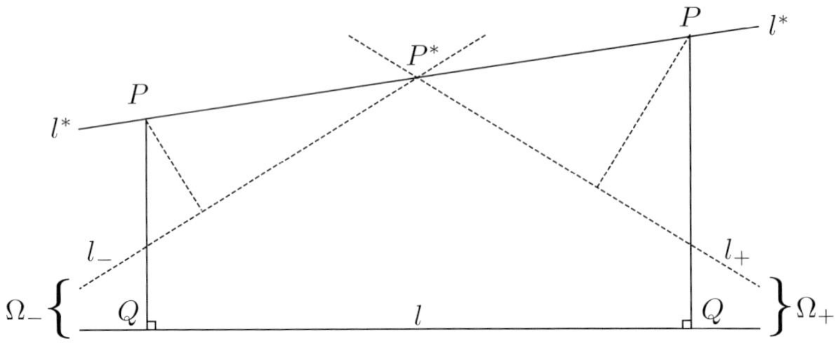

Fig. 13. Figures for Theorem 2.4

Hint: Otherwise there would be a rectangle; which is impossible under the hyperbolic angle-sum postulate.

Theorem 2.15. *For a pair of ultra-parallel straight lines, their common perpendicular segment realizes the shortest distance between the two straight lines.*

Hint: Try to apply Theorem 1.32 in one case, and the other case is easy.

Theorem 2.16. *Suppose a quadrilateral $ABCD$ has consecutive right angles at A and B. From a point F within the side CD draw perpendicular FE to AB with foot E. Then $|EF| < \max\{|AD|, |BC|\}$.*

Hint: Try to establish Corollary 2.17 bellow first.

Corollary 2.17. *In a Saccheri quadrilateral $ABCD$ with base AB, from a point F within the side CD draw perpendicular FE to AB with foot E. Then segment EF is less than each of the equal legs of $\square ABCD$.*

Theorem 2.18. *Suppose a quadrilateral $ABCD$ has consecutive right angles at A and B. From the midpoint F of CD draw perpendicular FE to AB with foot E. If E is the midpoint of AB then $|AD| = |BC|$.*

Hint: Otherwise, AD and BC would have two common perpendiculars.

Theorem 2.19. *In a right triangle the median segment joining the vertex of the right angle to the midpoint of the hypotenuse is less than half the hypotenuse.*

Hint: It follows from the configuration in §1.11.

Theorem 2.20. *In a triangle any midline (straight line segment joining the midpoints of two sides) segment is less than half the opposite side.*

Hint: It follows from the configuration in §1.11.

Theorem 2.21. *In a triangle ABC let X, Y and Z be the midpoints of the sides BC, CA and AB, respectively, and let P be the intersection point of AX and YZ. Then $|AP| > |PX|$. Furthermore, if $|AB| > |AC|$ then $|ZP| < |PY|$. (See Figure 14.)*

Hint: The first part follows from the important configuration in §1.11. The second part can be established with a bit calculation, using the same configuration.

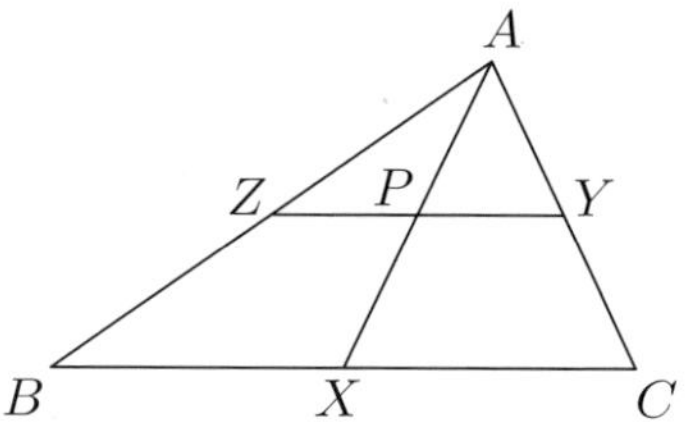

Fig. 14. Figure for Theorem 2.21

Theorem 2.22 (Exercise 36.4 in [12]). *Let $ABCD$ be a Saccheri quadrilateral with base AB (see Figure 15(a)). Then $\measuredangle DCA < \measuredangle CAB$ and* $\text{Area}(\triangle DCA) > \text{Area}(\triangle CAB)$.

Hint: To prove the first inequality, draw diagonal AC and the midline joining midpoints of AB and CD and try to apply Theorem 1.19. For the second inequality, draw diagonal BD and apply area variation as in Theorem 1.18(6).

Theorem 2.23 (Exercise 36.5 in [12]). *Let X be the midpoint of side BC of $\triangle ABC$ (see Figure 15(b)). If $|AB| > |AC|$, then* $\text{Area}(\triangle ABX) < \text{Area}(\triangle ACX)$.

Hint: Try to use the important configuration in §1.11 and apply area variation as in Theorem 1.18(6).

Exercise. Prove that in a Saccheri quadrilateral the straight line joining the midpoints of the equal legs is perpendicular to the straight line joining the midpoints of the base and summit, and that it bisects the diagonals.

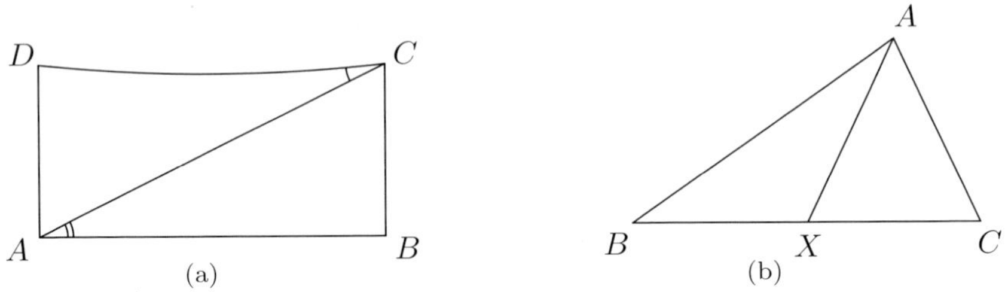

Fig. 15. Figures for Theorems 2.22 and 2.23

In a hyperbolic plane the angles of a triangle determine the sides.

Theorem 2.24 (AAA Criterion). *If in a hyperbolic plane two triangles have the three angles equal to the three angles, respectively, then they have the three sides equal to the three sides, those opposite the equal angles.*

Hint: We may assume the two triangles have a pair of equal angles coincide. Then prove that the two triangles coincide.

2.6. *Construction of the common perpendicular to two ultra-parallel straight lines*

Hilbert obtained in [15] the following nice ruler-and-compass construction of the common perpendicular for two ultra-parallel straight lines.

Hilbert's construction. Let a and b be two ultra-parallel straight lines in a hyperbolic plane. (See Figure 16.) Take two points A and A_1 on a and draw $AB \perp b$ and $A_1B_1 \perp b$ with feet B and B_1 on b. If $|AB| = |A_1B_1|$ then the straight line joining the midpoints of AA_1 and BB_1 is the common perpendicular to a and b. Otherwise, $|AB| \neq |A_1B_1|$. We may assume $|AB| > |A_1B_1|$. Mark point A' within AB so that $|A'B| = |A_1B_1|$. Draw straight line a' through A' so that a' makes with straight line AB an angle equal to that made by a and straight line A_1B_1, as Figure 16 illustrates. (We claim that a' will cut a; see Proposition 2.25 below.) Suppose a and a' meet in A_2 and draw $A_2B_2 \perp b$ with foot B_2 on b. Mark point A_3 in ray AA_1 so that $|A_1A_3| = |A'A_2|$, and draw $A_3B_3 \perp b$ with foot B_3 on b. Then we have $|A_2B_2| = |A_3B_3|$ since $\triangle BA'A_2 \cong \triangle B_1A_1A_3$ and $\triangle A_2BB_2 \cong \triangle A_3B_1B_3$. Therefore the straight line joining the midpoints of A_2A_3 and B_2B_3 is the common perpendicular to a and b.

Proposition 2.25. *In Hilbert's construction, a and a' do intersect.*

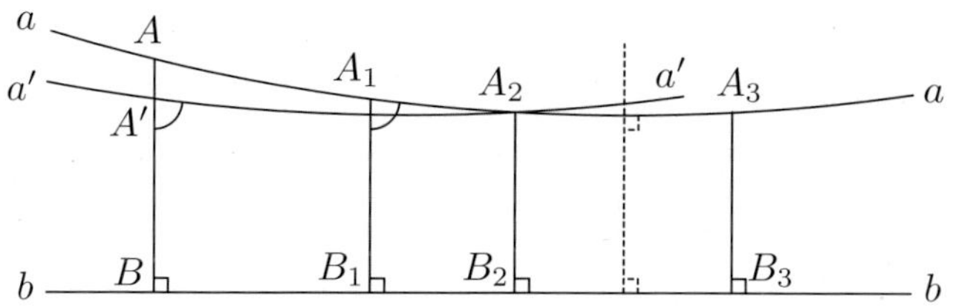

Fig. 16. Hilbert's construction of common perpendicular

Proof. Let A^* and B^* be the symmetric points of A and B, respectively, with respect to the common perpendicular of a and b, and similarly for A_1^* and B_1^* (not drawn in Figure 16; please construct the figure yourself). Thus $A^*B^* \perp b$ and $A_1^*B_1^* \perp b$. Produce $B_1^*A_1^*$ to $A_1^{*\prime}$ so that $|B_1^*A_1^{*\prime}| = |B^*A^*|$. It is easy to show that quadrilaterals $A^*B^*B_1A_1$ and $A^{*\prime}B^*B_1A_1$ are congruent. Then $\measuredangle A^*A_1B_1 = \measuredangle A^{*\prime}A'B$, and hence $A^{*\prime}$ lies on a' (thus above a). Therefore a' must cut a. □

2.7. *Construction of asymptotic parallels*

In §34 of his Appendix [4], J. Bolyai gave a very simple ruler-and-compass construction of the asymptotic parallel to a given directed straight line through a given point. The following is a well-known variant of Bolyai's construction, with exactly the same configuration. An elementary proof of this construction that uses neither trigonometry nor solid geometry was first given by Liebmann; see, for example, Wolfe [21] (page 95). A very simple proof within projective geometry is presented in [9] (page 204).

Construction. Given a directed straight line $\overline{BC}$ and a point A, with A not on straight line BC, draw $AB \perp BC$ with foot B and draw $AD \perp AB$. Take any point C on ray BC and draw $CD \perp AD$ with foot D. With center A and radius equal to $|BC|$, draw a circle cutting segment CD in E. (See Figure 17.) Then $\overline{AE} \parallel^{\text{a}} \overline{BC}$.

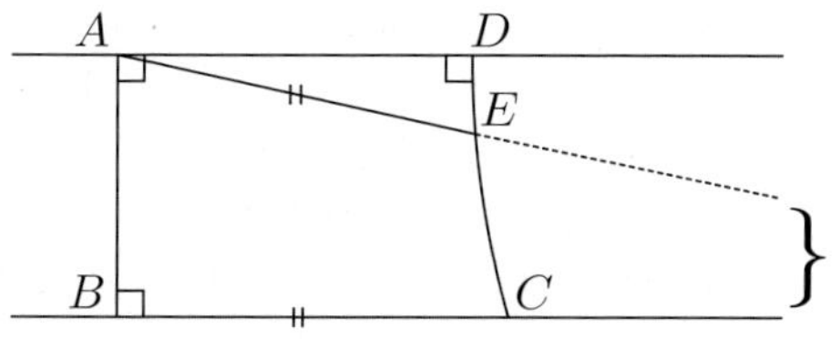

Fig. 17. Construction of asymptotic parallels

Exercise. Show that the circle in the construction above does cut the segment CD exactly once.

The construction described above is in fact a byproduct of Bolyai's construction in which he constructed for a given straight line segment its angle of parallelism, as the following theorem asserts.

Theorem 2.26 (Bolyai). *In a Lambert quadrilateral $ABCD$ with the angle at C acute, let E be within CD so that $|AE| = |BC|$. Then $\measuredangle AED = \Pi(|CD|)$.*

In his proof J. Bolyai made use of trigonometry, or explicitly, his version of the Law of Sines for hyperbolic triangles; see Bonola [3] (page 105).

2.8. *Ideal points*

In a hyperbolic plane two asymptotic straight lines will be regarded as to intersect in a point at infinity, an *ideal* point. Thus corresponding to a directed straight line is an ideal point which is not contained in the plane, and two ideal points coincide if the corresponding directed straight lines are asymptotically parallel. To distinguish from ideal points, the points in the hyperbolic plane are occasionally called *proper* points. The set of all ideal points is topologically a circle, called the *ideal boundary* of the plane.

An ideal point behaves like a proper point in many aspects, as the following theorem shows.

Theorem 2.27 (Ideal vs. Proper Points). *In a hyperbolic plane,*

(a) a proper point and an ideal point are joined by a unique straight line, and
(b) two distinct ideal points are joined by a unique straight line.

Proof. (a) is a direct consequence of the definition of asymptotic parallelism. To prove (b), let Ω_1 and Ω_2 be two distinct ideal points. Take any proper point P and draw rays $P\Omega_1$ and $P\Omega_2$. If the two straight lines $P\Omega_1$ and $P\Omega_2$ coincide then it is a straight line joining Ω_1 and Ω_2. Otherwise, the angle $\measuredangle\Omega_1 P\Omega_2$ is less then two right angles. Draw ray PQ to bisect the angle $\Omega_1 P\Omega_2$. Then there exists a point Q in ray PQ so that the perpendicular l to PQ through Q is asymptotically parallel to ray $P\Omega_1$. By symmetry, it is easy to conclude that l is also asymptotically parallel to ray $P\Omega_2$. The straight line l thus joins Ω_1 and Ω_2. To show the uniqueness of such a straight line, suppose a different straight line l' joins Ω_1. Then

l' is asymptotically parallel to $\overline{\Omega_2 Q \Omega_1}$ and hence cannot be asymptotically parallel to $\overline{\Omega_1 Q \Omega_2}$. □

We can consider "triangles" with one or more vertices ideal points.

Definition 2.28 (Asymptotic Triangles). Given A, B and C, each of which is a proper point in a hyperbolic plane or an ideal point associated to the plane, we shall call the configuration formed by the three straight lines BC, CA and AB a *singly asymptotic triangle*, a *doubly asymptotic triangle*, or a *triply asymptotic triangle* and denote it by $\triangle ABC$, according as exactly one, two, or three of A, B and C are ideal points.

Thus a singly asymptotic triangle has one finite side and two proper angles, a doubly asymptotic triangle has no finite side and one proper angle, while a triply asymptotic triangle has neither finite sides nor proper angles.

Some of the congruence theorems for proper triangles extend to asymptotic triangles. In the following congruence theorems for asymptotic triangles, the letter "I" in "AAI" and "AIS" stands for "Ideal angle".

Theorem 2.29 (AAI Congruence Criterion). *If two singly asymptotic triangles have the proper angles equal to the proper angles respectively, then they have the finite side equal to the finite side.*

Theorem 2.30 (AIS Congruence Criterion). *If two singly asymptotic triangles have the finite side equal to the finite side, and one proper angle equal to one proper angle, then they have the remaining proper angle equal to the remaining proper angle.*

2.9. *Horocycles*

By the definition of a circle, if two points are equidistant from a third point then the two points lie on a circle centered at the third point. Note that the condition that the three points form an isosceles triangle is equivalent to that the two base angles be equal. Considering singly asymptotic isosceles triangles, we get the notion of a "circle" centered at an ideal point.

Definition 2.31 (Corresponding Points Relative to an Ideal Point). Two proper points, A and B, are said to *correspond* to each other relative to a given ideal point Ω if $\measuredangle \Omega AB = \measuredangle \Omega BA$.

Theorem 2.32. *Suppose A and B are two distinct proper points and Ω is an ideal point. Then A and B are corresponding points relative to Ω if and only if the perpendicular bisector of the segment AB ends in Ω.*

Theorem 2.33. *Suppose A and B, and also B and C, are corresponding points relative to the same ideal point Ω. Then so are A and C.*

Thus we can speak of three or more corresponding points relative to the same ideal point.

Theorem 2.34. *Three distinct corresponding points relative to the same ideal point are not collinear.*

Proof. Let A, B and C be three corresponding points relative to ideal point Ω. We may assume that A and C lie on different sides of straight line $B\Omega$. Then we have $\measuredangle \Omega AB = \measuredangle \Omega BA$ and $\measuredangle \Omega BC = \measuredangle \Omega CB$, equalities of acute angles. It follows that $\measuredangle ABC$ is less than two right angles. Hence A, B and C are not collinear. □

Definition 2.35 (Horocycle). Given a point P and an ideal point Ω, the curve formed by P and all the points which correspond to P relative to Ω is called the *horocycle* through P centered at Ω.

For each point Q on the horocycle, the straight line $Q\Omega$ is called the radius through Q of the horocycle.

It follows from Theorem 2.34 that a horocycle is not a straight line. Note that historically horocycles were called "boundary lines" by Lobachevsky in [17] and "L-lines" by J. Bolyai in [4].

Theorem 2.36. *Two concentric horocycles are everywhere equidistant, namely, they cut out segments of equal length in all common radii.*

Proof. Let h_1 and h_2 be two concentric horocycles centered at Ω, with h_1 the inner, and let A_1 and B_1 be two points on h_1. Let the intersection points of radii $A_1\Omega$ and $B_1\Omega$ with h_2 be A_2 and B_2, respectively. Then in the quadrilateral $A_1A_2B_2B_1$ there hold $\measuredangle A_1 = \measuredangle B_1$ and $\measuredangle A_2 = \measuredangle B_2$. It is then not hard to obtain that $|A_1A_2| = |B_1B_2|$. □

2.10. *Construction of the straight line joining two given ideal points*

Let Ω_1 and Ω_2 be two ideal points, represented by rays $P_1\Omega_1$ and $P_2\Omega_2$, respectively, where P_1 and P_2 are two distinct proper points.

Case 1. If the rays $P_1\Omega_1$ and $P_2\Omega_2$ are collinear then we are done.

Case 2. Otherwise, $P_1\Omega_1$ and $P_2\Omega_2$ are not collinear. We draw rays $P_1\Omega_2$ and $P_2\Omega_1$. Let P_1X_1 and P_2X_2 be the internal bisectors of angles

$\measuredangle\Omega_1 P_1 \Omega_2$ and $\measuredangle\Omega_1 P_2 \Omega_2$, respectively. Then it is easy to show that the straight line joining Ω_1 and Ω_2 is perpendicular to both P_1X_1 and P_2X_2.

Case 2a. If the two lines P_1X_1 and P_2X_2 do not coincide, then constructing the common perpendicular of P_1X_1 and P_2X_2 will give the unique straight line joining ideal points Ω_1 and Ω_2, as desired. See Figure 18.

Case 2b. Otherwise, we replace P_1 by an arbitrary point P_1' within ray $P_1\Omega_1$ and proceed as in Case 2a, therein the construction will finish.

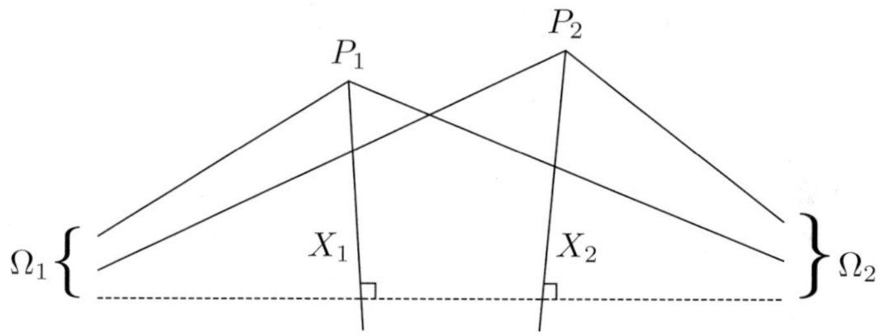

Fig. 18. Construction of straight line joining two ideal points

2.11. *Ultra-ideal points*

To further extend the idea of adjoining ideal points on the boundary of the hyperbolic plane, two ultra-parallel straight lines will be regarded as to intersect in an *ultra-ideal* point, a point further outside the hyperbolic plane and its ideal boundary. Explicitly, corresponding to each ultra-ideal point is a straight line, its representative straight line, such that all the straight lines that "pass through" the ultra-ideal point are exactly those which are perpendicular to its representative straight line, and, conversely, corresponding to each straight line is the ultra-ideal point through which pass all the perpendiculars to the given straight line. In other words, the set of all ultra-ideal points and the set of all straight lines in the hyperbolic plane are in one-to-one correspondence.

Exercise. Prove that the set of all the ultra-ideal points associated to a hyperbolic plane is topologically an open Möbius strip, namely, the surface obtained from the top-and-bottom-open square $[0,1] \times (0,1)$ by identifying points $(0,t)$ and $(1,1-t)$ for all $t \in (0,1)$.

2.12. *The projective plane associated to a hyperbolic plane*

So far we have associated to a hyperbolic plane points of three types, namely, the proper points, the ideal points, and the ultra-ideal points. They can be thought of, in a unified viewpoint, as corresponding to elliptic, parabolic, and hyperbolic pencils of straight lines in the hyperbolic plane. All these points together form an extended "plane", the *projective plane* associated to the hyperbolic plane. (Exercise: Show that the projective plane so named is indeed a topological, real projective plane.)

Projective lines are defined as certain subsets of the projective plane as follows. A projective line corresponding to a proper point P consists of all the ultra-ideal points that correspond to the straight lines passing through P. A projective line corresponding to an ideal point Ω consists of Ω and all the ultra-ideal points that correspond to the straight lines ending in Ω. A projective line corresponding to a straight line l consists of all the proper points on l, the two ideal endpoints of l, and all the ultra-ideal points that correspond to the straight lines perpendicular to l. Thus the projective lines are in one-to-one correspondence with the (elliptic, parabolic, or hyperbolic) pencils of straight lines in the hyperbolic plane.

The projective lines so defined satisfy the projective parallel postulate, namely, there are *no* parallels in the projective plane.

Proposition 2.37. *In the projective plane associated to a hyperbolic plane, any two distinct projective lines intersect in a unique point (proper, ideal or ultra-ideal).*

Hint: There are nine cases to consider which include all of the basic configurations consisting of a pair of (ideal) points/straight lines in the hyperbolic plane.

One is thus ready to develop the classical theory of real projective plane geometry, in particular, define the projective transformations, which we shall not pursue here; the reader is referred to, for example, Coxeter [8].

Conversely, as an alternative approach, hyperbolic geometry can be studied as a subgeometry of projective geometry; see Coxeter [9] for a systematic treatment.

2.13. *Center-pencils of a hyperbolic triangle*

We have seen that the incenter and centroid of a hyperbolic triangle are proper points. The analogs of circumcenter, orthocenter and excenters for a hyperbolic triangle will, however, be points in the projective plane (or pencils of straight lines in the hyperbolic plane).

Theorem 2.38 (Circumcenter). *Given a triangle in a hyperbolic plane, the perpendicular bisectors of the three sides of the triangle belong to a unique pencil of straight lines, called the circumcenter pencil of the triangle.*

Theorem 2.39 (Orthocenter). *Given a triangle in a hyperbolic plane, the three altitude straight lines of the triangle belong to a unique pencil of straight lines, called the orthocenter pencil of the triangle.*

Theorem 2.40 (Excenters). *Given a triangle in a hyperbolic plane, the internal bisector of one of the angles and the external bisectors of the remaining two angles of the triangle belong to a unique pencil of straight lines, called the excenter pencil associated with that internal angle.*

2.14. *Equidistant curves*

In a hyperbolic plane an *equidistant curve* of distance d away from a straight line l is the curve formed by all the points on one side of l that are distance $d(> 0)$ away from l. Thus on the other side of l there is an equidistant curve of the same distance away from l (this is different from what Coxeter uses in [9] where he calls the two curves *one* equidistant curve).

Theorem 2.41. *In a hyperbolic plane an equidistant curve is not a straight line.*

Hint: The proof is similar to that for a horocycle or a circle.

Definition 2.42. A *generalized circle* in a hyperbolic plane is either a circle, a horocycle, or an equidistant curve. The pencil consisting of all the straight lines perpendicular to a generalized circle is called its *radii pencil.*

With this terminology, we have the following

Theorem 2.43 (Circumcircle Theorem). *Through the three vertices of a triangle in a hyperbolic plane passes a unique generalized circle, called the generalized circumcircle of the triangle.*

Exercise. Show that the circumcenter pencil of a triangle in a hyperbolic plane coincides with the radii pencil of its generalized circumcircle.

We say that a simple quadrilateral in the hyperbolic plane is *generalized cyclic* if all the vertices of the quadrilateral lie on the same generalized circle.

Theorem 2.44. *A simple quadrilateral in a hyperbolic plane is generalized cyclic if and only if it has the sum of two opposite angles equal to that of the remaining two angles.*

Hint: The proof is similar to that for the Euclidean case.

2.15. *Positions of proper points relative to an ideal point*

Regarding the positions of two proper points relative to an ideal point, we make the following definition.

Definition 2.45. Given an ideal point Ω and two distinct points A and B in H^2, we say that A and B are *vertical* relative to Ω, if the straight line through A and B ends in Ω, *horizontal* if A and B lie on a horocycle centered at Ω, and *slant* if A and B lie on an equidistant curve (based on some straight line) ending in Ω.

Theorem 2.46. *Relative to a given ideal point, any two distinct points in a hyperbolic plane are either vertical, horizontal or slant, and exactly one of the three cases holds.*

Proof. Let Ω be an ideal point and let A and B be distinct points in H^2. Let l be the straight line through A and B. If l ends in Ω, then A and B are vertical relative to Ω. Otherwise, Ω is not an ideal endpoint of l. Let $l_\perp$ be the perpendicular bisector of the segment AB. If $l_\perp$ ends in Ω, then A and B lie on a horocycle centered at Ω, and they are horizontal relative to Ω. Otherwise, Ω is not an ideal endpoint of $l_\perp$. Let l' be the perpendicular drawn from Ω to $l_\perp$. Then it is easy to show that A and B are of equal distance away from l'. Hence A and B lie on an equidistant curve based on l', namely, they are slant relative to Ω. □

2.16. *Hyperbolic areas via equivalence of triangles*

Throughout this subsection we shall consider only simple polygons in a hyperbolic plane.

A polygon can be *partitioned* into two new polygons by introducing a straight line segment or a broken line within the polygon which connects two points on the perimeter of the polygon. When this process is performed a finite number of times, a polygon will be partitioned into a finite collection of polygons. We say that two polygons are *equivalent* if they can be partitioned into the same finite number of triangles and a one-to-one correspondence between the two sets of triangles can be established so that each pair of the corresponding triangles are congruent. It follows that two polygons are equivalent if they are each equivalent to a third polygon.

By the additivity of the defect of polygons, we have

Theorem 2.47. *Any two equivalent polygons have the same defect.*

One can establish the converse of Theorem 2.47 for triangles.

Theorem 2.48. *Two triangles with a side of one equal to a side of the other and having the same defect are equivalent.*

Hint: By the midline configuration in §1.11, every triangle is equivalent to a Saccheri quadrilateral with its base any side of the triangle and with the sum of its summit angles equal to the angle-sum of the triangle.

Theorem 2.49. *Any two triangles with the same defect are equivalent.*

Hint: One can construct a new triangle such that it has the same defect as the two triangles, has a side equal to a side of the first triangle, and has another side equal to a side of the second triangle; see [21] (page 125).

Without giving a specific definition, we assume the area of a polygon in a hyperbolic plane is defined in a reasonable way.

Theorem 2.50. *The area of a polygon in a hyperbolic plane equals its defect multiplied by a positive constant of the plane.*

Thus we have obtained

Theorem 2.51. *Two triangles in a hyperbolic plane have the same area if and only if they are equivalent.*

The area of an asymptotic triangle is defined naturally as the sum of the areas of an infinite collection of triangles into which the asymptotic triangle can be partitioned in any way.

Exercise. Prove that the area of an asymptotic triangle is well defined, namely, it is independent of the way the asymptotic triangle is partitioned into triangles.

We need to confirm the finiteness of the area of an asymptotic triangle.

Theorem 2.52. *Any asymptotic triangle has finite area.*

Proof. It is easy to see that we only need to establish the conclusion for a singly asymptotic right triangle. Let $\triangle P_0Q_0\Omega$ be a singly asymptotic triangle with Ω an ideal point and with the angle at Q_0 right. We shall partition $\triangle P_0Q_0\Omega$ into a sequence of quadrilaterals $\square_n$, $n = 1, 2, \cdots$, so that the area of each $\square_{n+1}$ is less than half the area of $\square_n$. It will follow that the area of $\triangle P_0Q_0\Omega$, which equals the sum of the areas of all the $\square_n$, $n = 1, 2, \cdots$, must be finite.

For this, let Q_0' be the symmetric point of Q_0 with respect to straight line $P_0\Omega$. In straight line $Q_0'\Omega$ there exists a unique point P_1' such that the distance from P_1' to straight line $Q_0\Omega$ equals $|P_0Q_0|$; the perpendicular foot Q_1 of P_1' on straight line $Q_0\Omega$ will fall within ray $Q_0\Omega$. Suppose $P_1'Q_1$ cuts $P_0\Omega$ in P_1. See Figure 19

Produce Q_0Q_1 successively to $Q_2, \cdots, Q_n, \cdots$ so that

$$|Q_0Q_1| = |Q_1Q_2| = |Q_2Q_3| = \cdots = |Q_{n-1}Q_n| = \cdots.$$

For each $n = 2, 3, \cdots$, draw $Q_nP_n \perp Q_0\Omega$; then Q_nP_n must cut $P_0\Omega$ (in P_n, say). Thus $\triangle P_0Q_0\Omega$ is partitioned into an infinite sequence of quadrilaterals

$$\square Q_0Q_1P_1P_0,\ \square Q_1Q_2P_2P_1,\ \cdots,\ \square Q_{n-1}Q_nP_nP_{n-1},\ \cdots.$$

We claim (Exercise!) that, for each $n = 2, 3, \cdots$, the area of $\square Q_{n-1}Q_nP_nP_{n-1}$ is less than half the area of $\square Q_{n-2}Q_{n-1}P_{n-1}P_{n-2}$. This proves Theorem 2.52. $\square$

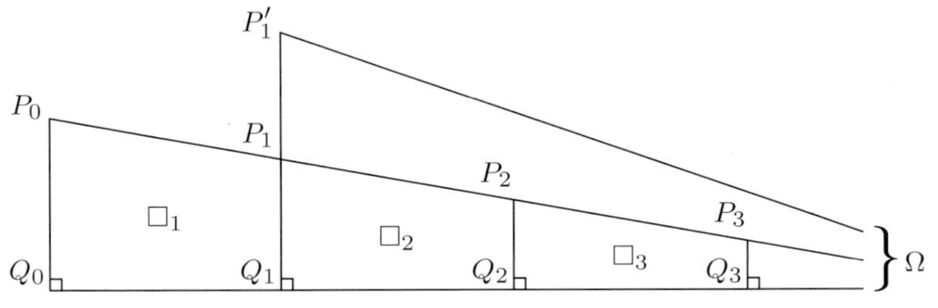

Fig. 19. Proof of Theorem 2.52: $|P_0Q_0| = |P_1'Q_1|$

In [16] (page 43) Liebmann obtained a beautiful proof of Theorem 2.52 by partitioning a singly asymptotic triangle into an infinite sequence of triangles and then rearranging them to form a finite quadrilateral.

Theorem 2.53. *All triply asymptotic triangles are equivalent to each other and hence have the same area.*

Proof. Dropping perpendicular from one of the ideal vertices of a triply asymptotic triangle to the opposite side partitions the asymptotic triangle into two doubly asymptotic right triangles. Since any two doubly asymptotic right triangles are obviously congruent, all triply asymptotic triangles are equivalent. □

Therefore every triply asymptotic triangle has area equal to $k^2\pi$ for some constant $k > 0$. We shall call the positive constant k^2 the *area factor* of the hyperbolic plane.

Theorem 2.54. *In a hyperbolic plane with area factor k^2, the area of a proper or asymptotic triangle with defect δ in radian equals $k^2\delta$.*

Proof. We need to prove that the common area factor for all proper triangles and the area factor of all triply asymptotic triangles agree. This can be done easily by partitioning an asymptotic triangle into an infinite sequence of proper triangles. □

Note that Gauss gave his proof of the above theorem in a letter to W. Bolyai in response to the latter's sending him a copy of the book containing J. Bolyai's famous Appendix. For rephrases of Gauss's proof (assuming the finiteness of the areas of asymptotic triangles), see Wolfe [21] (page 128), or Coxeter[7] (page 297).

2.17. *Metric relations of corresponding arcs in concentric horocycles*

Note that we may choose the unit of length and that of area independently for a hyperbolic plane. But in accordance with the convention in Euclidean geometry or Riemannian geometry, we always choose the units of length and of area so that when passing to infinitesimal regions of the plane the Euclidean length-area relations will be satisfied.

In a hyperbolic plane with area factor $k^2 (k > 0)$, a fundamental relation of any two corresponding arcs in concentric horocycles is: if the horocycles are unit distance apart, then the length of the outer arc equals that of the inner one multiplied by $e^{1/k}$.

Theorem 2.55. *In a hyperbolic plane with area factor $k^2 (k > 0)$, if s_1 and s_2 are respectively the lengths of the inner and the outer of two corresponding arcs in concentric horocycles, then*

$$s_2 = s_1 e^{x/k}, \tag{2.2}$$

where x is (the measure of) the distance between the arcs.

We remark that the laws of cosines and of sines for triangles, as well as the explicit formula for the angle-of-parallelism function Π (see §4.2 for the formulas mentioned) can all be derived step by step starting from (2.2), as presented in, for example, [21] (pages 135–156), in a very enlightening way which we shall not pursue here; we shall choose another way instead.

3. Isometries of the Hyperbolic Plane

Throughout this section we shall fix a hyperbolic plane H^2 with ideal boundary ∂H^2, and assume that everything we discuss will be about this fixed hyperbolic plane, unless otherwise clearly stated.

According to Kline's Erlangen Program, the study of Euclidean or hyperbolic geometry lies essentially in the search of properties left invariant by the isometries (and similarities). In this section we analyze the structures of the group of isometries of a hyperbolic plane. We shall see that an isometry of a hyperbolic plane resembles that of a Euclidean plane in many aspects: in both cases it is generated by three or fewer reflections in straight lines, and the differences lie in the different natures of two straight lines in respective planes. For detailed expositions of structures of isometries of the Euclidean plane, see Barker-Howe [1] or Martin [18].

3.1. *Isometries and reflections in straight lines*

We begin with the self-explaining definition of an isometry of the plane.

Definition 3.1 (Isometry). An *isometry* of H^2 is a bijective self-mapping of H^2 that preserves the distance between each pair of points in H^2.

In other words, a bijective mapping $f : H^2 \to H^2$ is an isometry if and only if for each pair of points P and Q in H^2, $|f(P)f(Q)| = |PQ|$.

As in the Euclidean case, the following theorem is easy to establish.

Theorem 3.2. *An isometry of a hyperbolic plane sends any straight line in the plane onto a straight line.*

The simplest isometry is, of course, the identity mapping. The inverse of any isometry and the composition of two isometries are isometries; thus the set of all the isometries of the plane is a group under composition.

The next simplest family of isometries are reflections in straight lines.

Definition 3.3 (Mirror Point). Given a straight line l and a point P in H^2, the *mirror point* of P with respect to l is the point P' in H^2 obtained by dropping perpendicular straight line segment PM from P to l, with foot M on l, and producing PM to P' so that $|PM| = |MP'|$.

Definition 3.4 (Reflection in a Straight Line). Given a straight line l in H^2, the *reflection in* l is the mapping $\sigma_l : H^2 \to H^2$ that sends each point in H^2 to its mirror point with respect to l.

It is easy to see that the reflection in a straight line is an *involution*, namely, $\sigma_l\sigma_l = \mathrm{id}_{H^2}$. We also observe that σ_l fixes the straight line l pointwise and fixes no other points in H^2. It follows that, for straight lines l_1 and l_2 in H^2, $\sigma_{l_1} = \sigma_{l_2}$ if and only if $l_1 = l_2$.

By applying the SAS congruence criterion, one easily shows that reflections in straight lines are isometries of the plane.

Theorem 3.5. *The reflection in any straight line is an isometry.*

The reflections in straight lines are important in that they generate all the isometries of the plane. We leave the proofs of Theorems 3.6 and 3.7 below as exercises.

Theorem 3.6. *If an isometry of the plane fixes each of some two distinct points, then it fixes pointwise the straight line joining the two points.*

Theorem 3.7. *If an isometry of the plane fixes each of some three non-collinear points, then it is the identity mapping.*

As an immediate corollary of Theorem 3.7, we have

Theorem 3.8. *If two isometries of the plane agree at each of some three non-collinear points, then they agree everywhere in the plane.*

Same as in Euclidean case, we have the following structure theorem for isometries of the hyperbolic plane.

Theorem 3.9 (Three or Fewer Reflections). *Any isometry of H^2 is the composition of at most three reflections in straight lines in H^2.*

Proof. Let ϕ be an isometry of H^2 and consider any triangle ABC in H^2. Let A^*, B^* and C^* be the image points of A, B and C, respectively, under mapping ϕ.

If A^* does not coincide with A, then let l_1 be the perpendicular bisector of segment AA^*. Let A_1, B_1 and C_1 be the mirror points of A, B and C, respectively, with respect to l_1. Then A_1 coincides with A^*.

If further B_1 does not coincide with B^*, then let l_2 be the perpendicular bisector of segment B_1B^*. Let A_2, B_2 and C_2 be the mirror points of A_1, B_1 and C_1, respectively, with respect to l_2. Then $A_2 = A^*$ and $B_2 = B^*$.

If furthermore C_2 does not coincide with C^*, then let l_3 be the perpendicular bisector of segment C_2C^*. Let A_3, B_3 and C_3 be the mirror points of A_2, B_2 and C_2, respectively, with respect to l_3. Then $A_3 = A^*$, $B_3 = B^*$ and $C_3 = B^*$. This means the given isometry ϕ and the composition isometry $\sigma_{l_3}\sigma_{l_2}\sigma_{l_1}$ agree at each of A, B and C. Therefore $\phi = \sigma_{l_3}\sigma_{l_2}\sigma_{l_1}$ in the case that all the three "if"s hold.

If one or more of the three "if"s do not hold, then we delete accordingly one or more σ's from the composition of σ_{l_1}, σ_{l_2} and σ_{l_3}. It follows that every isometry of H^2 is the composition, in certain order, of at most three reflections in straight lines in H^2. □

We follow the natural convention that the composition of *zero* reflections in straight lines in the plane is the identity mapping.

3.2. *Orientation preserving/reversing isometries*

A triangle in the plane can be given a clockwise or anticlockwise orientation by cyclically ordering its vertices. Thus if triangle ABC with vertices in this ordering is anticlockwise, then so are BCA and CAB, whereas CBA, BAC and ACB are clockwise.

Any isometry of the plane sends an anticlockwise triangle onto an anticlockwise or a clockwise triangle. It is easy to prove that if an isometry sends one anticlockwise triangle onto an anticlockwise triangle then it does so for every anticlockwise triangle.

Definition 3.10 (Orientation Preserving/Reversing Isometries). An isometry of H^2 is said to be *orientation preserving* or *orientation reversing* according as it sends an anticlockwise triangle in H^2 onto an anticlockwise or a clockwise triangle.

The following two theorems can be easily verified.

Theorem 3.11. *The reflection in any straight line is orientation reversing.*

Theorem 3.12. *An isometry of H^2 is orientation preserving or reversing according as it is the composition of an even or odd number of reflections in straight lines.*

In other words, an orientation preserving isometry is the composition of *zero* or *two* reflections in straight lines, while an orientation reversing isometry is the composition of *one* or *three* reflections in straight lines.

This leads to the following classification of orientation preserving isometries.

Definition 3.13 (Types of Isometries). An orientation preserving non-identity isometry of H^2 is said to be of *elliptic type*, of *hyperbolic type*, or of *parabolic type*, according as, in its representation as the composition of two reflections in straight lines, whether the two straight lines intersect, are ultra-parallel, or are asymptotically parallel.

In the following three subsections we shall reveal the natures of all orientation preserving isometries of H^2: one of elliptic type is a rotation about a point, that of hyperbolic type is a translation along a straight line, and that of parabolic type can be thought of as the common limit of isometries of the other two types, with center/axis degenerate to an ideal point.

3.3. *Rotations*

The same as in a Euclidean plane, a rotation about a point O in H^2 fixes O and rotates any point P other than O through a fixed angle along the circle centered at O and through P; it therefore sends any straight line through O to a straight line through O. We call O the *center* of the rotation.

Theorem 3.14. *Suppose that l_1 and l_2 are straight lines which meet in O, and that θ is the measure of the anticlockwise angle from l_1 to l_2. Then the isometry $\sigma_{l_2}\sigma_{l_1}$ is the rotation about O through an anticlockwise angle of measure $2\theta \bmod 2\pi$.*

In particular, the composition of two reflections in perpendicular straight lines is a rotation of angle π which is an orientation preserving involution (in fact, this gives the only involutions among all orientation preserving isometries of the plane).

Definition 3.15 (Half Turn). For a point P in H^2, the *half turn* about P, denoted by σ_P, is the isometry of H^2 given by the π rotation about P.

Thus for any two perpendicular straight lines l_1 and l_2 meeting in P,

$$\sigma_{l_2}\sigma_{l_1} = \sigma_{l_1}\sigma_{l_2} = \sigma_P.$$

Theorem 3.16. *Suppose l_1 and l_2 are distinct straight lines in H^2. Then $\sigma_{l_1}\sigma_{l_2} = \sigma_{l_2}\sigma_{l_1}$ if and only if l_1 and l_2 are perpendicular to each other.*

Corollary 3.17. *Suppose l_1, l_2 and l_3 straight lines in H^2, where l_1 and l_2 are both perpendicular to l_3. Then $\sigma_{l_3}\sigma_{l_2}\sigma_{l_1} = \sigma_{l_2}\sigma_{l_1}\sigma_{l_3}$.*

3.4. *Translations*

Similar to the case in Euclidean geometry, a translation τ along a straight line l in H^2 shifts points on l through the same distance and in the same direction, and shifts any point P not on l along the equidistant curve based on l and through P. We shall call l the *axis* of τ, and call the fixed distance through which τ shifts points on its axis the *translation distance* of τ.

Note that a translation sends any straight line perpendicular to its axis onto another such straight line and hence fixes no points in the plane. (Exercise: Show that a translation moves points on its axis the shortest distance.)

Theorem 3.18. *Suppose l_1 and l_2 are ultra-parallel straight lines in H^2 with common perpendicular l. Then the isometry $\sigma_{l_2}\sigma_{l_1}$ is the translation along l through distance $2d$ in the direction from l_1 to l_2, where d is the distance between l_1 and l_2 measured along l.*

Theorem 3.19. *Suppose l_1 and l_2 are ultra-parallel straight lines in H^2 with common perpendicular l and let P_1 and P_2 be the intersection points of l with l_1 and l_2, respectively. Then $\sigma_{l_2}\sigma_{l_1} = \sigma_{P_2}\sigma_{P_1}$.*

Proof. As $l_1 \perp l$ and $l \perp l_2$, we have $\sigma_{P_1} = \sigma_l\sigma_{l_1}$ and $\sigma_{P_2} = \sigma_{l_2}\sigma_l$; hence $\sigma_{P_2}\sigma_{P_1} = (\sigma_{l_2}\sigma_l)(\sigma_l\sigma_{l_1}) = \sigma_{l_2}(\sigma_l\sigma_l)\sigma_{l_1} = \sigma_{l_2}\sigma_{l_1}$. □

Remark. A translation τ sends any horocycle centered at Ω, an ideal endpoint of the axis of τ, onto a different horocycle centered at Ω; precisely, τ sends the horocycle centered at Ω and through a point P to the horocycle centered at Ω and through the point $\tau(P)$. It follows that the translation distance and the direction of translation of τ are determined by the action of τ on the set of horocycles centered at Ω.

3.5. *Isometries of parabolic type*

We may think of an isometry of H^2 of parabolic type as an infinitesimal rotation or translation, with center or axis being degenerate to an ideal point, which we call the *ideal center* of the isometry.

Theorem 3.20. *Suppose l_1 and l_2 are asymptotically parallel straight lines in H^2, both ending in ideal point Ω. Then $\sigma_{l_2}\sigma_{l_1}$ sends any straight line ending in Ω to another such straight line and moves any point P along the horocycle centered at Ω and through P, in the direction from l_1 to l_2.*

In particular, an isometry of parabolic type fixes no points in H^2.

3.6. *Redundancy of two reflections*

The representation of an orientation preserving isometry of H^2 as the composition of two reflections in straight lines is redundant in a useful way: one of the two straight lines can be chosen arbitrarily in the pencil.

Theorem 3.21. *Suppose l_1, l_2, l_3 and l_4 are straight lines in H^2 passing through the same point such that the directed angle from l_1 to l_2 equals the directed angle from l_3 to l_4. Then $\sigma_{l_2}\sigma_{l_1} = \sigma_{l_4}\sigma_{l_3}$.*

Theorem 3.22. *Suppose l_1, l_2, l_3 and l_4 are straight lines in H^2 perpendicular to the same straight line such that the directed distance from l_1 to l_2 equals that from l_3 to l_4. Then $\sigma_{l_2}\sigma_{l_1} = \sigma_{l_4}\sigma_{l_3}$.*

Theorem 3.23. *Suppose l_1, l_2, l_3 and l_4 are straight lines in H^2 ending in an ideal point Ω such that along a horocycle centered at Ω the directed arclength from l_1 to l_2 equals that from l_3 to l_4. Then $\sigma_{l_2}\sigma_{l_1} = \sigma_{l_4}\sigma_{l_3}$.*

The three theorems above can be unified as the following single theorem.

Theorem 3.24. *Suppose l_1, l_2, and l_3 (resp. l_4) are straight lines belonging to the same pencil of straight lines in H^2. Then there exists a unique straight line l_4 (resp. l_3) in the pencil such that $\sigma_{l_2}\sigma_{l_1} = \sigma_{l_4}\sigma_{l_3}$*

We leave the proofs of Theorems 3.21–3.24 as exercises.

3.7. *Orientation reversing isometries as reflections and glide reflections*

In this subsection we show that all orientation reversing isometries of H^2 are reflections in straight lines and the so-called glide reflections.

Definition 3.25 (Glide Reflection). An isometry of H^2 is called a *glide reflection* if it is the composition of a translation along a straight line l followed by the reflection in l. We call l the *axis* of the glide reflection.

By decomposing the translation in the definition, a glide reflection is obtained as the composition of reflections in three straight lines where the last straight line is perpendicular to the first two.

Proposition 3.26. *If l_1, l_2 and l_3 are straight lines in H^2 such that l_1 and l_2 are perpendicular to l_3, then $\sigma_{l_3}\sigma_{l_2}\sigma_{l_1} = \sigma_{l_2}\sigma_{l_3}\sigma_{l_1} = \sigma_{l_2}\sigma_{l_1}\sigma_{l_3}$.*

Proof. The identities of isometries follow from Corollary 3.17. □

Thus a glide reflection is also a "reflection glide" with the same axis.

Theorem 3.27 (When Three Equals One). *Given straight lines l_1, l_2 and l_3 in H^2, the isometry $\sigma_{l_3}\sigma_{l_2}\sigma_{l_1}$ is the reflection in a straight line if and only if l_1, l_2 and l_3 belong to the same pencil of straight lines in H^2.*

Proof. Suppose l_1, l_2 and l_3 belong to the same pencil of straight lines in H^2. By Theorem 3.24, there exists straight line l_0 in the same pencil such that $\sigma_{l_3}\sigma_{l_0} = \sigma_{l_2}\sigma_{l_1}$. Then post-composing with σ_{l_3} gives $\sigma_{l_0} = \sigma_{l_3}\sigma_{l_2}\sigma_{l_1}$.

Conversely, suppose $\sigma_{l_3}\sigma_{l_2}\sigma_{l_1} = \sigma_{l_0}$ for some straight line l_0 in H^2. Then $\sigma_{l_2}\sigma_{l_1} = \sigma_{l_3}\sigma_{l_0}$. We may assume l_1 and l_2 are distinct; hence so are l_0 and l_3. It is easy to convince that the pencil of straight lines in H^2 containing l_1 and l_2 is the same as that containing l_0 and l_3. Thus l_1, l_2 and l_3 belong to the same pencil of straight lines. □

Theorem 3.28. *Any orientation reversing isometry of H^2 is a reflection in straight line or a glide reflection.*

Proof. Suppose l_1, l_2 and l_3 are distinct straight lines in H^2. We need to prove that the composition $\sigma_{l_3}\sigma_{l_2}\sigma_{l_1}$ is a glide reflection.

Case 1. Suppose l_1 and l_2 either intersect in P or are asymptotically parallel, ending in the same ideal point Ω. Let l_2' be the straight line through P or joining Ω and perpendicular to l_3, with foot Q on l_3. Then there exists a unique straight line l_1' through P or joining Ω such that $\sigma_{l_2}\sigma_{l_1} = \sigma_{l_2'}\sigma_{l_1'}$. Further, let l_3' be the straight line through Q and perpendicular to l_1' and let l_2'' be the straight line through Q and perpendicular to l_3'. See the left in Figure 20 for the case where l_1 and l_2 intersect. Then $\sigma_{l_3}\sigma_{l_2'} = \sigma_{l_3'}\sigma_{l_2''}$ since $l_2' \perp l_3$ and $l_2'' \perp l_3'$. It follows that

$$\sigma_{l_3}\sigma_{l_2}\sigma_{l_1} = \sigma_{l_3}\sigma_{l_2'}\sigma_{l_1'} = \sigma_{l_3'}\sigma_{l_2''}\sigma_{l_1'} \tag{3.1}$$

is a glide reflection since we have $l_1' \perp l_3'$ and $l_2'' \perp l_3'$.

Case 2. Suppose l_1 and l_2 are ultra-parallel, with common perpendicular l. Let P_1 and P_2 be the intersection points of l_1 and l_2, respectively, with l.

Then $\sigma_{l_2}\sigma_{l_1} = \sigma_{P_2}\sigma_{P_1}$. Let l'_2 be the straight line through P_2 and perpendicular to l_3 and let l''_2 be the straight line through P_2 and perpendicular to l'_2. Further, let l'_1 be the straight line through P_1 and perpendicular to l'_2 and let l''_1 be the straight line through P_1 and perpendicular to l'_1. See the right in Figure 20. Then we have

$$\sigma_{l_3}\sigma_{l_2}\sigma_{l_1} = \sigma_{l_3}\sigma_{P_2}\sigma_{P_1} = \sigma_{l'_2}(\sigma_{l_3}\sigma_{l''_2}\sigma_{l'_1})\sigma_{l''_1} = \sigma_{l'_2}\sigma_{l'_3}\sigma_{l''_1} \tag{3.2}$$

for some straight lines l'_3 which is perpendicular to l'_2 with foot Q on l'_2. Now let l''_3 be the straight line through Q and perpendicular to l''_1 and let l'''_2 be the the straight line through Q and perpendicular to l''_3. Then

$$\sigma_{l_3}\sigma_{l_2}\sigma_{l_1} = \sigma_{l'_2}\sigma_{l'_3}\sigma_{l''_1} = \sigma_{l''_3}\sigma_{l'''_2}\sigma_{l''_1} \tag{3.3}$$

is a glide reflection with axis $\sigma_{l''_3}$. This completes the proof of Theorem 3.28. □

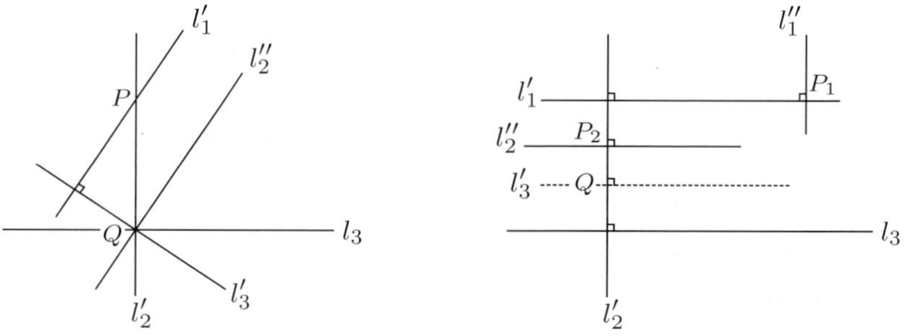

Fig. 20. Figures for the proof of Theorem 3.28

3.8. *Isometries as projective transformations*

Viewed in projective geometry, an isometry ϕ of H^2 extends naturally to a projective transformation Φ of the projective plane P^2 associated to H^2, namely, a bijective self-mapping Φ of P^2 that sends any projective line to a projective line. It is easy to see that Φ sends ideal points to ideal points, thus Φ leaves invariant the ideal boundary ∂H^2 of H^2. Conversely, isometries of H^2 are characterized as those projective transformations of P^2 that leaves invariant both the hyperbolic plane and its ideal boundary.

Given an isometry ϕ of H^2, we find the set of fixed points of its extension Φ in P^2, as follows.

a) If ϕ is a half turn about O, then the set of fixed points of Φ consists of the proper point O and the projective line corresponding to O.

b) If ϕ is a rotation, but not a half turn, about O, then the only fixed point of Φ is the proper point O.

c) If ϕ is a translation with axis l, then the set of fixed points of Φ consists of the two ideal endpoints of l and the ultra-ideal point corresponding to l.

d) If ϕ is a parabolic isometry with ideal center Ω, then the only fixed point of Φ is the ideal point Ω.

e) If ϕ is the reflection in straight line l, then the set of fixed points of Φ consists of the ultra-ideal point corresponding to l and the projective line corresponding to l.

f) If ϕ is a glide reflection with axis l, then the set of fixed points of Φ consists of the two ideal endpoints of l and the ultra-ideal point corresponding to l.

3.9. *Invariant projective lines of* Φ

Let Φ be the extension in P^2 of an isometry ϕ of H^2. To find the set of invariant projective lines of Φ, we resort to the following duality between the fixed points and the invariant projective lines of Φ.

Theorem 3.29. *Let Φ be the extension in P^2 of an isometry ϕ of H^2. Then Φ fixes $Q \in P^2$ if and only if Φ leaves invariant the projective line corresponding to Q.*

Proof. Recall that the projective line $\mathcal{L}$ corresponding to a point Q in P^2 consists of points in P^2 which are incident to Q.

If Φ fixes Q, then Φ sends each point incident to Q to a point incident to Q. Hence Φ leaves $\mathcal{L}$ invariant.

Now suppose Φ leaves invariant a projective line $\mathcal{L}$ which corresponds to point Q in P^2. If Q is a proper point, then ϕ sends each straight line through Q onto a straight line through Q. It follows that $\phi^{-1}(Q)$ lies on each straight line through Q. Hence $\phi^{-1}(Q)$ coincides with Q. If Q is an ideal point, then ϕ sends each straight line ending in Q onto a straight line ending in Q. It follows that $\Phi^{-1}(Q)$ is an ideal endpoint of each straight line ending in Q. Hence $\Phi^{-1}(Q)$ coincides with Q. If Q is an ultra-ideal point corresponding to a straight line l, then ϕ sends each straight line perpendicular to l onto a straight line perpendicular to l. It follows that $\phi^{-1}(l)$ cuts each straight line perpendicular to l. Hence $\phi^{-1}(l)$ coincides with l. $\square$

3.10. *Composition of two orientation preserving isometries other than two translations*

To obtain the composition of two such isometries of H^2, we shall need a common straight line in their so-called principal invariant pencils.

Theorem 3.30. *Any two distinct pencils of straight lines in H^2, which are not both hyperbolic, share a unique common straight line.*

Proof. Suppose $\mathcal{P}_1$ and $\mathcal{P}_2$ are two distinct pencils of straight lines in H^2 where $\mathcal{P}_1$ is either an elliptic pencil with center O_1 or a parabolic one with ideal center Ω_1.

If $\mathcal{P}_2$ is either elliptic with center O_2 or parabolic with ideal center Ω_2, then the desired straight line is that joins O_1 or Ω_1 to O_2 or Ω_2.

If $\mathcal{P}_2$ is a hyperbolic pencil with axis l_2, then the desired straight line is that passes through O_1 or joins Ω_1, and is perpendicular to l_2. □

Definition 3.31 (Principal Invariant Pencil). For a non-identity isometry ϕ of H^2, we define the *principal invariant pencil* of ϕ to be the pencil of straight lines in H^2 that corresponds to
a) the center of rotation, if ϕ is a rotation,
b) the ideal center, if ϕ is an isometry of parabolic type,
c) the axis of translation, if ϕ is a translation, or
d) the axis of glide reflection, if ϕ is a glide reflection.

As an immediate corollary of Theorem 3.30, we have

Theorem 3.32. *Given any two non-identity orientation preserving isometries of H^2 which are not two translations, their principal invariant pencils either coincide or share a unique common straight line.*

By virtue of Theorem 3.32, the composition of two such isometries of H^2 can be performed with ease. For this, suppose ϕ_1 and ϕ_2 are orientation preserving isometries of H^2 which are not two translations. Let l be a common straight line of the two principal invariant pencils, $\mathcal{P}_1$ of ϕ_1 and $\mathcal{P}_2$ of ϕ_2. Then there exist straight lines l_1 in $\mathcal{P}_1$ and l_2 in $\mathcal{P}_2$ such that $\phi_1 = \sigma_l\sigma_{l_1}$ and $\phi_2 = \sigma_{l_2}\sigma_l$. It follows that

$$\phi_2\phi_1 = (\sigma_{l_2}\sigma_l)(\sigma_l\sigma_{l_1}) = \sigma_{l_2}(\sigma_l\sigma_l)\sigma_{l_1} = \sigma_{l_2}\sigma_{l_1}. \tag{3.4}$$

Therefore the isometry $\phi_2\phi_1$ will be a rotation, a translation, or a parabolic isometry, according as whether l_1 and l_2 intersect, are ultra-parallel, or are asymptotically parallel.

3.11. *Composition of two translations*

Suppose τ_1 and τ_2 are translations of H^2 with axes a_1 and a_2, respectively. According to whether a_1 and a_2 intersect, are ultra-parallel, or are asymptotically parallel, there are three cases to consider.

Case 1. The axes a_1 and a_2 are ultra-parallel.

Let l be the common perpendicular of a_1 and a_2. Then there exist straight line $l_1 \perp a_1$ and straight line $l_2 \perp a_2$ such that $\tau_1 = \sigma_l \sigma_{l_1}$ and $\tau_2 = \sigma_{l_2} \sigma_l$. It follows that

$$\tau_2 \tau_1 = \sigma_{l_2} \sigma_{l_1}. \tag{3.5}$$

Therefore the isometry $\tau_2 \tau_1$ will be a rotation, a translation, or a parabolic isometry, according as whether l_1 and l_2 intersect, are ultra-parallel, or are asymptotically parallel.

Case 2. The axes a_1 and a_2 intersect.

Suppose a_1 and a_2 intersect in O. Then there exist point O_1 in a_1 and point O_2 in a_2 such that $\tau_1 = \sigma_O \sigma_{O_1}$ and $\tau_2 = \sigma_{O_2} \sigma_O$. It follows that

$$\tau_2 \tau_1 = \sigma_{O_2} \sigma_{O_1}, \tag{3.6}$$

which is a translation with axis the straight line through O_1 and O_2.

Case 3. The axes a_1 and a_2 are asymptotically parallel.

Let Ω be the common ideal endpoint of a_1 and a_2. Then $\tau_2 \tau_1$ fixes Ω since each of τ_1 and τ_2 does. Hence $\tau_2 \tau_1$ is either a parabolic isometry with Ω its ideal center, or a translation with its axis some straight line ending in Ω. The two subcases are distinguished by whether the sum of the *signed* translation distances of τ_1 and τ_2 vanishes, here we sign the translation distance of τ_1 (resp. τ_2) as positive or negative according as it moves points on its axis towards or away from Ω.

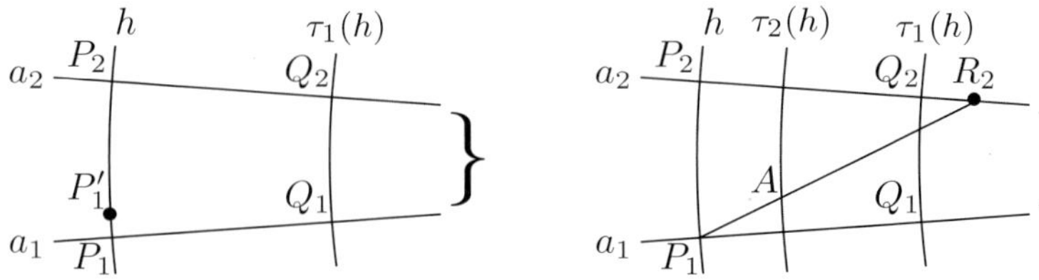

Fig. 21. Case 3: Composition of two translations

Case 3a. The sum of the signed translation distances of τ_1 and τ_2 vanishes.

In this case $\tau_2\tau_1$ sends any horocycle h centered at Ω back to itself. Hence $\tau_2\tau_1$ is parabolic and is determined by its action restricted to h.

To describe the action of the parabolic isometry $\tau_2\tau_1$ on h, let P_1 and P_2 be the intersection points of h with a_1 and a_2, respectively. Set $Q_1 = \tau_1(P_1)$, $Q_2 = \tau_2^{-1}(P_2)$ and $P_1' = \tau_2(Q_1) = \tau_2\tau_1(P_1)$. (See the left in Figure 21.) Then Q_1 and Q_2 lie on horocycle $\tau_1(h)$ while P_1' lies on h with $P_1' \neq P_1$. In fact, P_1' is the unique point on h which is located on the same side of a_2 as Q_1 and is equidistant from a_2 as Q_1. In particular, $\tau_2\tau_1$ shifts points on h through an horocycle arclength which equals the difference of the lengths of the horocycle arcs P_1P_2 and Q_1Q_2.

Case 3b. The sum of the signed translation distances of τ_1 and τ_2 does not vanish.

In this case $\tau_2\tau_1$ sends any horocycle centered at Ω to a different horocycle centered at Ω. Hence $\tau_2\tau_1$ is a translation with axis some straight line ending in Ω and with signed translation distance equal to the sum of those of τ_1 and τ_2.

To find the axis of $\tau_2\tau_1$, let P_1 be the intersection point of h with a_1 and set $Q_1 = \tau_1(P_1)$. (See the right in Figure 21.) Let Q_2 be the intersection point of $\tau_1(h)$ with a_2 and set $R_2 = \tau_2(Q_2)$. Let A be the intersection point of $\tau_2(h)$ with the equidistant curve through P_1 and R_2 and ending in Ω. Let a be the straight line through A and ending in Ω. We claim that a is the axis of translation $\tau_2\tau_1$.

Exercise. Prove that a is indeed the axis of $\tau_2\tau_1$, as claimed.

Thus we have finished composing two translations in all cases.

3.12. *Conjugates of isometries*

Given isometries ϕ and ψ of the plane, the isometry $\psi\phi\psi^{-1}$ is called the *conjugate of ϕ by ψ*. A conjugate of an isometry has the same nature as that of the isometry, as the following theorem establishes.

Theorem 3.33. *Given isometries ϕ and ψ of H^2, the isometries $\psi\phi\psi^{-1}$ and ϕ are of the same nature, namely, they are of the same type and have equal rotation angles or translation distances.*

In particular, a conjugate of a reflection is a reflection (and similarly for a half turn). With this proved first, the rest of the proof of Theorem 3.33 is an exercise in group theory.

Given an isometry ϕ of H^2, the set of fixed points of ϕ in H^2 will be denoted by $\mathrm{Fix}(\phi)$. Furthermore, the set of fixed points of (the extension of) ϕ in $H^2 \cup \partial H^2$ will be denoted by $\overline{\mathrm{Fix}}(\phi)$.

Theorem 3.34. *Given isometries ϕ and ψ of H^2, the sets of fixed points of $\psi\phi\psi^{-1}$ in H^2 and in $H^2 \cup \partial H^2$ equal respectively the images under ψ of those of ϕ, namely,*

$$\mathrm{Fix}(\psi\phi\psi^{-1}) = \psi(\mathrm{Fix}(\phi)), \quad \overline{\mathrm{Fix}}(\psi\phi\psi^{-1}) = \psi(\overline{\mathrm{Fix}}(\phi)). \tag{3.7}$$

Theorem 3.35. *Given two non-identity isometries ϕ and ψ of H^2, they commute, namely $\psi\phi = \psi\phi$, if and only if either they are reflections in perpendicular axes, or they are a reflection and a half-turn about a point on the axis of the reflection, or they are rotations about the same center, or they have the same nonempty set of fixed points in ∂H^2.*

Proof. The "if" part is easy to verify; we prove the "only if" part only. Suppose ϕ and ψ are non-identity isometries of H^2 such that $\psi\phi = \phi\psi$. This is equivalent to that $\psi\phi\psi^{-1} = \phi$. By Theorem 3.34, $\psi(\mathrm{Fix}(\phi)) = \mathrm{Fix}(\phi)$ and $\psi(\overline{\mathrm{Fix}}(\phi)) = \overline{\mathrm{Fix}}(\phi)$. And the same as above is true with ϕ and ψ interchanged. A careful type-by-type discussion (Exercise!) will give the list of cases as stated in the conclusion of the theorem. □

3.13. *The orthic triangle*

Given an acute triangle, the triangle formed by the feet of its three altitudes is called its *orthic triangle*. We leave the proof of the following theorem as a challenging exercise.

Theorem 3.36 (Orthocenter-Incenter). *The orthocenter of an acute triangle coincides with the incenter of its orthic triangle.*

Exercise. Try to obtain a theorem similar to Theorem 3.36 for obtuse triangles.

4. Hyperbolic Trigonometry Derived from Isometries

Throughout this section we shall fix a hyperbolic plane $H^2(k)$ which has area factor k^2 where $k > 0$, and take as angle measure the usual radian measure. We shall first obtain several identities of isometries associated with a triangle in a neutral plane, and then evaluate these identities in the upper half-plane model $\mathbf{U}^2$ of $H^2(1)$ as identities of 2×2 real matrices of determinant 1 and derive as consequence certain trigonometric laws. Some of the basic facts about the upper half-plane model of $H^2(1)$ will be outlined, with details left as exercises.

Recall that by a directed angle we mean an angle with one of its two arms specified as initial and the other terminal. For a rotation of the plane to be fully determined, we need to specify the center of rotation as well as the directed angle through which the rotation rotates points in the plane. For measuring directed angles we assume the plane is oriented, so we are able to speak of anticlockwise measure. Thus a directed angle in the plane has a well-defined anticlockwise radian measure $\theta \in \mathbf{R}/2\pi\mathbb{Z}$.

Notation. For a point A in the plane and $\theta \in \mathbf{R}/2\pi\mathbb{Z}$, we denote by $\rho_A(\theta)$ the rotation about A through directed angle of anticlockwise radian measure θ. For a directed straight line l in the plane and $\lambda \in \mathbf{R}$, we denote by $\tau_l(\lambda)$ the translation along directed axis l through signed distance λ. For a triangle ABC in the plane, we always write $a = |BC|$, $b = |CA|$, $c = |AB|$ and denote the anticlockwise radian measures of the directed interior angles at A, B and C by α, β and γ, respectively; that is, $\alpha = \measuredangle(AB, AC)$, $\beta = \measuredangle(BC, BA)$ and $\gamma = \measuredangle(CA, CB)$. By this convention, if ABC is a *clockwise* triangle in the plane then $\alpha, \beta, \gamma \in (-\pi, 0)$ modulo 2π.

We continue to use our notation for reflections in straight lines and half-turns: for a straight line l in the plane, the reflection in l is denoted by σ_l, and for a point P in the plane, the half-turn about P is denoted by σ_P; thus $\sigma_P = \rho_P(\pi)$.

4.1. *Some identities of isometries of a neutral plane*

We first obtain the following conjugation identities of isometries of a neutral plane.

Lemma 4.1 (Conjugation of Rotations). *For a directed straight line AB with $|AB| = \lambda$ and $\theta \in \mathbf{R}/2\pi\mathbb{Z}$, we have*

$$\rho_B(\theta) = \tau_{AB}(\lambda)\,\rho_A(\theta)\,\tau_{AB}(-\lambda), \tag{4.1}$$

or, in an equivalent form,

$$\rho_B(\theta)\,\tau_{AB}(\lambda) = \tau_{AB}(\lambda)\,\rho_A(\theta). \tag{4.2}$$

Lemma 4.2 (Conjugation of Translations). *For a directed angle $\angle(BA, BC)$ of anticlockwise radian measure $\theta \in \mathbf{R}/2\pi\mathbb{Z}$ and $\lambda \in \mathbf{R}$, we have*

$$\tau_{BC}(\lambda) = \rho_B(\theta)\,\tau_{BA}(\lambda)\,\rho_B(-\theta), \tag{4.3}$$

or, in an equivalent form,

$$\tau_{BC}(\lambda)\,\rho_B(\theta) = \rho_B(\theta)\,\tau_{BA}(\lambda). \tag{4.4}$$

The following identity is obtained by tracing a triangle with an arrow.

Theorem 4.3 (Tracing Identity). *For a triangle ABC in the plane, we have*

$$\rho_A(\pi-\alpha)\,\tau_{CA}(b)\,\rho_C(\pi-\gamma)\,\tau_{BC}(a)\,\rho_B(\pi-\beta)\,\tau_{AB}(c) = \mathrm{id}. \tag{4.5}$$

Proof. It is easily checked that the ray AB is pointwise fixed by the composition in that order of the six orientation preserving isometries of the plane appeared on the left side of equation (4.5). Therefore the composition of the six isometries in that order gives the identity isometry. □

We obtain identity (4.6) below from the tracing identity (4.5) by performing appropriate conjugations so that the resulting translations have the same axis AB and the resulting rotations have the same center B.

Theorem 4.4 (Conjugated Tracing Identity). *For a triangle ABC in the plane, we have*

$$\rho_B(\pi-\beta)\,\tau_{AB}(a)\,\rho_B(\pi-\gamma)\,\tau_{AB}(b)\,\rho_B(\pi-\alpha)\,\tau_{AB}(c) = \mathrm{id}. \tag{4.6}$$

Proof. First, we have

$$\begin{aligned}
\rho_C(\pi-\gamma)\,\tau_{BC}(a) &= \tau_{BC}(a)\,\rho_B(\pi-\gamma)\\
&= \rho_B(\pi-\beta)\,\tau_{AB}(a)\,\rho_B(\beta-\pi)\,\rho_B(\pi-\gamma)\\
&= \rho_B(\pi-\beta)\,\tau_{AB}(a)\,\rho_B(\beta-\gamma); \qquad (4.7)\\
\rho_A(\pi-\alpha)\,\tau_{CA}(b) &= \tau_{AB}(b)\,\rho_A(\pi-\alpha)\\
&= \tau_{AB}(b)\,\tau_{AB}(-c)\,\rho_B(\pi-\alpha)\,\tau_{AB}(c)\\
&= \tau_{AB}(b-c)\,\rho_B(\pi-\alpha)\,\tau_{AB}(c). \qquad (4.8)
\end{aligned}$$

Substituting (4.7) and (4.8) into (4.5), we obtain

$$\tau_{AB}(b-c)\,\rho_B(\pi-\alpha)\,\tau_{AB}(c)\rho_B(\pi-\beta)\,\tau_{AB}(a)\,\rho_B(\beta-\gamma)\rho_B(\pi-\beta)\,\tau_{AB}(c) = \mathrm{id},$$

from which follows (4.6), the desired identity. □

With similar proofs, we obtain identities (4.9)–(4.11) below.

Theorem 4.5 (Holonomy Identity). *For a triangle ABC in the plane, we have*

$$\tau_{CA}(b)\,\tau_{BC}(a)\,\tau_{AB}(c) \;=\; \rho_A(\alpha+\beta+\gamma-\pi). \tag{4.9}$$

Theorem 4.6. *For a triangle ABC in the plane with X, Y and Z the midpoints of sides BC, CA and AB, respectively, we have*

$$\sigma_Z\sigma_Y\sigma_X \;=\; \rho_B(\alpha+\beta+\gamma). \tag{4.10}$$

Theorem 4.7. *For a triangle ABC in the plane, we have*

$$\rho_C(\pi-\gamma)\,\rho_B(\pi-\beta)\,\rho_A(\pi-\alpha) \;=\; \tau_{AC}(a+b+c). \tag{4.11}$$

The identities (4.12) and (4.13) below come relatively easy.

Theorem 4.8. *For a triangle ABC in the plane, we have*

$$\rho_A(2\alpha)\,\rho_B(2\beta)\,\rho_C(2\gamma) = \mathrm{id}. \tag{4.12}$$

Proof. The desired identity follows from the obvious identity

$$\sigma_{CA}\sigma_{AB}\sigma_{AB}\sigma_{BC}\sigma_{BC}\sigma_{CA} = \mathrm{id}$$

by observing that $\sigma_{CA}\sigma_{AB} = \rho_A(2\alpha)$ and so on. □

Theorem 4.9. *For a triangle ABC in the plane, we have*

$$\tau_{AB}(2c)\,\tau_{CA}(2b)\,\tau_{BC}(2a) = \mathrm{id}. \tag{4.13}$$

Proof. The desired identity follows from the obvious identity

$$\sigma_B\sigma_A\sigma_A\sigma_C\sigma_C\sigma_B = \mathrm{id}$$

by observing that $\sigma_C\sigma_B = \tau_{BC}(2a)$ and so on. □

4.2. *Some trigonometric formulas in $H^2(k)$*

In this subsection we simply collect some trigonometric formulas in $H^2(k)$ and include no proofs.

Angle of parallelism. In $H^2(k)$, the fundamental formula of Bolyai-Lobachevsky for the angle of parallelism function $\Pi : (0,\infty) \to (0,\pi/2)$ takes the form

$$\Pi(x) = 2\arctan(e^{x/k}), \tag{4.14}$$

from which follow the formulas (4.15)–(4.17) below relating the trigonometric functions of $\Pi(x)$ to the hyperbolic functions of x/k:

$$\sin\Pi(x) = \operatorname{sech}(x/k) = (\cosh(x/k))^{-1}; \tag{4.15}$$

$$\cos\Pi(x) = \tanh(x/k); \tag{4.16}$$

$$\tan\Pi(x) = \operatorname{csch}(x/k) = (\sinh(x/k))^{-1}. \tag{4.17}$$

Lengths of circular arcs and equidistant arcs. The length of equidistant arc distance d away and subtending a core segment of length l is

$$l' = l\cosh(d/k). \tag{4.18}$$

The length of the circular arc of radius r subtending an angle of radian measure θ is

$$\theta' = \theta\sinh(r/k). \tag{4.19}$$

Therefore the circumference of a circle of radius r is

$$C = 2\pi k\sinh(r/k). \tag{4.20}$$

Laws of Cosines, of Cotangents, and of Sines. For a triangle in $H^2(k)$ with side-lengths a, b, c and measures of respective opposite angles α, β, γ, there hold two Laws of Cosines, (LC1) and (LC2), Law of Cotangents (LCot), and Law of Sines, (LS), as follows.

(LC1) The First Law of Cosines:

$$\cosh(c/k) = \cosh(a/k)\cosh(b/k) - \sinh(a/k)\sinh(b/k)\cos\gamma. \tag{4.21}$$

(LC2) The Second Law of Cosines:

$$\cos\gamma = -\cos\alpha\cos\beta + \sin\alpha\sin\beta\cosh(c/k). \tag{4.22}$$

(LCot) The Law of Cotangents:

$$\cot\alpha\sin\gamma + \cosh(b/k)\cos\gamma = \coth(a/k)\sinh(b/k). \tag{4.23}$$

(LS) The Law of Sines:

$$\frac{\sinh(a/k)}{\sin\alpha} = \frac{\sinh(b/k)}{\sin\beta} = \frac{\sinh(c/k)}{\sin\gamma}. \tag{4.24}$$

Lobachevsky's trigonometric formulas. In (§36 of) [17], Lobachevsky made use of only trigonometric functions and expressed his trigonometric formulas for triangles in $H^2(k)$ as:

$$\sin\alpha\tan\Pi(a) = \sin\beta\tan\Pi(b), \tag{4.25}$$

$$\cos\alpha\cos\Pi(b)\cos\Pi(c) + \csc\Pi(a)\sin\Pi(b)\sin\Pi(c) = 1, \tag{4.26}$$

$$\cot\alpha\sin\gamma\sin\Pi(b) + \cos\gamma = \sec\Pi(a)\cos\Pi(b), \tag{4.27}$$

$$\cos\alpha + \cos\beta\cos\gamma = \sin\beta\sin\gamma\csc\Pi(a). \tag{4.28}$$

By formulas (4.15)–(4.17), it is easy to check that formulas (4.25)–(4.28) above are respectively equivalent to the Law of Sines, the First Law of Cosines, the Law of Cotangents and the Second Law of Cosines.

Trigonometric formulas for a right triangle. In the following theorem we recollect the trigonometric formulas Lobachevsky obtained in (§35 of) [17] for right triangles.

Theorem 4.10 (Lobachevsky). *For a right triangle in $H^2(1)$ with side-lengths a, b, c and measures of respective opposite angles $\alpha, \beta, \pi/2$, we have*

$$\cosh c = \cosh a\cosh b, \quad \cosh c = \cot\alpha\cot\beta; \tag{4.29}$$

$$\sin\alpha = \sinh a/\sinh c, \quad \sin\alpha = \cos\beta/\cosh b; \tag{4.30}$$

$$\sin\beta = \sinh b/\sinh c, \quad \sin\beta = \cos\alpha/\cosh a; \tag{4.31}$$

$$\tanh a = \tanh c\cos\beta, \quad \tanh a = \sinh b\tan\alpha; \tag{4.32}$$

$$\tanh b = \tanh c\cos\alpha, \quad \tanh b = \sinh a\tan\beta. \tag{4.33}$$

4.3. *Upper half-plane model* $\mathbf{U}^2$ *for hyperbolic plane* $H^2(1)$

We shall need to evaluate some of the identities of isometries obtained in the previous subsection in an analytic model of a hyperbolic plane. The upper half-plane model $\mathbf{U}^2$ of the hyperbolic plane $H^2(1)$ is a convenient choice. This is the upper half-plane

$$\mathbf{U}^2 = \{(x, y) \in \mathbf{R}^2 \mid y > 0\} = \{z \in \mathbf{C} \mid \operatorname{Im} z > 0\} \tag{4.34}$$

equipped with Riemannian metric of constant Gaussian curvature -1:

$$ds^2 = y^{-2}(dx^2 + dy^2) = (\operatorname{Im} z)^{-2}|dz|^2. \tag{4.35}$$

The ideal boundary $\partial\mathbf{U}^2$, namely, the set of all the ideal points, consists of the x-axis and ∞ (with the usual operations as used in complex analysis): $\partial\mathbf{U}^2 = \mathbf{R} \cup \{\infty\}$. It is easy to observe the fact that the Riemannian metric is conformal, that is, the angle between any two intersecting hyperbolic curves equals the angle between their Euclidean representatives.

The hyperbolic straight lines (h-lines, for short) in $\mathbf{U}^2$ are the complete geodesics in the Riemannian metric, which are the Euclidean vertical upper half-lines $l^+(x_0)$: for $x_0 \in \mathbf{R}$,

$$l^+(x_0) = \{(x_0, y) \in \mathbf{R}^2 \mid y > 0\}, \tag{4.36}$$

$l^+(0)$ being the upper y-axis, and the Euclidean upper half-circles $C^+(x_0; r)$ centered in x-axis: for $x_0 \in \mathbf{R}$ and $r > 0$,

$$C^+(x_0; r) = \{(x, y) \in \mathbf{R}^2 \mid (x - x_0)^2 + y^2 = r^2,\ y > 0\}. \tag{4.37}$$

The hyperbolic distance between two points in h-line $l^+(0)$ is given by

$$d_{\mathbf{U}^2}(y_1 i, y_2 i) = |\ln(y_2/y_1)| = |\ln y_2 - \ln y_1|. \tag{4.38}$$

Thus the h-line $l^+(0)$ as a geodesic of unit speed is parametrized as:

$$z(t) = e^t i, \quad t \in \mathbf{R}, \tag{4.39}$$

with $z(0) = i$. The reflection σ_0 in h-line $l^+(0)$ is given by:

$$\sigma_0(z) = \sigma_0(x + yi) = -x + yi = -\bar{z}, \qquad z = x + yi \in \mathbf{U}^2. \tag{4.40}$$

In general, the reflection σ_{x_0} in h-line $l^+(x_0)$, $x_0 \in \mathbf{R}$ is given by:

$$\sigma_{x_0}(x + yi) = -x + 2x_0 + yi, \quad x + yi \in \mathbf{U}^2. \tag{4.41}$$

Therefore the parabolic isometries with ideal center ∞ are of the form $\hbar_{2x_0}$: for $x_0 \in \mathbf{R}$,

$$\hbar_{2x_0}(z) = \sigma_{x_0}\sigma_0(z) = z + 2x_0, \quad z \in \mathbf{U}^2. \tag{4.42}$$

The reflection $\sigma_{0;r}$ in h-line $C^+(0; r)$ is given by

$$\sigma_{0;r}(z) = r^2 z|z|^{-2} = r^2(\bar{z})^{-1}, \quad z \in \mathbf{U}^2 \tag{4.43}$$

as can be checked for points in the upper y-axis. Thus translations along h-line $l^+(0)$ are of the form: for $r > 0$ with $r \neq 1$,

$$\phi(z) = \sigma_{0;r}\sigma_{0;1}(z) = r^2 z, \quad z \in \mathbf{U}^2, \tag{4.44}$$

with signed translation distance (towards ∞) $\ln(r^2) = 2\ln r \in \mathbf{R}$.

The reflection $\sigma_{x_0;r}$ in h-line $C^+(x_0; r)$ is obtained by conjugation $\sigma_{x_0;r} = \hbar_{x_0}\sigma_{0;r}\hbar_{-x_0}$:

$$\sigma_{x_0;r}(z) = \hbar_{x_0}\sigma_{0;r}\hbar_{-x_0}(z) = r^2/(\bar{z} - x_0) + x_0, \quad z \in \mathbf{U}^2. \tag{4.45}$$

Notation of h-lines. The directed h-line with ideal endpoints $x_1, x_2 \in \mathbf{R} \cup \{\infty\}$ and with direction from x_1 to x_2 will be denoted by $\vec{l}_{[x_1,x_2]}$, whereas the undirected h-line will be denoted by either $l_{[x_1,x_2]}$ or $l_{[x_2,x_1]}$.

The directed h-line through i obtained from the directed h-line $\vec{l}_{[\infty,0]}$ by rotating about i through directed angle θ mod 2π is

$$\vec{l}_{[-\cot(\theta/2),\tan(\theta/2)]}, \tag{4.46}$$

represented by the upper half-circle $C^+(-\cot\theta; |\csc\theta|)$. See Figure 22.

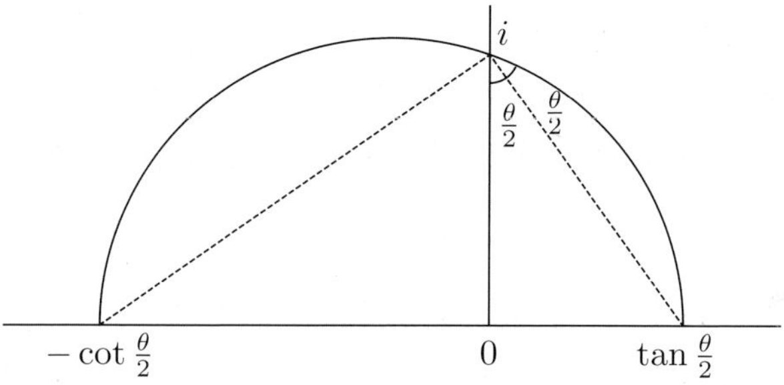

Fig. 22. Geodesic $l_{[-\cot(\theta/2),\tan(\theta/2)]}$

Fractional linear transformations of the complex projective line $\mathbf{C}P^1 = \mathbf{C}\cup\{\infty\}$ (or the extended complex plane) are the projective transformations of $\mathbf{C}P^1$, which are of the form (where $a, b, c, d \in \mathbf{C}$ with $ad - bc \neq 0$)

$$\begin{bmatrix} z_1 \\ z_2 \end{bmatrix} \mapsto \begin{bmatrix} a & b \\ c & d \end{bmatrix} \begin{bmatrix} z_1 \\ z_2 \end{bmatrix} = \begin{bmatrix} az_1 + bz_2 \\ cz_1 + dz_2 \end{bmatrix}, \tag{4.47}$$

using homogeneous coordinates $[z_1, z_2]$ where $(z_1, z_2) \in \mathbf{C}^2\backslash\{(0,0)\}$. Here we use the bracket matrix to mean ± 1 times the parenthesis matrix. Writing $z = z_1/z_2$, the above projective transformations give rise to the fractional linear transformations of the form

$$z = \frac{z_1}{z_2} \mapsto \frac{az_1 + bz_2}{cz_1 + dz_2} = \frac{az + b}{cz + d}, \quad z \in \mathbf{U}^2. \tag{4.48}$$

The fractional linear transformations that leave both of $\mathbf{U}^2$ and $\partial\mathbf{U}^2$ invariant are those such that $a, b, c, d \in \mathbf{R}$ and $ad - bc > 0$ (Exercise). Normalized by requiring $ad - bc = 1$, all such fractional linear transformations are in one-to-one correspondence with pairs of $\mathrm{SL}(2, \mathbf{R})$ matrices:

$$\left\{ \begin{bmatrix} a & b \\ c & d \end{bmatrix} = \pm \begin{pmatrix} a & b \\ c & d \end{pmatrix} \;\middle|\; a, b, c, d \in \mathbf{R},\ ad - bc = 1 \right\}. \tag{4.49}$$

This gives the quotient group $\mathrm{PSL}(2, \mathbf{R}) = \mathrm{SL}(2, \mathbf{R})/\{\pm I\}$, where I is the identity 2×2 matrix and

$$\mathrm{SL}(2, \mathbf{R}) = \{A \in \mathrm{Mat}(2, \mathbf{R}) \mid \det(A) = 1\}. \tag{4.50}$$

Let us denote the full group of isometries of $\mathbf{U}^2$ by $\mathrm{Isom}(\mathbf{U}^2)$ and the group of all orientation preserving isometries of $\mathbf{U}^2$ by $\mathrm{Isom}^+(\mathbf{U}^2)$. It follows from the above discussions that $\mathrm{Isom}^+(\mathbf{U}^2)$ is isomorphic to $\mathrm{PSL}(2,\mathbf{R})$.

The orientation reversing isometries of $\mathbf{U}^2$ are of the form

$$z \mapsto \frac{a(-\bar{z})+b}{c(-\bar{z})+d}, \quad z \in \mathbf{U}^2, \tag{4.51}$$

where $a,b,c,d \in \mathbf{R}$ and $ad-bc=1$, obtained from the orientation preserving isometries in (4.48) composed with σ_0.

Exercise. The full group $\mathrm{Isom}(\mathbf{U}^2)$ of isometries of $\mathbf{U}^2$ is isomorphic to

$$\mathrm{PGL}(2,\mathbf{R}) \cong \{A \in \mathrm{Mat}(2,\mathbf{R}) \mid \det(A) = \pm 1\}/\{\pm I\}. \tag{4.52}$$

4.4. *Matrices of certain isometries of* $\mathbf{U}^2$

The bracket matrices representing translations along $l^+(0)$ and rotations about points in $l^+(0)$ are found as follows.

Lemma 4.11. *The translation $\tau_{[0,\infty]}(d)$ along the directed geodesic $\vec{l}_{[0,\infty]}$ in $\mathbf{U}^2$ through signed distance $d \in \mathbf{R}$ is given by bracket matrix*

$$\tau_{[0,\infty]}(d) = \begin{bmatrix} e^{d/2} & 0 \\ 0 & e^{-d/2} \end{bmatrix}. \tag{4.53}$$

Lemma 4.12. *The anticlockwise rotation $\rho_i(\theta)$ about $i \in \mathbf{U}^2$ through directed angle $\theta \in \mathbf{R}/2\pi\mathbb{Z}$ is given by bracket matrix*

$$\rho_i(\theta) = \begin{bmatrix} \cos(\theta/2) & \sin(\theta/2) \\ -\sin(\theta/2) & \cos(\theta/2) \end{bmatrix}. \tag{4.54}$$

We then obtain the bracket matrix of the rotation $\rho_{ie^l}(\theta)$ from the conjugation relation $\rho_{ie^l}(\theta) = \tau_{[0,\infty]}(l)\,\rho_i(\theta)\,\tau_{[0,\infty]}(-l)$.

Lemma 4.13. *The anticlockwise rotation $\rho_{ie^l}(\theta)$ about $ie^l \in \mathbf{U}^2$, $l \in \mathbf{R}$ through directed angle $\theta \in \mathbf{R}/2\pi\mathbb{Z}$ is given by bracket matrix*

$$\rho_{ie^l}(\theta) = \begin{bmatrix} \cos\frac{\theta}{2} & e^l \sin\frac{\theta}{2} \\ -e^{-l}\sin\frac{\theta}{2} & \cos\frac{\theta}{2} \end{bmatrix}. \tag{4.55}$$

4.5. *Trigonometric laws via identities of isometries*

In the upper half-plane model $\mathbf{U}^2$ of hyperbolic plane $H^2(1)$, an identity of isometries of $H^2(1)$ will become an identity of $\mathrm{SL}(2,\mathbf{R})$ matrices, with a resulting plus or minus sign, which can be explicitly determined.

Let us take identity (4.12), which is $\rho_A(2\alpha)\,\rho_B(2\beta)\,\rho_C(2\gamma) = \mathrm{id}$, as an example and work out the detail. For this, let us consider an anticlockwise triangle ABC. We may assume that $A = i$ and $C = ie^b$. Then, by conjugation relations, we find $\rho_B(2\beta)$ as

$$\rho_B(2\beta) = \rho_i(-\alpha)\rho_{ie^c}(2\beta)\rho_i(\alpha). \tag{4.56}$$

Thus identity (4.12) becomes

$$\rho_i(\alpha)\rho_{ie^c}(2\beta)\rho_i(\alpha)\rho_{ie^b}(2\gamma) = \mathrm{id}, \tag{4.57}$$

or, in an equivalent form,

$$\rho_i(\alpha)\rho_{ie^c}(2\beta) = \rho_{ie^b}(-2\gamma)\rho_i(-\alpha). \tag{4.58}$$

Replacing isometries in (4.58) by $\mathrm{SL}(2,\mathbf{R})$ matrices, we get

$$\begin{pmatrix} \cos\frac{\alpha}{2} & \sin\frac{\alpha}{2} \\ -\sin\frac{\alpha}{2} & \cos\frac{\alpha}{2} \end{pmatrix} \begin{pmatrix} \cos\beta & e^c\sin\beta \\ -e^{-c}\sin\beta & \cos\beta \end{pmatrix} = \pm \begin{pmatrix} \cos\gamma & -e^b\sin\gamma \\ e^{-b}\sin\gamma & \cos\gamma \end{pmatrix} \begin{pmatrix} \cos\frac{\alpha}{2} & -\sin\frac{\alpha}{2} \\ \sin\frac{\alpha}{2} & \cos\frac{\alpha}{2} \end{pmatrix}. \tag{4.59}$$

The $\pm$ sign on the right side of (4.59) must be "$-$" by the requirement of equality (4.61) below, or by simply considering $\alpha \to 0$ and $\beta \to 0$.

Working out the product matrices on both sides of (4.59) and equating respectively the sum/difference of the corresponding entries $(1,1) \pm (2,2)$ and $(1,2) \pm (2,1)$, we get the following four identities:

$$\cos\frac{\alpha}{2}\cos\beta - \sin\frac{\alpha}{2}\sin\beta\cosh c = \sin\frac{\alpha}{2}\sin\gamma\cosh b - \cos\frac{\alpha}{2}\cos\gamma, \tag{4.60}$$

$$\sin\frac{\alpha}{2}\sin\beta\sinh c = \sin\frac{\alpha}{2}\sin\gamma\sinh b, \tag{4.61}$$

$$\cos\frac{\alpha}{2}\sin\beta\sinh c = \cos\frac{\alpha}{2}\sin\gamma\sinh b, \tag{4.62}$$

$$\sin\frac{\alpha}{2}\cos\beta + \cos\frac{\alpha}{2}\sin\beta\cosh c = \sin\frac{\alpha}{2}\cos\gamma + \cos\frac{\alpha}{2}\sin\gamma\cosh b. \tag{4.63}$$

Note that each of (4.61) and (4.62) gives the Law of Sines, in the form $\sin\beta\sinh c = \sin\gamma\sinh b$. From (4.60) and (4.63) combined, one can easily obtain the Second Law of Cosines, in the form

$$\sin\alpha\sin\beta\cosh c = \cos\gamma + \cos\alpha\cos\beta, \tag{4.64}$$

as well as the Law of Cotangents, in the equivalent form

$$\sin\alpha\cos\gamma = \sin\beta\cosh c - \cos\alpha\sin\gamma\cosh b. \tag{4.65}$$

From identity (4.11) by evaluating isometries of $H^2(1)$ by $\mathrm{SL}(2,\mathbf{R})$ matrices and manipulating similarly, we obtain (4.66) and (4.67) below.

Proposition 4.14. *For a triangle ABC in $H^2(1)$ we have*

$$\sinh c \sin\frac{\alpha}{2}\cos\frac{\beta}{2} = \cos\frac{\gamma}{2}\sinh(s-b), \tag{4.66}$$

$$\sinh c \cos\frac{\alpha}{2}\cos\frac{\beta}{2} = \sin\frac{\gamma}{2}\sinh s, \tag{4.67}$$

where $s = \frac{1}{2}(a+b+c)$ is the semi-perimeter of the triangle.

As a corollary of (4.67), we obtain an extended law of sines as follows.

Theorem 4.15. *For a triangle ABC in $H^2(1)$ we have*

$$\frac{\sinh a}{\sin\alpha} = \frac{\sinh b}{\sin\beta} = \frac{\sinh c}{\sin\gamma} = \frac{\sinh\frac{1}{2}(a+b+c)}{2\cos\frac{1}{2}\alpha\,\cos\frac{1}{2}\beta\,\cos\frac{1}{2}\gamma}. \tag{4.68}$$

One can derive from Proposition 4.14 the half-angle formulas below.

Theorem 4.16. *For a triangle ABC in $H^2(1)$ we have*

$$\sin\frac{\alpha}{2} = \sqrt{\frac{\sinh(s-b)\sinh(s-c)}{\sinh b\,\sinh c}}, \tag{4.69}$$

$$\cos\frac{\alpha}{2} = \sqrt{\frac{\sinh s\,\sinh(s-a)}{\sinh b\,\sinh c}}, \tag{4.70}$$

$$\tan\frac{\alpha}{2} = \sqrt{\frac{\sinh(s-b)\sinh(s-c)}{\sinh s\,\sinh(s-a)}}, \tag{4.71}$$

where $s = \frac{1}{2}(a+b+c)$ is the semi-perimeter of the triangle.

From identity (4.6) by evaluating isometries of $H^2(1)$ by $\mathrm{SL}(2,\mathbf{R})$ matrices, we obtain the Delambre-Gauss formulas for triangles in $H^2(1)$.

Theorem 4.17 (Delambre-Gauss Formulas). *For a triangle ABC in $H^2(1)$, we have the following Delambre-Gauss formulas:*

$$\cosh\tfrac{1}{2}(a+b)\,\sin\tfrac{1}{2}\gamma = \cos\tfrac{1}{2}(\alpha+\beta)\,\cosh\tfrac{1}{2}c, \tag{4.72}$$

$$\sinh\tfrac{1}{2}(a+b)\,\sin\tfrac{1}{2}\gamma = \cos\tfrac{1}{2}(\alpha-\beta)\,\sinh\tfrac{1}{2}c, \tag{4.73}$$

$$\cosh\tfrac{1}{2}(a-b)\,\cos\tfrac{1}{2}\gamma = \sin\tfrac{1}{2}(\alpha+\beta)\,\cosh\tfrac{1}{2}c, \tag{4.74}$$

$$\sinh\tfrac{1}{2}(a-b)\,\cos\tfrac{1}{2}\gamma = \sin\tfrac{1}{2}(\alpha-\beta)\,\sinh\tfrac{1}{2}c. \tag{4.75}$$

Note that the original Delambre-Gauss formulas for spherical triangles can be recovered from formulas (4.72)–(4.75) above by changing all the hyperbolic functions appeared to the corresponding trigonometric ones.

We leave it to the reader to evaluate (4.9), (4.10) and (4.13) as identities of $\mathrm{SL}(2,\mathbf{R})$ matrices and obtain more trigonometric formulas.

4.6. *Suggested further readings*

We refer the interested reader to Casey [6] for hundreds of trigonometric formulas for spherical triangles. Following a well-known philosophy, one may translate these formulas to obtain formulas for hyperbolic triangles in $H^2(1)$ by changing all the trigonometric functions of side-lengths appeared in the formulas to the corresponding hyperbolic functions.

Elucidating ideas of Schilling, Fenchel [10] obtained formulas, such as the Laws of Cosines, of Sines, and of Cotangents, in complex arguments for oriented right-angled hexagons in the hyperbolic space $H^3(1)$. In doing so he derived various formulas in a unified way for a variety of polygons, convex or self-intersecting, in $H^2(1)$, such as triangles, quadrilaterals with two right angles, pentagons with four right angles, and right-angled hexagons.

For accounts and studies of the history of discovery and developments of non-Euclidean geometry, we recommend Bonola [3], Sommerville [20], Greenberg [11] and Rosenfeld [19]. For axiomatic treatments see Hilbert [14] and Hartshorne [12]. For many facts on hyperbolic geometry obtained using the upper half-space model see Beardon [2].

Acknowledgments

These notes grow out of a series of four lectures delivered by the author in June, 2010 in the summer school "Geometry, Topology and Dynamics of Character Varieties" held by the Institute for Mathematical Sciences of the National University of Singapore. The author would like to thank Bill Goldman, Caroline Series and Ser Peow Tan, the chief organizers of the summer school, to give him the chance to deliver the lectures. Thanks also go to the graduate participants of the summer school who gave the author useful feedbacks, and to the anonymous referee who read the first version of the notes carefully and gave helpful suggestions for revision.

The author was supported by NSFC grant no. 10871139 while preparing and revising the notes.

References

1. W. H. Barker, R. Howe, *Continuous Symmetry: From Euclid to Kline*, Amer. Math. Society, Providence, RI, 2007.
2. A. F. Beardon, *The Geometry of Discrete Groups*, Springer-Verlag, New York, 1983.
3. R. Bonola, *Non-Euclidean Geometry: A Critical and Historical Study of Its Development*, Translated from the Italian by H. S. Carslaw, with a supplement containing the Dr. G. B. Halstead translation of *The Science of Absolute*

Space by John Bolyai and *The Theory of Parallels* by Nicholas Lobachevski, Dover Publications, New York, 1955.

4. J. Bolyai, *Appendix to Tantamen: The Science of Absolute Space*, Translated by G. B. Halstead, reprinted in R. Bonola's book [3].
5. H. S. Carslaw, *Elements of Non-Euclidean Plane Geometry and Trigonometry*, Longmans-Green, London, 1916.
6. J. Casey, *A Treatise on Spherical Trigonometry*, Longmans-Green, London, 1889.
7. H. S. M. Coxeter, *Introduction to Geometry*, Second edition, Wiley, New York, 1969.
8. H. S. M. Coxeter, *The Real Projective Plane*, McGraw-Hill, New York, 1949.
9. H. S. M. Coxeter, *Non-Euclidean Geometry*, Sixth edition, Math. Assoc. Amer., 1998.
10. W. Fenchel, *Elementary Geometry in Hyperbolic Space*, Walter de Gruyter, Berlin, 1989.
11. M. J. Greenberg, *Euclidean and Non-Euclidean Geometries: Development and History*, Third edition, W. H. Freeman, New York, 1993.
12. R. Hartshorne, *Geometry: Euclid and Beyond*, Springer-Verlag, New York, 2000.
13. T. L. Heath, *The Thirteen Books of Euclid's Elements, Volume I: Introduction and Books I, II*, Cambridge University Press, Cambridge, 1908.
14. D. Hilbert, *Foundations of Geometry*, Translated from the German by L. Unger, Open Court Publishing, Chicago, 1902.
15. D. Hilbert, *Neue Begründung der Bolyai-Lobatschefskyschen Geometrie*, Math. Ann., **57** (1903), 137–150.
16. H. Liebmann, *Nichteuklidische Geometrie*, Third edition, de Gruyter, Berlin-Leipzig, 1923.
17. N. Lobachevsky, *Geometric Researches on the Theory of Parallels*, Translated by G. B. Halstead, Open Court Publishing, Chicago, 1914.
18. G. E. Martin, *Transformation Geometry*, Springer-Verlag, New York, 1982.
19. B. A. Rosenfeld, *A History of Non-Euclidean Geometry: Evolution of the Concept of a Geometric Space*, Translated by A. Shenitzer, Springer-Verlag, New York, 1988.
20. D. M. Y. Sommerville, *Elements of Non-Euclidean Geometry*, Open Court Publishing, Chicago, 1919.
21. W. E. Wolfe, *Introduction to Non-Euclidean Geometry*, The Dryden Press, New York, 1945.

HYPERBOLIC STRUCTURES ON SURFACES

Javier Aramayona
School of Mathematics, Statistics and Applied Mathematics
National University of Ireland, Galway, Ireland
Javier.Aramayona@nuigalway.ie

We give a brief introduction to hyperbolic structures on surfaces. Using the concepts of developing map and holonomy, we sketch a proof that every surface equipped with a complete hyperbolic metric is isometric to a quotient of $\mathbb{H}$ by a Fuchsian group. We then define Teichmüller spaces and explain Fenchel-Nielsen coordinates. Finally, we introduce mapping class groups and show that they act properly discontinuously on Teichmüller space.

Keywords: Geometric structure, Teichmüller space, mapping class group

Mathematics Subject Classification 2000: 57M50, 20F65

1. Introduction

This paper is intended as a brief introduction to hyperbolic structures on surfaces, Teichmüller spaces and mapping class groups. It is based on the first half of the course "Hyperbolic structures on surfaces", given by C. Leininger and the author during the programme "Geometry, Topology and Dynamics of Character Varieties" at the Institute for Mathematical Sciences of Singapore in July 2010. It accompanies the article [23], also in this volume, which discusses degenerations of hyperbolic structures.

In order to keep the exposition as concise and self-contained as possible, we have narrowed our attention to three particular strands. First, that a surface S equipped with a complete hyperbolic structure may be identified with a quotient of $\mathbb{H}$ by a torsion-free Fuchsian group, via the developing map. Second, that the Teichmüller space $\mathcal{T}(S)$, that is, the space of complete hyperbolic structures on S, is homeomorphic to some $\mathbb{R}^n$, where n depends only on the topology of S. Finally, that the mapping class group $\text{Mod}(S)$ of S, that is, the group of self-homeomorphisms of S up to homotopy, acts properly discontinuously on $\mathcal{T}(S)$.

The plan of the paper is as follows. In Section 2 we recall some basic facts about plane hyperbolic geometry. In Section 3 we introduce the notion of a hyperbolic structure on a surface, and explain why every complete hyperbolic surface is isometric to $\mathbb{H}/\Gamma$, where Γ is a torsion-free Fuchsian group. In Section 4 we define Teichmüller spaces and describe Fenchel-Nielsen coordinates. Finally, in Section 5 we introduce the mapping class group and prove that it acts properly discontinuously on Teichmüller space.

2. Plane Hyperbolic Geometry

We refer the reader to [2, 4, 5, 8, 19, 20, 24, 26] for a detailed discussion of the topics presented in this section.

2.1. *Möbius transformations*

Let $\bar{\mathbb{C}} = \mathbb{C} \cup \{\infty\}$ be the Riemann sphere. A Möbius transformation is a map $T : \bar{\mathbb{C}} \to \bar{\mathbb{C}}$ of the form

$$T(z) = \frac{az+b}{cz+d},$$

where $a, b, c, d \in \mathbb{C}$ and $ad - bc \neq 0$. Here, $T(\infty) = \frac{a}{c}$ and $T\left(\frac{-c}{d}\right) = \infty$.

Denote by $\text{Möb}(\bar{\mathbb{C}})$ the set of all Möbius transformations. Every element of $\text{Möb}(\bar{\mathbb{C}})$ is a bijection; the inverse of T is

$$T^{-1}(z) = \frac{dz-b}{-cz+a} \in \text{Möb}(\bar{\mathbb{C}}) \,.$$

Moreover, the composition of two Möbius transformations is a Möbius transformation, and thus $\text{Möb}(\bar{\mathbb{C}})$ is a group under composition.

To every Möbius transformation

$$T(z) = \frac{az+b}{cz+d}.$$

we may associate a matrix of non-zero determinant, namely

$$\begin{pmatrix} a & b \\ c & d \end{pmatrix}.$$

Multiplying the matrix by a non-zero complex number does not change the Möbius transformation it represents, and thus there is a surjective map $\text{SL}(2, \mathbb{C}) \to \text{Möb}(\bar{\mathbb{C}})$. It is easy to verify that this map is in fact a homomorphism, with kernel $\{\pm I\}$. Therefore,

$$\text{Möb}(\bar{\mathbb{C}}) \cong \text{PSL}(2, \mathbb{C}).$$

We will make use of the following important properties of Möbius maps; for a proof, see for instance ([19], Thm. 2.4.1 and 2.11.3).

Proposition 2.1. *(1) Every element of* $\text{Möb}(\bar{\mathbb{C}})$ *is conformal.*
(2) Let L *be either a Euclidean circle or a Euclidean line in* $\mathbb{C}$*, and let* $T \in \text{Möb}(\bar{\mathbb{C}})$*. Then* $T(L)$ *is either a Euclidean circle or a Euclidean line in* $\mathbb{C}$*.*

2.1.1. *Classification in terms of trace and fixed points*

The *trace* of the Möbius transformation $T(z) = \frac{az+b}{cz+d}$ is $\text{tr}(T) := a + d$. Observe that $\text{tr}(T)$ is only defined up to sign; however, $\text{tr}^2(T) = (a + d)^2$ is well-defined and thus yields a function $\text{tr}^2 : \text{PSL}(2,\mathbb{C}) \to \mathbb{C}$ that is continuous with respect to the natural topology on $\text{PSL}(2,\mathbb{C})$, and is constant on each conjugacy class.

If $T \in \text{Möb}(\bar{\mathbb{C}})$ is not the identity, then its fixed points are given by

$$z = \frac{(a-d) \pm \sqrt{\text{tr}^2(T) - 4}}{2c}.$$

Therefore, T has exactly one fixed point if and only if $\text{tr}^2(T) = 4$; otherwise it has two.

If T has exactly one fixed point, then it is called *parabolic*. A parabolic transformation is conjugate in $\text{Möb}(\bar{\mathbb{C}})$ to $z \to z + 1$.

If T has two fixed points then, up to conjugation in $\text{Möb}(\bar{\mathbb{C}})$, $T(z) = \lambda z$ for some $\lambda \in \mathbb{C} \setminus \{0, 1\}$. The number λ is called the *multiplier* of T; note that the multiplier is also a conjugacy invariant, for

$$\text{tr}^2(T) = \lambda + \lambda^{-1} + 2. \tag{2.1}$$

If $|\lambda| = 1$ then T is called *elliptic*; observe that T is elliptic if and only if $\text{tr}^2(T) \in [0, 4)$. Otherwise, T is called *loxodromic*. In the special case that $\lambda \in \mathbb{R}$, T is called *hyperbolic*; observe that T is hyperbolic if and only if $\text{tr}^2(T) > 4$.

2.2. *Models for hyperbolic geometry*

We will consider two (equivalent) models for plane hyperbolic geometry. The first is the *upper half-plane*

$$\mathbb{H} = \{x + iy \in \mathbb{C} \mid y > 0\},$$

equipped with the Riemannian metric

$$ds^2_{\mathbb{H}} = \frac{dx^2 + dy^2}{y^2}.$$

The second is the *Poincaré disc*, namely the open unit disc

$$\mathbb{D} = \{x + iy \in \mathbb{C} \mid x^2 + y^2 < 1\}$$

in the complex plane, equipped with the Riemannian metric

$$ds^2_{\mathbb{D}} = \frac{4(dx^2 + dy^2)}{(1 - (x^2 + y^2))^2}.$$

2.2.1. *Hyperbolic distance*

Let $\gamma : [a, b] \to \mathbb{H}$ be a piecewise differentiable path. The *hyperbolic length* of γ is defined as

$$l_{\mathbb{H}}(\gamma) = \int_\gamma ds_{\mathbb{H}}.$$

Given $z, w \in \mathbb{H}$, define the *hyperbolic distance* between z and w by

$$d_{\mathbb{H}}(z, w) = \inf\{l_{\mathbb{H}}(\gamma) \mid \gamma \text{ is a piecewise differentiable path from } z \text{ to } w\}.$$

Hyperbolic distance in $d_{\mathbb{D}}$ in $\mathbb{D}$ is defined in an analogous way. It is an easy exercise to verify that $(\mathbb{H}, d_{\mathbb{H}})$ and $(\mathbb{D}, d_{\mathbb{D}})$ are metric spaces.

2.2.2. *Möbius transformations act by isometries*

Let Möb($\mathbb{H}$) (resp. Möb($\mathbb{D}$)) be the subgroup of Möb($\bar{\mathbb{C}}$) consiting of all Möbius transformations that preserve $\mathbb{H}$ (resp. $\mathbb{D}$). As one may easily verify,

$$\text{Möb}(\mathbb{H}) \cong \text{PSL}(2, \mathbb{R}) \quad \text{and} \quad \text{Möb}(\mathbb{D}) \cong \text{PSU}(2, \mathbb{C}).$$

Lemma 2.2. *Every element of* Möb($\mathbb{H}$) *(resp.* Möb($\mathbb{D}$)*) is an isometry of* $\mathbb{H}$ *(resp.* $\mathbb{D}$*).*

Proof. We prove the result for $\mathbb{H}$, as the one for $\mathbb{D}$ is obtained in an analogous way. Let $\gamma : I \to \mathbb{H}$ be a piecewise differentiable path, and let

$$T(z) = \frac{az + b}{cz + d} \in \text{Möb}(\mathbb{H}).$$

Write $w = T(z)$, and observe that

$$\text{Im}(w) = \frac{\text{Im}(z)}{|cz + d|^2},$$

where $\text{Im}(z)$ denotes the imaginary part of z. Then:

$$l_{\mathbb{H}}(T(\gamma)) = \int_{T(\gamma)} \frac{|dw|}{\text{Im}(w)} = \int_\gamma \frac{|cz + d|^2}{\text{Im}(z)} \cdot \frac{|dz|}{|cz + d|^2} = l_{\mathbb{H}}(\gamma),$$

as desired. □

2.2.3. *The Cayley transformation*

The *Cayley transformation* is the Möbius map

$$C(z) = \frac{z-i}{z+i} \in \text{Möb}(\bar{\mathbb{C}}).$$

It is easy to check that $C(\mathbb{H}) = \mathbb{D}$. Using a similar calculation to that of Lemma 2.2, we obtain:

Lemma 2.3. *The Cayley transformation $C : \mathbb{H} \to \mathbb{D}$ is an isometry.*

2.2.4. *Hyperbolic geodesics*

A piecewise differentiable path in $\mathbb{H}$ or $\mathbb{D}$ is said to be *geodesic* if the length of any of its segments realizes the distance between the endpoints. The following gives a full description of geodesics in $\mathbb{H}$ and $\mathbb{D}$.

Proposition 2.4. *(i) The geodesics in $\mathbb{H}$ are either vertical Euclidean lines or Euclidean semicircles perpendicular to $\mathbb{R}$.*
(ii) The geodesics in $\mathbb{D}$ are either diameters of $\mathbb{D}$ or arcs of Euclidean semicircles perpendicular to $\mathbb{S}^1$.

Proof. (i) Let $z, w \in \mathbb{H}$. Suppose first that $z, w \in i\mathbb{R}$; thus, up to relabelling, $z = ip$ and $w = iq$ where $p < q$. Let $\gamma : [a, b] \to \mathbb{H}$ be a piecewise differentiable path from z to w, where $\gamma(t) = (x(t), y(t))$. Then:

$$l_{\mathbb{H}}(\gamma) = \int_a^b \frac{1}{y(t)} \sqrt{\left(\frac{dx}{dt}\right)^2 + \left(\frac{dy}{dt}\right)^2} dt \geq \int_a^b \frac{1}{y(t)} \frac{dy}{dt} dt = \int_p^q \frac{dy}{y} = \log\left(\frac{q}{p}\right),$$

with equality if and only if $\gamma([a, b])$ is the vertical segment from ip to iq.

Now consider arbitrary $z, w \in \mathbb{H}$, and let L be either the vertical Euclidean line through z, w (if z, w have the same real part) or the Euclidean semicircle through z, w and with center in $\mathbb{R}$ (if z, w have different real parts). It is an easy exercise to check that there exists $T \in \text{Möb}(\mathbb{H})$ such that $T(L) = i\mathbb{R}$. The result now follows from the above paragraph and Lemma 2.2.

(ii) The proof for $\mathbb{D}$ is a direct consequence of (i), Lemma 2.3 and Proposition 2.1. □

2.2.5. *The boundary at infinity*

Let $p < 1$. Note that Proposition 2.4 gives that $d_{\mathbb{H}}(i, pi) = -\log(p)$. In particular, $d_{\mathbb{H}}(i, pi) \to \infty$ as $p \to 0$. For this reason, the set $\bar{\mathbb{R}} = \mathbb{R} \cup \{\infty\}$ is called the *boundary at infinity* of $\mathbb{H}$. Similarly, $\mathbb{S}^1$ is called the *boundary at infinity* of $\mathbb{D}$.

2.2.6. *The full isometry group*

Denote by $\mathrm{Isom}^+(\mathbb{H})$ (resp. $\mathrm{Isom}^+(\mathbb{D})$) the group of orientation-preserving isometries of $\mathbb{H}$ (resp. $\mathbb{D}$). We have:

Proposition 2.5. $\mathrm{Isom}^+(\mathbb{H}) \cong \mathrm{PSL}(2, \mathbb{R})$ *and* $\mathrm{Isom}^+(\mathbb{D}) \cong \mathrm{PSU}(2, \mathbb{C})$.

Proof. Again, we prove the result only for $\mathbb{H}$. In view of Lemma 2.2, we must show that every element of $\mathrm{Isom}^+(\mathbb{H})$ is a Möbius transformation. Let $F \in \mathrm{Isom}^+(\mathbb{H})$. Composing with an element of $\mathrm{PSL}(2, \mathbb{R})$ if necessary, we may assume that F fixes two distinct points $z_1, z_2 \in i\mathbb{R}$. Choose a point $w \notin i\mathbb{R}$, noting that $T(w)$ lies on the hyperbolic circle C_i of centre z_i and radius $d_{\mathbb{H}}(z_i, w)$, for $i = 1, 2$.

Now, it is not difficult to verify (see [24], Ch. 2) that every hyperbolic circle is also a Euclidean circle. Therefore, C_1 and C_2 intersect at two points: one of them is w, and the other one is on the other side of $i\mathbb{R}$ from w. Since F is orientation-preserving, we get that $F(w) = w$. Therefore, F is an isometry fixing three points, and hence the identity. □

2.2.7. *Dynamics of elements of* $\mathrm{Isom}^+(\mathbb{H})$

Recall the classification of Möbius transformations into *elliptic, parabolic* and *loxodromic*. Since $\mathrm{Isom}^+(\mathbb{H}) \cong \mathrm{PSL}(2, \mathbb{R})$, loxodromic isometries of $\mathbb{H}$ are all *hyperbolic*. We now make a few comments on the dynamics of the different types of isometries:

(i) If $T \in \mathrm{Isom}^+(\mathbb{H})$ is parabolic, then it has exactly one fixed point on $\bar{\mathbb{R}}$. Thus, up to conjugation in $\mathrm{Isom}^+(\mathbb{H})$, $T(z) = z + a$ for some $a \in \mathbb{R}$; observe that T leaves invariant every Euclidean line of the form $y =$ constant.

(ii) If $T \in \mathrm{Isom}^+(\mathbb{H})$ is hyperbolic, then it has two fixed points on $\bar{\mathbb{R}}$; the geodesic between them is called the *axis* of T. Up to conjugation, $T(z) = \lambda z$ for some $\lambda \in \mathbb{R}$. The map T acts on its axis as a hyperbolic translation, with translation distance $l = \log \lambda$; observe that (2.1) gives that $\mathrm{tr}^2(T) = 4\cosh^2(l/2)$.

(iii) Finally, if $T \in \mathrm{Isom}^+(\mathbb{H})$ is elliptic, then it has exactly one fixed point in $\mathbb{H}$. Up to conjugation, $T(z) = \frac{\cos(t)z+\sin(t)}{\sin(t)z+\cos(t)}$, for some $t \in \mathbb{R}$.

We refer the interested reader to ([24], Chapter 3) for pictures showing the dynamics of the different types of elements of $\mathrm{Isom}^+(\mathbb{H})$.

2.3. *Fuchsian groups and fundamental domains*

2.3.1. *Fuchsian groups*

Let Γ be a group acting by homeomorphisms on a metric space $\mathbb{X}$. We say that Γ acts *properly discontinously* on $\mathbb{X}$ if, for all compact subsets $K \subset \mathbb{X}$, the set

$$\{\gamma \in \Gamma \mid \gamma(K) \cap K \neq \emptyset\}$$

is finite. We say that Γ acts *freely* on $\mathbb{X}$ if every non-identity element of Γ acts without fixed points.

We will be interested mainly in the case where $\mathbb{X} = \mathbb{H}$ and Γ is subgroup of $\mathrm{PSL}(2,\mathbb{R})$ that is *discrete* with respect to the natural topology inherited from $\mathrm{PSL}(2,\mathbb{R})$. Discrete subgroups of $\mathrm{PSL}(2,\mathbb{R})$ are called *Fuchsian groups.* We will need the following well-known result; for a proof see ([5], Prop. B.1.6), for instance.

Proposition 2.6. *Let Γ be a subgroup of* $\mathrm{PSL}(2,\mathbb{R})$. *The following conditions are equivalent:*

(1) Γ acts freely and properly discontinuously on $\mathbb{H}$.
(2) $\mathbb{X}/\Gamma$ is Hausdorff and the projection $\mathbb{X} \to \mathbb{X}/\Gamma$ is a covering map.
(3) Γ is a torsion-free Fuchsian group.

2.3.2. *Fundamental domains*

We now introduce the concept of fundamental domain for the action of a group on a metric space.

Definition 2.7 (Fundamental Domain). Let Γ be a group acting properly discontinously by homeomorphisms on a metric space $\mathbb{X}$. A *fundamental domain* for the action of Γ on $\mathbb{X}$ is a closed subset $C \subset \mathbb{X}$ such that:

(1) The interior $\mathrm{int}(C)$ of C is not empty.
(2) If $T \neq \mathrm{id}$ then $T(\mathrm{int}(C)) \cap \mathrm{int}(C) = \emptyset$.

(3) The Γ-translates of C *tessellate* $\mathbb{X}$; that is, $\bigcup_{T\in\Gamma} T(C) = \mathbb{X}$.

As it turns out, every Fuchsian group admits a particularly nice type of fundamental domain, known as the *Dirichlet domain*, which we now describe.

Definition 2.8 (Dirichlet Domain). Let Γ be a group acting properly discontinously by isometries on a metric space $\mathbb{X}$, and let $z_0 \in X$ be a point not fixed by any non-trivial element of Γ. The *Dirichlet domain* of Γ centered at z_0 is

$$\mathcal{D}_\Gamma(z_0) = \{x \in \mathbb{X} \mid d(x, z_0) \leq d(x, T(z_0)), \forall T \in \Gamma\}.$$

Proposition 2.9. *Let Γ be a Fuchsian group, and let z_0 be a point not fixed by any non-trivial element of Γ. Then $\mathcal{D}_\Gamma(z_0)$ is a convex fundamental domain for the action of Γ on $\mathbb{H}$.*

Proof. First, $\mathcal{D}_\Gamma(z_0)$ is closed and convex since it is the intersection of closed half-planes of $\mathbb{H}$. Moreover, $z_0 \in \text{int}(\mathcal{D}_\Gamma(z_0))$ since Γ is discrete.

We now claim that the Γ-translates of $\mathcal{D}_\Gamma(z_0)$ tessellate $\mathbb{H}$. Let $z \in \mathbb{H}$. Since Γ is discrete, there exists $T \in \Gamma$ such that

$$d(z, T(z_0)) = \min_{S\in\Gamma}\{d(z, S(z_0))\}.$$

Thus $T^{-1}(z) \in \mathcal{D}_\Gamma(z_0)$, and therefore $z \in T(\mathcal{D}_\Gamma(z_0))$, as desired.

Finally, suppose for contradiction that there exists $z \in \text{int}(\mathcal{D}_\Gamma(z_0))$ and $T \neq \text{Id}$ such that $T(z) \in \text{int}(\mathcal{D}_\Gamma(z_0))$. In particular,

$$d(z, z_0) < d(z, T^{-1}(z_0)) = d(T(z), z_0)$$

and

$$d(T(z), z_0) < d(T(z), T(z_0)) = d(z, z_0),$$

which is impossible. □

Observe that the fact that Γ acts properly discontinously on $\mathbb{H}$ implies that $\mathcal{D}_\Gamma(z_0)$ is *locally finite*, that is, for every compact set $K \subset \mathbb{H}$, there are only finitely many Γ-translates of $\mathcal{D}_\Gamma(z_0)$ that intersect K; see ([20], Thm. 3.5.1) for details.

Example 2.10. (a) Let $T \in \text{Möb}(\mathbb{H})$ be a parabolic isometry so, up to conjugation, $T(z) = z + a$ for some $a \in \mathbb{R} \setminus \{0\}$. Let $\Gamma = \langle T \rangle$ and $z_0 \in \mathbb{H}$. Then $\mathcal{D}_\Gamma(z_0) = \{z \in \mathbb{C} \mid \text{Re}(z_0) - \frac{a}{2} \leq \text{Re}(z) \leq \text{Re}(z_0) + \frac{a}{2}\}$.

(b) Let $\Gamma = \mathrm{SL}(2,\mathbb{Z})$ and let $z_0 = 2i$, which is not fixed by any element of Γ. Then $\mathcal{D}_\Gamma(2i) = \{z \in \mathbb{H} \mid -1/2 \leq \mathrm{Re}(z) \leq 1/2, |z| \geq 1\}$.

(c) Consider the Euclidean isometries $A : (x,y) \to (x+1,y)$ and $B : (x,y) \to (x, y+b)$, where $b > 0$. Let $G \cong \mathbb{Z}^2$ be the group generated by A and B. Then the Dirichlet domain for the action of Γ on Euclidean plane $\mathbb{E}^2$ is generically a hexagon; in the special case when $b = 1$, it is a square.

2.3.3. *The action of a group on a Dirichlet domain*

Let Γ be a discrete group acting properly discontinuously by isometries on $\mathbb{X} = \mathbb{H}$ or $\mathbb{E}$, and let D be a Dirichlet domain for the action of Γ. We will assume, for simplicity, that Γ acts freely on $\mathbb{X}$, so that $\mathbb{X}/\Gamma$ is a smooth surface, and that D is compact, so that D is a finite sided polygon. In the case when Γ is a Fuchsian group, then the fact that D is compact implies that Γ has no parabolic elements; see ([20], Thm. 4.2.1). The group Γ identifies the sides of D in pairs and, in fact, Γ is generated by the (finite) collection of all side pairings; see ([20], Thm. 3.5.4). Each Γ-orbit of vertices of D is called a *cycle*, and the sum of the internal angles at the vertices of a cycle is always equal to 2π; see ([20], Thm. 3.5.3).

A converse to this situation is described in *Poincaré's Polygon Theorem*, which we now state; for a proof, see ([24], Ch. 7). Again, $\mathbb{X} = \mathbb{H}$ or $\mathbb{E}$.

Theorem 2.11 (Poincaré). *Let $P \subset \mathbb{X}$ be a compact polygon whose sides are identified in pairs by isometries of $\mathbb{X}$, and let Γ be the group generated by those isometries. Suppose that, for each Γ-orbit of vertices of P, the internal angles at the vertices in that orbit add up to 2π. Then Γ is a discrete group acting freely and properly discontinously on $\mathbb{X}$; moreover, D is a fundamental domain for the action.*

3. Hyperbolic Structures on Surfaces

We refer the reader to the texts [5, 10, 11, 24, 26] for a more detailed discussion on the material presented in this section.

3.1. *Definition and examples*

Let $\mathbb{X}$ denote the hyperbolic plane $\mathbb{H}$, the Euclidean plane $\mathbb{E}$ or the unit sphere $\mathbb{S}^2$ in $\mathbb{R}^3$.

Definition 3.1 (Geometric Structure). Let S be a topological surface.

A *geometric structure* on S consists of an open cover $\{U_i\}_{i\in I}$ of S and a collection $\{\phi_i\}_{i\in I}$ of maps, with $\phi_i : U_i \to \mathbb{X}$, such that

(1) ϕ_i is a homeomorphism onto its image, for each $i \in I$, and
(2) if $U_i \cap U_j \neq \emptyset$, the restriction of the transition map

$$\phi_i \circ \phi_j^{-1} : \phi_j(U_i \cap U_j) \to \phi_i(U_i \cap U_j)$$

to each connected component of $\phi_j(U_i \cap U_j)$ is an orientation-preserving isometry of $\mathbb{X}$.

In the case where $\mathbb{X} = \mathbb{H}$ (resp. $\mathbb{X} = \mathbb{E}$ or $\mathbb{X} = \mathbb{S}^2$), we say that the surface S is equipped with a *hyperbolic structure* (resp. *Euclidean or spherical structure*). In the definition above, each pair (U_i, ϕ_i) is called a *chart*. The set of all charts is called an *atlas* of S; note that every atlas is contained in a unique maximal atlas. Finally, observe that a surface equipped with a geometric structure supports a natural path-metric, obtained by deeming each chart map to be an isometry.

Remark 3.2 (Geometric Structure on a Covering Space). Suppose that S is equipped with a geometric structure $\{(U_i, \phi_i)\}_{i\in I}$ and let $\pi : \tilde{S} \to S$ be a covering map; without loss of generality, we assume that U_i is evenly covered for all $i \in I$. Then $\tilde{S}$ comes equipped with a natural geometric structure, where the open sets are the preimages of the U_i under π and the chart maps are the restrictions of $\phi_i \circ \pi$ to each of these sets.

A geometric structure on a covering space also induces a geometric structure on the quotient space; see Examples 3.4 and 3.5 below. From now on, we will focus our attention mainly on the case $\mathbb{X} = \mathbb{H}$.

Remark 3.3 (Hyperbolic Structure on a Surface with Boundary). If S has boundary, we define a *hyperbolic structure with geodesic boundary* on S by requiring that each of the sets U_i in Definition 3.1 be an open set of a closed half-plane in $\mathbb{H}$.

Example 3.4 (Hyperbolic Structure on a Surface of Genus $g \geq 2$). Let $g \geq 2$ and P be a regular hyperbolic $4g$-gon in $\mathbb{D}$ with internal angles $\pi/2g$. To see that such polygon exists, consider $4g$ equispaced geodesic rays in $\mathbb{D}$ emanating from the origin O, as in Figure 1. Consider the hyperbolic polygon P_t whose vertices are the points of intersection between these rays and the hyperbolic circle of center O and hyperbolic radius $t > 0$. As t increases, the internal angle of P_t decreases from the Euclidean value

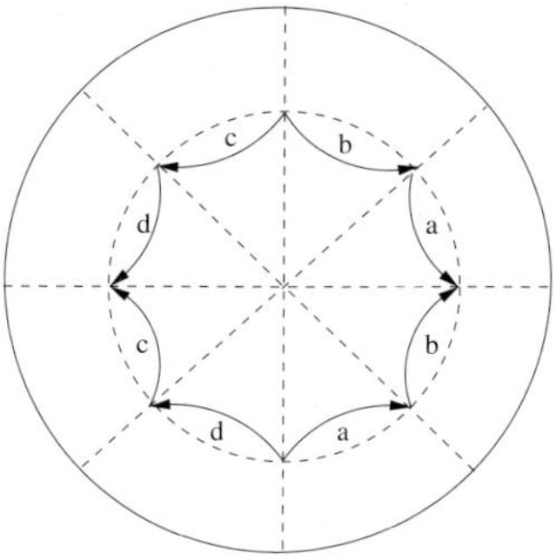

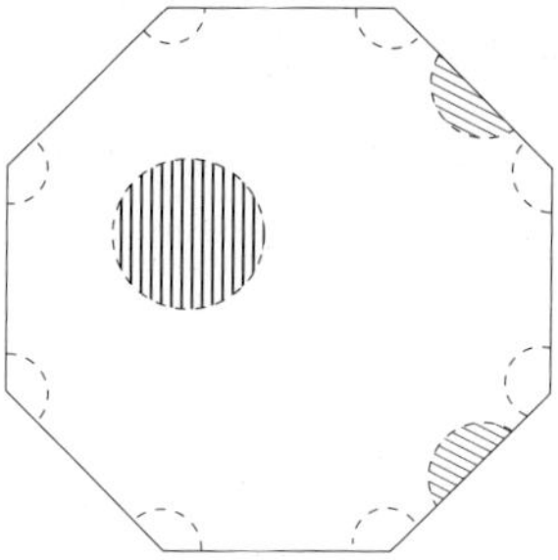

Fig. 1. The left figure shows a regular hyperbolic octagon; for exactly one value of the radius of the shaded circle, the internal angles will be $\pi/4$. The sides are identified by isometries $a, b, c, d \in \mathrm{PSL}(2, \mathbb{R})$ according to the labelling shown. The quotient surface $\mathbb{H}/\Gamma$, where $\Gamma = \langle a, b, c, d \rangle$, is homeomorphic to a closed surface of genus 2, and is equipped with a complete hyperbolic structure. The schematics of the charts around a point p are shown in the right figure, depending on whether a lift of p lies in the interior of the polygon, or in the interior of a side, or is a vertex.

$(4g - 2)\pi$, down to 0. By continuity, there is a value of t for which the internal angle is equal to $\pi/2g$.

Suppose that the sides of P are identified in pairs by elements of $\mathrm{PSL}(2, \mathbb{R})$ according to the labelling outlined in Figure 1. We see that the hypotheses of Poincaré's Polygon Theorem are satisfied, and thus the group Γ generated by the side pairings is a Fuchsian group acting freely on $\mathbb{H}$; observe there is only one Γ-orbit of vertices. The quotient space $\bar{P} = \mathbb{H}/\Gamma$ is homeomorphic to a closed surface of genus g. We define a hyperbolic structure on $\bar{P}$ by a specifying a chart around each point in $\bar{P}$; such charts are schematically shown in Figure 1 for $g = 2$, depending on whether a lift of the point is in the interior of P, on one of the sides of P, or is a vertex of P. Observe that, since the angle around any vertex of P is $\pi/2g$, then the angle around the corresponding point in $\bar{P}$ is 2π and thus the chart is well-defined. Finally, note the natural path-metric on $\bar{P}$ is complete.

In fact, the previous example is a special case of a more general situation, as we now explain.

Example 3.5 (Quotient of $\mathbb{H}$ by a Fuchsian Group). Let Γ be a Fuchsian group acting freely on $\mathbb{H}$. Let $S = \mathbb{H}/\Gamma$ and let $\pi : \mathbb{H} \to S$ be the natural covering map. We endow $S = \mathbb{H}/\Gamma$ with a hyperbolic structure by specifying a chart around each point $p \in S$, as follows. Let U_p be an evenly covered open neighbourhood of p, and let f_p be a homeomorphism identifying U_p with any of the open sets in $\mathbb{H}$ covering U_p. The collection $\{(U_p, f_p)\}_{p \in S}$ gives a hyperbolic structure on S; again, the natural path-metric on the surface is complete.

It is easy to see that the hyperbolic structures on $\mathbb{H}/\Gamma$ given in Examples 3.4 and 3.5 are in fact equivalent.

Remark 3.6 (Euclidean Structure on a Torus). By applying the same reasoning as above, we obtain a Euclidean structure on a surface of genus $g = 1$ by identifying opposite sides of a rectangle in the Euclidean plane $\mathbb{E}$. More generally, the quotient of $\mathbb{E}$ by a discrete group of Euclidean isometries is naturally equipped with a Euclidean structure.

Remark 3.7 (Geometric Structures on Closed Surfaces). Observe that a surface equipped with a geometric structure has constant Gaussian curvature. Therefore, a closed surface of genus $g \geq 1$ admits a hyperbolic (resp. Euclidean) structure if and only if $g \geq 2$ (resp. $g = 1$), as follows from Example 3.4, Remark 3.6 and the Gauss-Bonnet theorem.

3.2. *The Cartan-Hadamard Theorem. Developing map and holonomy*

The next result, a special case of the celebrated *Cartan-Hadamard Theorem*, asserts that Example 3.5 is the only way of obtaining a surface equipped with a hyperbolic structure, provided we restrict our attention to *complete* structures. We refer the reader to [3, 9, 12] for more general versions of the Cartan-Hadamard Theorem, and to ([8], Ch. 6) for a discussion on *incomplete* hyperbolic structures on surfaces.

Theorem 3.8 (Cartan-Hadamard). *Let X be a connected surface equipped with a hyperbolic structure, and suppose that the natural path-metric on X is complete. Then X is isometric to $\mathbb{H}/\Gamma$, where Γ is a Fuchsian group acting freely on $\mathbb{H}$.*

The rest of this section is devoted to give a sketch of the proof of Theorem 3.8. The strategy is as follows. First, we will construct an isometry

$$\mathrm{Dev} : \tilde{\mathrm{X}} \to \mathbb{H},$$

called the *developing map*; here $\tilde{X}$ denotes the universal cover of X. The map Dev will induce an isomorphism

$$\mathrm{Hol} : \pi_1(\mathrm{X}) \to \Gamma,$$

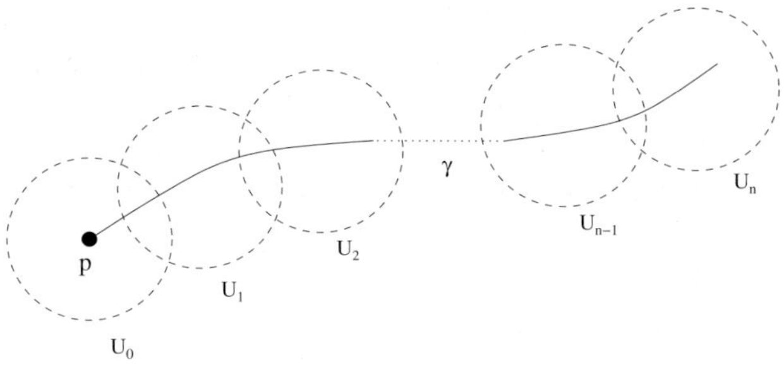

Fig. 2. Constructing the developing map.

where Γ is a torsion-free Fuchsian group; the map Hol is called the *holonomy representation* of $\pi_1(X)$. Once all this has been established, it will easily follow that X is isometric to $\mathbb{H}/\Gamma$.

Next, we explain some of the details, and refer the reader to [5, 11, 24] for a more thorough discussion; we remark that one obtains a Cartan-Hadamard Theorem for Euclidean surfaces using the same ideas as below, with the obvious modifications.

3.2.1. *The developing map*

Let $\{(U_i, \phi_i)\}_{i \in I}$ be an atlas defining the hyperbolic structure on X. Fix, once and for all, a basepoint $p \in U_0$. Let $\tilde{X}$ be the universal cover of X, namely the set of homotopy classes of paths in X that start at p; recall that $\tilde{X}$ has a natural hyperbolic structure coming from that of X, by Remark 3.2.

Let $[\gamma] \in \tilde{X}$ and choose a representative $\gamma : [0, 1] \to X$ of $[\gamma]$. We cover $\gamma([0, 1])$ with a finite collection $\{(U_i, \phi_i)\}_{i=0}^{n}$ of charts as shown in Figure 2; in particular $U_i \cap U_{i+1}$ is connected. We define the map Dev successively, as follows. First, set

$$\mathrm{Dev}|_{(\mathrm{U}_0 \cap \gamma)} = \phi_0|_{(\mathrm{U}_0 \cap \gamma)}.$$

Since $U_0 \cap U_1$, is connected, the definition of hyperbolic structure yields that $\phi_0 \circ \phi_1^{-1}|_{(U_0 \cap U_1)} = T_1 \in \mathrm{PSL}(2, \mathbb{R})$. Set

$$\mathrm{Dev}|_{(\mathrm{U}_1 \cap \gamma)} = \mathrm{T}_1 \circ \phi_1|_{(\mathrm{U}_1 \cap \gamma)},$$

and note Dev is now defined on $(U_0 \cup U_1) \cap \gamma$. Repeating this process, we obtain maps $T_2, \ldots, T_n \in \mathrm{PSL}(2, \mathbb{R})$, and define

$$\mathrm{Dev}|_{(\mathrm{U}_n \cap \gamma)} = \mathrm{T}_1 \circ \mathrm{T}_2 \circ \cdots \circ \mathrm{T}_n \circ \phi_n|_{(\mathrm{U}_n \cap \gamma)}, \tag{3.1}$$

noting that Dev is now well-defined on $(U_0 \cup \cdots \cup U_n) \cap \gamma$, and thus on the whole of γ. We set

$$\mathrm{Dev}(\gamma) = \mathrm{Dev}(\gamma(1)) \in \mathbb{H}.$$

At this point, it is straightforward, although not terribly amusing, to show that $\mathrm{Dev}(\gamma)$ depends only on the initial chart $\phi_0 : U_0 \to \mathbb{H}$ and the homotopy class of γ; this is carefully explained in ([5], Prop. B.1.3), for instance. Thus we have obtained a well-defined map

$$\mathrm{Dev} : \tilde{\mathrm{X}} \to \mathbb{H},$$

which is a local isometry with respect to the natural hyperbolic structure on $\tilde{X}$ (as we will see, if X is complete then Dev will be a global isometry). As a consequence, we obtain that any two choices of initial chart produce developing maps which differ by an element of $\mathrm{PSL}(2, \mathbb{R})$.

3.2.2. *Two technical lemmas*

Having introduced the developing map, we continue towards a proof of Theorem 3.8. Following the strategy of [5, 24], the proof is based on the two results we now present.

Lemma 3.9. *Suppose X is equipped with a complete hyperbolic structure. Then its universal cover $\tilde{X}$ is also complete.*

Proof. Let $(\tilde{z}_n)_n \subset \tilde{X}$ be a Cauchy sequence. As the covering map $\pi : \tilde{X} \to X$ does not increase distances, then $(\pi(\tilde{z}_n))_n$ is a Cauchy sequence and thus converges to a point $z \in X$, since X is complete. Let U be an evenly covered open neighbourhood of z. Since $(\tilde{z}_n)_n$ is Cauchy, all but finitely many elements of $(\tilde{z}_n)$ belong to exactly one of the preimages $\tilde{U}$ of U and thus converge to the preimage of z contained in $\tilde{U}$. □

Lemma 3.10. *Let X be a surface equipped with a complete hyperbolic structure. Then, the developing map* $\mathrm{Dev} : \tilde{\mathrm{X}} \to \mathbb{H}$ *is a surjective covering map.*

Proof. Since Dev is a local homeomorphism by construction, it suffices to prove that Dev satisfies the *path-lifting property*. This is, we want to establish that, for all $z_0 \in \mathrm{Dev}(\tilde{\mathrm{X}})$, all $\tilde{z}_0 \in \mathrm{Dev}^{-1}(\mathrm{z}_0)$, and all piecewise differentiable paths $\gamma : [0, 1] \to \mathbb{H}$ with $\gamma(0) = z_0$, there exists a path $\tilde{\gamma} : [0, 1] \to \tilde{X}$ such that $\tilde{\gamma}(0) = \tilde{z}_0$ and $\mathrm{Dev} \circ \tilde{\gamma} = \gamma$.

Let $z_0 \in \mathrm{Dev}(\tilde{\mathrm{X}})$, $\tilde{z}_0 \in \mathrm{Dev}^{-1}(\mathrm{z}_0)$, and $\gamma : [0,1] \to \mathbb{H}$ a piecewise differentiable path with $\gamma(0) = z_0$. Consider

$$t_0 = \sup \left\{ t \in [0,1] \mid \exists \tilde{\gamma} : [0,t] \to \tilde{X} \text{ with } \tilde{\gamma}(0) = \tilde{z}_0 \text{ and } \mathrm{Dev} \circ \tilde{\gamma} = \gamma|_{[0,\mathrm{t}]} \right\}.$$

We want to show that $t_0 = 1$. First, note that, since Dev is a local isometry, then $t_0 > 0$. Consider, for all $t < t_0$, the lift $\tilde{\gamma} : [0,t] \to \tilde{X}$ of $\gamma : [0,t] \to \mathbb{H}$ and observe that $\tilde{\gamma}$ is unique, again because Dev is a local isometry. Let t_n be an increasing sequence converging to t_0. Then $(\tilde{\gamma}(t_n))_n$ is a Cauchy sequence in $\tilde{X}$; otherwise $\tilde{\gamma}([0,t_0))$ would have infinite length, which is impossible; see ([5], Prop. B.1.3) for details. Therefore $(\tilde{\gamma}(t_n))_n$ converges, by Lemma 3.9, and thus we define $\tilde{\gamma}(t_0)$ to be this limit. Finally, since Dev is an isometry in a neighbourhood of $\tilde{\gamma}(t_0)$, it follows that $t_0 = 1$, as claimed. Therefore, Dev is a covering map.

The fact that we can lift paths from $\mathbb{H}$ to $\tilde{X}$ quickly implies that Dev is surjective. Indeed, let $z \in \mathbb{H}$, and choose $z_0 \in \mathrm{Dev}(\tilde{\mathrm{X}})$ and a path $\gamma : [0,1] \to \mathbb{H}$ with $\gamma(0) = z_0$. Denote by $\tilde{\gamma} : [0,1] \to \tilde{X}$ the lift of γ. Then $\mathrm{Dev}(\tilde{\gamma}(1)) = \mathrm{z}$, as desired. □

Since Dev is a surjective covering map and $\mathbb{H}$ is simply-connected, we deduce that $\tilde{X}$ is homeomorphic to $\mathbb{H}$. This, together with the fact that Dev is a local isometry, implies:

Corollary 3.11. *Let X be a surface equipped with a complete hyperbolic structure. Then, then universal cover $\tilde{X}$ of X is isometric to $\mathbb{H}$.*

3.2.3. *Holonomy*

Let X be a surface equipped with a hyperbolic structure, and choose a basepoint p on X. If we consider closed paths based at p in the construction (3.1) of the developing map above, we obtain a map

$$\mathrm{Hol} : \pi_1(\mathrm{X}, \mathrm{p}) \to \mathrm{PSL}(2, \mathbb{R}),$$

defined by $\mathrm{Hol}([\gamma]) = \mathrm{T}_1 \circ \mathrm{T}_2 \circ \cdots \circ \mathrm{T}_{\mathrm{n}} \in \mathrm{PSL}(2,\mathbb{R})$. By definition, the map Hol is a homomorphism, and is commonly referred to as the *holonomy representation* of $\pi(X,p)$. Again, Hol only depends on the choice of initial chart $\phi_0 : U_0 \to \mathbb{H}$, and any two choices produce conjugate homomorphisms. Moreover, we have:

Lemma 3.12. $\mathrm{Hol} : \pi_1(\mathrm{X}, \mathrm{p}) \to \mathrm{PSL}(2, \mathbb{R})$ *is injective.*

Proof. Let $[\gamma] \in \pi_1(X, p)$ and suppose that $\mathrm{Hol}([\gamma]) = \mathrm{Id} \in \mathrm{PSL}(2, \mathbb{R})$. Then, the developing image of γ is a loop based at p. Since $\mathbb{H}$ is simply-connected, we can find a homotopy between this loop and the trivial loop. Finally, since Dev is a covering map, we may lift such homotopy to a homotopy between γ and the trivial loop, as desired. □

Therefore, the holonomy representation gives an identification of $\pi_1(X, p)$ with a Fuchsian group $\Gamma = \mathrm{Hol}(\pi_1(\mathrm{X}, \mathrm{p}))$. Since $\pi_1(X, p)$ acts on $\tilde{X}$ freely and properly discontinously, the same holds for the action of Γ on $\mathbb{H}$, by Proposition 2.6. Therefore, Γ is a torsion-free Fuchsian group, again by Proposition 2.6.

Proof of Theorem 3.8. Let X be a surface equipped with a complete hyperbolic structure. By Corollary 3.11, its universal cover $\tilde{X}$ is isometric to $\mathbb{H}$ via the developing map. Moreover, the holonomy map gives an identification $\pi_1(X, p)$ with a torsion-free Fuchsian group Γ, and thus the result follows. □

4. Teichmüller Space

In this section we introduce the Teichmüller space $\mathcal{T}(S)$ of an orientable surface S of genus $g \geq 1$, the space of distinct geometric structures on S. In order to keep the exposition as simple as possible, we restrict our attention to closed surfaces only. In this case, Remark 3.7 gives that S carries a hyperbolic (resp. Euclidean) structure if and only if $g \geq 2$ (resp. $g = 1$).

In addition, we will focus solely on topological aspects of Teichmüller space, with the ultimate goal of proving, in Section 5, that the natural action of the *mapping class group* on Teichmüller space is properly discontinuous. In particular, we will not make reference to the various different metrics on Teichmüller space. We refer the reader to [1, 5, 13, 14, 16, 17], and the references therein, for a detailed exposition of Teichmüller spaces.

4.1. *Two definitions*

We now give two equivalent definitions of the Teichmüller space of a surface S of genus $g \geq 2$, one as the set of distinct hyperbolic structures on S and the other as the set of conjugacy classes of discrete faithful representations of $\pi_1(S)$ into $\mathrm{PSL}(2, \mathbb{R})$.

Definition 4.1 (Teichmüller Space of a Hyperbolic Surface, I). Let S be a closed topological surface of genus $g \geq 2$. The Teichmüller space $\mathcal{T}(S)$ of S is

$$\mathcal{T}(S) = \{(X, f)\}/\sim,$$

where

- X is S equipped with a hyperbolic structure,
- $f : S \to X$ is a homeomorphism, called the *marking*, and
- $(X, f) \sim (Y, g)$ if and only if there is an isometry $\iota : X \to Y$ such that $\iota \circ f$ is homotopic to g.

In order to reduce notation, we will denote points $[(X, f)] \in \mathcal{T}(S)$ simply by X whenever we do not need to make explicit reference to the marking. We now present an equivalent definition of Teichmüller space which, in particular, will allow us to define a natural topology on $\mathcal{T}(S)$.

Definition 4.2 (Teichmüller Space of a Hyperbolic Surface, II). Let S be a closed surface of genus $g \geq 2$. The Teichmüller space of S is

$$\mathcal{T}(S) = \mathcal{DF}(\pi_1(S), \mathrm{PSL}(2, \mathbb{R}))/\,\mathrm{PSL}(2, \mathbb{R}),$$

the set of discrete, faithful representations of $\pi_1(S)$ into $\mathrm{PSL}(2, \mathbb{R})$, up to conjugation.

The set $\mathcal{DF}(\pi_1(S), \mathrm{PSL}(2, \mathbb{R}))/\,\mathrm{PSL}(2, \mathbb{R})$ is called the $\mathrm{PSL}(2, \mathbb{R})$-*character variety* of $\pi_1(S)$. The equivalence of Definitions 4.1 and 4.2 is essentially contained in the statement of Theorem 3.8. Indeed, a point $[(X, f)] \in \mathcal{T}(S)$ determines a conjugacy class of faithful representations of $\pi_1(X) \cong \pi_1(S)$ into $\mathrm{PSL}(2, \mathbb{R})$ via the holonomy map. Conversely, given $\rho \in \mathcal{DF}(\pi_1(S), \mathrm{PSL}(2, \mathbb{R}))$, then $X = \mathbb{H}/\rho(\pi_1(S))$ comes equipped with a natural hyperbolic structure, by Example 3.5. Now, ρ induces a homotopy equivalence $h : S \to X$ which is then homotopic to a homeomorphism $f : S \to X$, the desired marking. Finally, any two conjugate representations produce isometric surfaces.

Remark 4.3 (Topology on $\mathcal{T}(S)$). Observe that, in light of Definition 4.2, $\mathcal{T}(S)$ carries a natural topology as a quotient of $\mathrm{PSL}(2, \mathbb{R})^{2g}$, since

$$\mathcal{DF}(\pi_1(S), \mathrm{PSL}(2, \mathbb{R})) \subset \mathrm{Hom}(\pi_1(S), \mathrm{PSL}(2, \mathbb{R})) = \mathrm{PSL}(2, \mathbb{R})^{2g}.$$

Example 4.4 (Teichmüller Space of the Torus). If S has genus 1, we define $\mathcal{T}(S)$ as the set of distinct Euclidean structures of *unit area* on S, by performing the obvious changes in Definition 4.1. By the same reasoning as above, we may identify $\mathcal{T}(S)$ with the set of *marked* torsion-free Euclidean lattices, modulo Euclidean isometries and scalings; the term "marked" means that every lattice has a specified ordered pair of generators. Up to isometry and scaling, we can arrange for one of the generators of the lattice to be 1, and the other one to lie above the x-axis. In this way have identified $\mathcal{T}(S)$ with the upper-half plane $\mathbb{H}$. For our purposes, this is just an identification as topological spaces; that said, Teichmüller spaces carry a natural metric, the so-called *Teichmüller metric*, for which the Teichmüller space of the torus, equipped with this metric, is isometric to the hyperbolic plane $(\mathbb{H}, d_{\mathbb{H}})$.

4.2. *Fenchel-Nielsen coordinates*

For a general surface S, the definition of Teichmüller space does not give a very clear insight on the structure of $\mathcal{T}(S)$. This will change once we introduce the so-called *Fenchel-Nielsen coordinates* for Teichmüller space. In terms of these coordinates, a point $X \in \mathcal{T}(S)$ will correspond to $6g-6$ real numbers; half of these correspond to the *lengths*, measured in X, of the curves in a fixed pants decomposition, and the other half correspond to the *twist* with which different pants have been glued to obtain the structure X. Before we define these coordinates, we need to introduce a few notions.

4.2.1. *Length functions*

Again, S denotes a closed surface of genus $g \geq 2$. Let γ be a homotopically non-trivial simple closed curve on S, and thus a non-trivial element of $\pi_1(S)$.

We claim that, given $X \in \mathcal{T}(S)$, there exists a unique simple closed geodesic in X that is homotopic to γ. To see this, we first regard $\pi_1(X) \cong \pi_1(S)$ as a subgroup of $\mathrm{PSL}(2, \mathbb{R})$, using the holonomy map. Under this identification, γ corresponds to a hyperbolic isometry $\bar{\gamma}$; otherwise it would be elliptic, which is impossible since X is a surface; or parabolic, which is also impossible since parabolic isometries have zero translation distance and X is compact. Now, γ is homotopic to the simple closed geodesic contained in projection of the axis of $\bar{\gamma}$. The uniqueness of the simple closed geodesic is obtained along similar lines; see Prop. 1.3 of [13].

Let $\mathcal{C}(S)$ be the set of homotopy classes of simple closed curves on S. For simplicity, we will refer to the elements of $\mathcal{C}(S)$ simply as *curves*, and we will often blur the distinction between a curve and any of its representatives.

Given $\gamma \in \mathcal{C}(S)$, the *length function* of γ is the function

$$l_{\cdot}(\gamma) : \mathcal{T}(S) \to \mathbb{R}_+$$

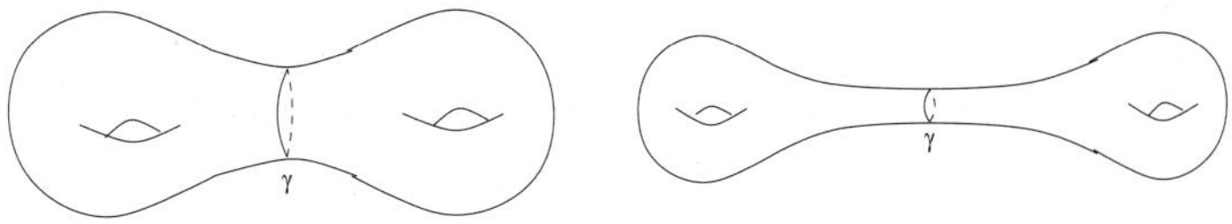

Fig. 3. Two different points in the Teichmüller space of a surface of genus 2.

given by

$$l_{[(X,f)]}(\gamma) = \text{length}_X(f(\gamma)),$$

where $\text{length}_X(f(\gamma))$ denotes the length of the unique geodesic representative of $f(\gamma)$ in X. For simplicity, we will denote $l_{[(X,f)]}(\gamma)$ simply by $l_X(\gamma)$.

Example 4.5. If X, $Y \in \mathcal{T}(S)$ are such that $\{l_X(\gamma)\}_{\gamma\in\mathcal{C}(S)} \neq \{l_Y(\gamma)\}_{\gamma\in\mathcal{C}(S)}$, then $X \neq Y$. In particular, the two surfaces in Figure 3 represent different different points in the Teichmüller space of the closed surface of genus 2.

As we will see, length functions are central to the definition of Fenchel-Nielsen coordinates; in addition, they are used to define the so-called *Thurston's compactification* of $\mathcal{T}(S)$, see the article [23] in this volume.

Let $X \in \mathcal{T}(S)$ and $\gamma \in \mathcal{C}(S)$. Let $\rho : \pi_1(X) \to \text{PSL}(2,\mathbb{R})$ be the holonomy representation of $\pi_1(X)$, noting that $\rho(\gamma)$ is a hyperbolic element of $\text{PSL}(2,\mathbb{R})$. Recall from (2.1) that the trace and translation distance of $\rho(\gamma)$ are related by $\text{tr}^2(\rho(\gamma)) = 4\cosh^2(\frac{l(\rho(\gamma))}{2})$. Also, note that $l(\rho(\gamma)) = l_X(\gamma)$, and so we deduce that length functions are continuous:

Lemma 4.6 (Length Functions are Continuous). *For every $\gamma \in \mathcal{C}(S)$, the function $l.(\gamma) : \mathcal{T}(S) \to \mathbb{R}_+$ is continuous.*

4.2.2. *Multicurves and pants decompositions*

Given two curves $\gamma, \gamma' \subset S$, the *intersection number* of γ and γ', denoted by $i(\gamma,\gamma')$, is the minimal cardinality of $\gamma \cap \gamma'$ among all representatives of γ and γ'. If $i(\gamma,\gamma') = 0$, we say that γ and γ' are *disjoint*. We say that two curves γ, γ' *fill* the surface if $S\backslash(\gamma \cup \gamma')$ is a union of topological disks; equivalently, if any non-trivial curve on S intersects at least one of γ or γ'.

A *multicurve* on S is a collection of pairwise distinct, pairwise disjoint curves; such a collection is necessarily finite, and consists of at most $3g-3$ curves. A multicurve μ that is maximal with respect to inclusion is called a *pants decomposition* of S; note that $S \setminus \mu$ has exactly $2g-2$ components, and that the closure of each of them is homeomorphic to a sphere with three boundary components, or *pair of pants.*

4.2.3. *The Teichmüller space of a pair of pants*

We start by stating a well-known result in hyperbolic geometry, namely that a right-angled hyperbolic hexagon is determined by the lengths of any three non-consecutive sides. By a *marked* hyperbolic hexagon H we mean a hexagon in $\mathbb{H}$, together with a distinguished vertex, and a labelling $s_1, \ldots, s_6$ of the sides of H, in such way that the sides occur in that order when travelling counterclockwise along H from the distinguised vertex. Denote by l_i the hyperbolic length of the side s_i. We have:

Lemma 4.7. *Let $a, b, c > 0$. There exists a marked right-angled hyperbolic hexagon $H \subset \mathbb{H}$ such that $l_1 = a$, $l_3 = b$, and $l_5 = c$. Moreover, any two such marked hexagons are isometric via an element of* $\mathrm{PSL}(2, \mathbb{R})$ *sending one distinguished vertex to the other.*

The proof of Lemma 4.7 is an exercise in hyperbolic geometry; see ([13], Prop. 10.4) for details. Armed with Lemma 4.7, we are now in a position to understand the Teichmüller space $\mathcal{T}(\mathcal{P})$ of a pair of pants $\mathcal{P}$; the definition of $\mathcal{T}(P)$ is analogous to Definition 4.1, now considering hyperbolic structures with geodesic boundary, and requiring the isometry and the homotopies to fix the boundary pointwise. Denoting the three boundary components of $\mathcal{P}$ by $\gamma_1, \gamma_2, \gamma_3$, we have:

Lemma 4.8. *The map*

$$F : \mathcal{T}(\mathcal{P}) \to \mathbb{R}^3_+,$$

given by $F(X) = (l_X(\gamma_1), l_X(\gamma_2), l_X(\gamma_3))$, is a homeomorphism.

Proof. (Sketch) *(i) F is onto:* Let $(a, b, c) \in \mathbb{R}^3_+$. By Lemma 4.7, up to the action of $\mathrm{PSL}(2, \mathbb{R})$ there exists a unique marked right-angled hyperbolic hexagon H such that s_1, s_3 and s_5 have length $a/2, b/2$ and $c/2$, respectively. Now, glue two copies of H along s_2, s_4, s_6, obtaining a hyperbolic structure with geodesic boundary on $\mathcal{P}$, such that the lengths of the three boundary components are equal to a, b, c, respectively.

(ii) F is injective: Consider $X \in \mathcal{T}(\mathcal{P})$ and let $F(X) = (a, b, c) \in \mathbb{R}^3_+$. For each $i \neq j$ there exists a unique geodesic arc A_{ij} from γ_i to γ_j, perpendicular to both γ_i and γ_j. Then $X \setminus (A_{12} \cup A_{23} \cup A_{13})$ has two connected components, and the closure of each is a right-angled hyperbolic hexagon.

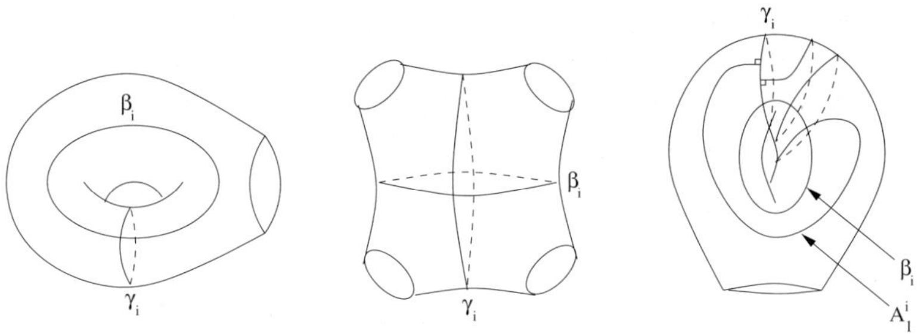

Fig. 4. Left and centre: A curve $\beta_i \subset S_i$ intersecting γ_i minimally, depending on the two possibilities for S_i. Right: The arc A_i when S_i is homeomorphic to a torus with one boundary component.

Since a, b, c are fixed, we know the lengths of three non-consecutive sides of each hexagon. Then, by Lemma 4.7, each hexagon is determined up to isometry and, therefore, so is the hyperbolic structure on X.

(iii) F is continuous. Finally, to see that F is continuous one first needs to modify Definition 4.2 to accommodate for surfaces with boundary. Once this is done, the continuity of F follows immediately from the definition; see ([13], Prop. 10.4) for details. □

4.2.4. *The coordinates*

Let S be a closed orientable surface of genus $g \geq 2$. We want to define a homeomorphism

$$F : \mathcal{T}(S) \to \mathbb{R}_+^{3g-3} \times \mathbb{R}^{3g-3}.$$

Fix a pants decomposition $\gamma_1, \ldots, \gamma_{3g-3}$ on S, and fix an orientation for each of the curves. Let $X \in \mathcal{T}(S)$. The first $3g-3$ coordinates of $F(X)$, known as the *length parameters* of X, are simply the lengths $l_X(\gamma_1), \ldots, l_X(\gamma_{3g-3})$ in X of the curves γ_i. The other $3g-3$ coordinates $\tau_1(X), \ldots, \tau_{3g-3}(X)$, known as the *twist parameters* of X, are slightly more complicated. There are many (equivalent) ways of defining them; see, for instance, [5, 6, 16, 17, 26]. One way to do it is as follows:

Each of the curves γ_i is contained in a unique component S_i of $X \backslash (\bigcup_{j \neq i} \gamma_j)$ whose closure is homeomorphic to either a torus with one boundary component, or a sphere with four boundary components. Choose a curve β_i that is contained in S_i and intersects γ_i minimally; see Figure 4. In addition, in each pair of pants of $S_i \backslash \gamma_i$ we consider the unique geodesic arc that is entirely contained in that pair of pants, has endpoints on γ_i

and is perpendicular to γ_i. Denote by $(A^i_j)_j$ the collection of arcs obtained in this way, observing that $(A^i_j)_j$ has exactly one element if S_i is homeomorphic to a torus with one boundary component, and that it has two otherwise; see Figure 4 for an example of the former case.

Choose a basepoint $p \in \gamma_i \cap A^i_1$, and observe that $\pi_1(S_i, p)$ is generated by elements that have representatives which are entirely contained in $\gamma_i \cup (\cup_j A^i_j)$. Therefore, we may homotope β_i onto a curve β'_i contained in $\gamma_i \cup (\cup_j A^i_j)$; moreover, by tightening β'_i if necessary, we may assume that β'_i does not backtrack along γ_i. Then define $\tau_i(X)$ as the signed length of the segment of β'_i that runs along γ_i and contains p; the sign is positive if β'_i runs along γ_i in the sense given by the fixed orientation on γ_i, and negative otherwise.

Once we have defined Fenchel-Nielsen coordinates, we may state our promised theorem:

Theorem 4.9. *The map*

$$F : \mathcal{T}(S) \to \mathbb{R}^{3g-3}_{+} \times \mathbb{R}^{3g-3},$$

given by

$$F(X) = (l_X(\gamma_1), \ldots, l_X(\gamma_{3g-3}), \tau_1(X), \ldots, \tau_{3g-3}(X)),$$

is a homeomorphism.

Proof. The map F is continous since it is defined in terms of length functions, which are continuous by Lemma 4.6. Also, F is bijective because it admits an inverse, which may intuitively be described as follows: given a tuple

$$(l_1, \ldots, l_{3g-3}, \tau_1, \ldots, \tau_{3g-3}) \in \mathbb{R}^{3g-3}_{+} \times \mathbb{R}^{3g-3},$$

one first constructs $2g - 2$ hyperbolic pairs of pants whose bounday components have length prescribed by the l_i, and then one glues the pairs of pants along the boundaries according to the twist parameters τ_i; see ([13], Thm. 10.6) for details. □

5. Mapping Class Groups

In this section we introduce the mapping class group of a surface and discuss some of its elements. We then describe how the mapping class group acts on Teichmüller space, and prove that this action is properly discontinuous. We refer the reader to [13, 14, 18] for a thorough discussion on mapping class groups.

5.1. *Definition and examples*

Let S be an orientable surface of genus $g \geq 1$. Again, for simplicity, we restrict our attention to the case where S is closed.

Definition 5.1. The *mapping class group* $\mathrm{Mod}(S)$ of S is the group of homotopy classes of orientation-preserving homeomorphisms of S; in other words,

$$\mathrm{Mod}(S) = \mathrm{Homeo}^+(S)/\,\mathrm{Homeo}_0(S),$$

where $\mathrm{Homeo}_0(S)$ denotes the connected component of $\mathrm{Homeo}^+(S)$ containing the identity. Elements of $\mathrm{Mod}(S)$ are called *mapping classes*. We will also need to consider the *extended mapping class group* $\mathrm{Mod}^{\pm}(S)$, that is, the group of all homeomorphisms of S up to homotopy.

Example 5.2 (Mapping Class Group of the Torus). If S is a torus, then $\mathrm{Mod}^{\pm}(S) \cong \mathrm{GL}(2,\mathbb{Z})$. Indeed, given a homeomorphism $g : S \to S$, let $g_* \in \mathrm{GL}(2,\mathbb{Z})$ be the induced automorphism of $\pi_1(S) \cong \mathbb{Z}^2$. Now, homotopic homeomorphisms induce conjugate automorphisms, and thus we have a homomorphism $G : \mathrm{Mod}^{\pm}(S) \to \mathrm{GL}(2,\mathbb{Z})$ given by $G([g]) = [g_*]$. The homomorphism G is clearly surjective; also, if $[g_*]$ is the identity, then g is homotopic to the identity and so G is also injective. Using the same reasoning, plus the fact that orientation-preserving homeomorphism must preserve algebraic intersection number, we obtain that $\mathrm{Mod}(S) \cong \mathrm{SL}(2,\mathbb{Z})$.

The example above is a particular instance of a general result, known as the *Dehn-Nielsen-Baer Theorem.* This result, which we state next, asserts that, if S is closed, the outer automorphism group $\mathrm{Out}(\pi_1(S))$ of $\pi_1(S)$ is isomorphic to the extended mapping class group $\mathrm{Mod}^{\pm}(S)$. The proof follows an argument similar to the one for the torus, but a substantial amount of extra work is required; see ([13], Thm. 8.1).

Theorem 5.3 (Dehn-Nielsen-Baer). *Let S be a closed surface of genus $g \geq 1$. Then* $\mathrm{Mod}^{\pm}(S) \cong \mathrm{Out}(\pi_1(S))$.

5.1.1. *Examples of mapping classes*

We now give some examples of mapping classes:

Example 5.4 (Finite Order). If $\psi : S \to S$ is a finite order homeomorphism, then its homotopy class $[\psi]$ is a finite order mapping class. Conversely, it is not difficult to see that every finite order mapping class is represented by a finite order homeomorphism; essentially, since Teichmüller space is contractible, every mapping class of finite order must have a fixed point, see ([13], Thm. 7.1) for details. More generally, a celebrated result of Kerkchoff [22] states that every finite subgroup of $\mathrm{Mod}(S)$ is realized by a finite group of surface homeomorphisms.

Example 5.5 (Dehn Twist). Consider the annulus $A = ([0,n] \times [0,1])/\sim$, where $(0,y) \sim (n,y)$. Let $T : A \to A$ be the affine homeomorphism of A that takes the vector $(0,1)$ to the vector $(n,1)$, so

$$T = \begin{pmatrix} 1 & n \\ 0 & 1 \end{pmatrix}.$$

Let $\gamma \in \mathcal{C}(S)$ and let N_γ be a regular neighbourhood of γ. Choose an orientation-preserving homeomorphism $h : A \to N_\gamma$. The *right Dehn twist* t_γ about γ is defined as

$$t_\gamma(x) = \begin{cases} hTh^{-1}(x), & x \in N_\gamma \\ x, & x \notin N_\gamma \end{cases}.$$

Observe that t_γ is only well-defined as a mapping class.

One of the many reasons why Dehn twists constitute an important type of mapping class is because they generate the mapping class group. In fact, one has more:

Theorem 5.6 (Dehn-Lickorish). $\mathrm{Mod}(S)$ *is generated by finitely many Dehn twists.*

We refer the reader to ([13], Ch. 4) for a proof of Theorem 5.6, and for explicit examples of Dehn twists that generate $\mathrm{Mod}(S)$.

We now introduce another important type of mapping classes, namely *pseudo-Anosov* mapping classes, by means of an example due to Thurston [25].

Example 5.7 (Pseudo-Anosov). Let α and β be two curves that fill S, and choose representatives of α and β that realize $i(\alpha,\beta)$. Since α and β fill S, every connected component of $S\backslash(\alpha \cup \beta)$ is a topological disk. We thus obtain a cell decomposition of S whose vertices are precisely the $i(\alpha,\beta)$ points of intersection between α and β, and whose 1-cells are contained in $\alpha \cup \beta$.

Now consider the dual cell complex $\mathcal{D}$ of this cell decomposition. The complex $\mathcal{D}$ is also a cell decomposition of S, whose 2-cells correspond precisely to the $i(\alpha,\beta)$ points of intersection of α and β. By deeming each 2-cell

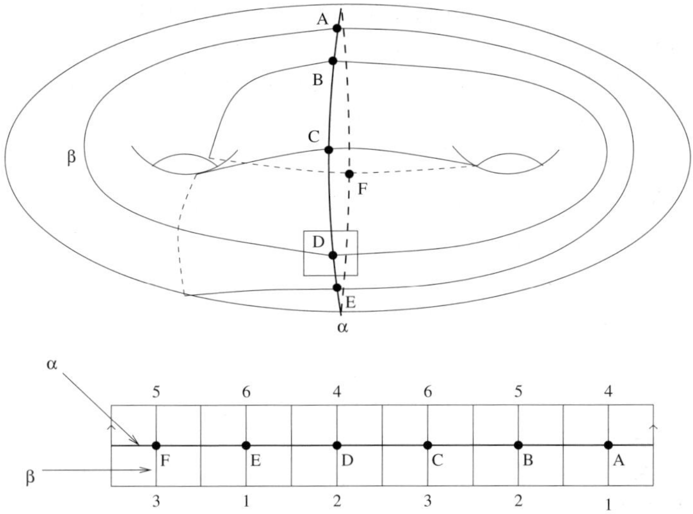

Fig. 5. Top: Two curves α and β that fill a genus 2 surface S; here $i(\alpha, \beta) = 6$. Bottom: The dual cell decomposition of S determined by α and β. We obtain S by gluing the top and bottom of the rectangle, and then the vertical sides according to the labelling given.

of $\mathcal{D}$ to be a Euclidean square, we obtain a *singular Euclidean structure* on S: away from the vertices of $\mathcal{D}$ the metric is locally Euclidean, and at the vertices of $\mathcal{D}$ there are *cone singularities*, each with cone angle $k\pi$ for some $k \geq 2$: this is explained in more detail in Leininger's article [23].

We choose geodesic representatives of α and β in this singular Euclidean structure, so that α and β bisect each square through the midpoints of parallel edges, and intersect each other at the centre of each square. See Figure 5 for a example on a closed surface of genus 2.

Now, the Dehn twists t_α and t_β act as affine transformations of this singular Euclidean structure, namely by the matrices

$$t_\alpha = \begin{pmatrix} 1 & n \\ 0 & 1 \end{pmatrix} \quad \text{and} \quad t_\beta = \begin{pmatrix} 1 & 0 \\ -n & 1 \end{pmatrix},$$

where $n = i(\alpha, \beta)$. Therefore

$$t_\alpha t_\beta^{-1} = \begin{pmatrix} 1 + n^2 & n \\ n & 1 \end{pmatrix}.$$

The matrix of $t_\alpha t_\beta^{-1}$ has two real eigenvalues $\lambda, \lambda^{-1} \neq 1$. The corresponding eigenvectors are orthogonal, and thus determine a pair of orthogonal *singular foliations* of the surface, with singularities at the vertices of $\mathcal{D}$; again, see Leininger's article [23] for a thorough explanation of this. Now, $t_\alpha t_\beta^{-1}$ preserves these foliations, and expands along one by a factor of λ and contracts along the other by a factor of λ^{-1}.

As mentioned before, the mapping class $t_\alpha t_\beta^{-1}$ is an example of a *pseudo-Anosov* mapping class. In general, a pseudo-Anosov mapping class comes equipped with a pair of orthogonal *singular foliations* of the surface (see Leininger's article [23] for a detailed exposition of foliations on surfaces), and expands along one by a fixed factor λ and contracts along the other by a factor of λ^{-1}. It is not difficult to see that, as a consequence, a pseudo-Anosov mapping class does not fix any non-trivial simple closed curves on S.

Nielsen-Thurston's Classification of Mapping Classes. A mapping class ϕ may fix a non-trivial multicurve on the surface (e.g. a Dehn twist) or it may not (e.g. a pseudo-Anosov mapping class). In the former case, ϕ is said to be *reducible*; in the latter case, ϕ is called *irreducible.* Observe that a finite order mapping class may be reducible or irreducible; see ([13], Ch. 13.2.2) for specific examples. The celebrated *Nielsen-Thurston classification* of elements of $\text{Mod}(S)$ asserts that every irreducible mapping class of infinite order is pseudo-Anosov. Namely:

Theorem 5.8 (Nielsen-Thurston Classification). *Let* $\phi \in \text{Mod}(S)$. *Then* ϕ *is either periodic, reducible or pseudo-Anosov.*

We point the reader to [13, 14] and the references therein for a detailed exposition on the Nielsen-Thurston classification of mapping classes.

Using the Nielsen-Thurston classification, one may give a complete description of the structure of a general mapping class, as we now briefly explain; see ([13], Ch. 13) for more details. A *reduction system* for $\phi \in \text{Mod}(S)$ is a multicurve $\mu \subset S$ such that $\phi(\mu) = \mu$. Now, ϕ fixes a canonically defined multicurve μ_ϕ on S, namely the intersection of all *maximal* (with respect to inclusion) reduction systems for ϕ; following [7], the multicurve μ_ϕ is called the *canonical reduction system* of ϕ. The mapping class ϕ may permute the elements of μ_ϕ, as well as the connected components of $S \setminus \mu_\phi$. However, there exists $n \in \mathbb{N}$ such that ϕ^n does not permute the components of μ_ϕ or $S \setminus \mu_\phi$; observe that n is uniformly bounded above in terms of the genus of S. Then, ϕ^n acts as a power of a Dehn twist about each component of μ_ϕ , and the restriction of ϕ^n to each connected component S' of $S \setminus \mu_\phi$ is either the identity or a pseudo-Anosov mapping class of $\text{Mod}(S')$.

5.2. *The action of* Mod(S) *on* $\mathcal{T}(S)$

The mapping class group acts naturally on Teichmüller space, namely if $\psi \in \mathrm{Mod}(S)$ and $[(X,f)] \in \mathcal{T}(S)$ then $\psi([(X,f)]) = [(X, f \circ g^{-1})]$, where g denotes any representative of ψ. In terms the character variety $\mathcal{DF}(\pi_1(S), \mathrm{PSL}(2,\mathbb{R}))$, the action of Mod($S$) on $\mathcal{T}(S)$ is given by Theorem 5.3. As mentioned above, the main goal of this section is to prove the following result:

Theorem 5.9. Mod(S) *acts on* $\mathcal{T}(S)$ *properly discontinuously.*

Proof. Suppose, for contradiction, that there exist a compact set $K \subset \mathcal{T}(S)$ and a sequence $(\psi_n)_n$ of *distinct* elements of Mod(S) such that

$$\psi_n(K) \cap K \neq \emptyset, \text{ for all } n \in \mathbb{N}.$$

Thus, there exists a sequence $(X_n)_n$ of elements of K such that $\psi_n(X_n) \in K$ for all $n \in \mathbb{N}$. Let α and β be two curves that fill S. On the one hand, Lemma 4.6 implies that there exists $R = R(K) > 0$ such that $l_\alpha(X) + l_\beta(X) \leq R$ for all $X \in K$. On the other hand, we will show that, up to relabelling α and β,

$$\lim_{n\to\infty} l_{\psi_n^{-1}(\alpha)}(X_n) = \infty.$$

Having showed this, we will obtain the desired contradiction since, by definition, $l_{\psi_n^{-1}(\alpha)}(X_n) = l_\alpha(\psi_n(X_n))$.

Claim 1. *At least one of* $(\psi_n^{-1}(\alpha))_n$ *and* $(\psi_n^{-1}(\beta))_n$ *has a non-constant subsequence.*

Proof. Indeed, suppose this were not the case. Then, up to taking a subsequence, there are simple closed curves α' and β' on S such that $\psi_n^{-1}(\alpha) = \alpha'$ and $\psi_n^{-1}(\beta) = \beta'$ for all n. Therefore, up to the action of Mod(S), we may assume that $\psi_n^{-1}(\alpha) = \alpha$ and $\psi_n^{-1}(\beta) = \beta$ for all n. Now, α and β fill S, and so each component of $S \setminus (\alpha \cup \beta)$ is a topological disk. Since ψ_n fixes both α and β, its action on S is determined by the induced permutation on the set of disks of $S \setminus (\alpha \cup \beta)$. As S is compact, there are only finitely many such disks, and we have a contradiction to the ψ_n being pairwise distinct. Thus our claim follows.

Hence, up to relabelling and extracting a subsequence, we may assume that $(\psi_n^{-1}(\alpha))_n$ is a sequence of distinct curves on S; in order to simplify notation, we will write $\alpha_n = \psi_n^{-1}(\alpha)$. Next, we claim:

Claim 2. *There exists a pants decomposition P of S such that $i(\alpha_n, \gamma) \to \infty$ for some $\gamma \in P$.*

Proof. Choose any pants decomposition Q and suppose that, for all $\gamma \in Q$, $i(\alpha_n, \gamma)$ is uniformly bounded. Therefore the number of arcs of α_n in the complement of Q is bounded independently of n. As a consequence, the curves α_n differ only up to Dehn twisting about some component of Q; more formally, again up to extracting a subsequence, there exists a simple closed curve α' on S such that $\alpha_n = T_n(\alpha')$, for some $T_n \in \mathbb{T}_Q$, the subgroup of $\mathrm{Mod}(S)$ generated by the Dehn twists on the components of Q. As the α_n are pairwise distinct, then $T_n \neq T_m$ for $n \neq m$; moreover, up to extracting a subsequence, there exists a curve $\gamma' \in Q$ such that every T_n is supported on a submulticurve of Q containing γ'. Let γ be a curve in $S - (Q - \gamma')$ such that either $i(\gamma, \gamma') = 1$ if $S - (Q - \gamma')$ contains a one-holed torus, or $i(\gamma, \gamma') = 2$ if $S - (Q - \gamma')$ contains a four-holed sphere; compare with Figure 4, where β_i and γ_i play the role of γ and γ' respectively. Setting $P = (Q - \gamma') \cup \gamma$, we obtain the desired result.

Continuing with the proof of the main result, we may choose a curve γ on S such that $i(\alpha_n, \gamma) \to \infty$, by Claim 2. Now, there exists $\epsilon = \epsilon(K) > 0$ such that, for all $X \in K$, the ϵ-neighborhood of γ in X is an embedded annulus in X; this may be seen explicitly by considering the construction of a hyperbolic pair of pants from hyperbolic hexagons and using that K is compact, and is also an easy consequence of the *Collar Lemma* of Keen [21] and Halpern [15]. Since $X_n \in K$ for all n, we have

$$l_{\psi_n^{-1}(\alpha)}(X_n) \geq \epsilon \cdot i(\psi_n^{-1}(\alpha), \gamma) \to \infty,$$

which gives the desired contradiction. This finishes the proof of Theorem 5.9. □

Acknowledgements

I would like to express my gratitude to the Institute for Mathematical Sciences of Singapore for their hospitality and financial support. The vast majority of this gratitude is targetted at Ser Peow Tan for all his help, support and friendship. I would like to say thanks to Chris Leininger for help and discussions. Thanks also to David Dumas, Jonah Gaster, Add Jaipong, Thomas Koberda, Brice Loustau and Sara Maloni for interesting conversations. Finally, I would like to thank whoever came up with the idea of a "pepper crab", for (s)he deserves a place in heaven.

References

1. W. Abikoff, *The real analytic theory of Teichmüller space.* Lecture Notes in Mathematics, 820. *Springer*, Berlin, 1980.
2. J.W. Anderson. *Hyperbolic geometry.* Springer Undergraduate Mathematics Series. Springer-Verlag London, Ltd., London, 1999.
3. W. Ballmann. *Lectures on spaces of nonpositive curvature. With an appendix by Misha Brin.* DMV Seminar, 25. Birkhuser Verlag, Basel, 1995.
4. A. F. Beardon, *The geometry of discrete groups.* Graduate Texts in Mathematics, 91. *Springer-Verlag*, New York, 1983.
5. R. Benedetti & C. Petronio. *Lectures on hyperbolic geometry.* Universitext. Springer-Verlag, Berlin, 1992. xiv+330 pp.
6. M. Bestvina, K. Bromberg, K. Fujiwara, J. Souto, Shearing coordinates and convexity of length functions. Preprint, 2009.
7. J. Birman, A. Lubotzky and J. McCarthy, *Abelian and solvable subgroups of the mapping class groups*, Duke Math. J. 50 (1983).
8. F. Bonahon, Low-dimensional geometry. *From Euclidean surfaces to hyperbolic knots.* Student Mathematical Library, 49. IAS/Park City Mathematical Subseries. American Mathematical Society, Providence, RI; Institute for Advanced Study (IAS), Princeton, NJ, 2009. xvi+384 pp.
9. M. R. Bridson & A. Haefliger. *Metric spaces of non-positive curvature.* Grundlehren der Mathematischen Wissenschaften [Fundamental Principles of Mathematical Sciences], 319. Springer-Verlag, Berlin.
10. R. D. Canary, D. B. A. Epstein & P. L. Green. *Notes on notes of Thurston With a new foreword by Canary.* London Math. Soc. Lecture Note Ser., 328, Fundamentals of hyperbolic geometry: selected expositions, 1115, Cambridge Univ. Press, Cambridge, 2006.
11. A. J. Casson & S. A. Bleiler. *Automorphisms of surfaces after Nielsen and Thurston.* London Mathematical Society Student Texts, 9. Cambridge University Press, Cambridge, 1988. iv+105 pp.
12. M. P. do Carmo. *Riemannian geometry.* Mathematics: Theory & Applications. Birkhäuser Boston, Inc., Boston, MA, 1992.
13. B. Farb & D. Margalit. *A primer on mapping class groups.* Princeton University Press, to appear.
14. A. Fathi, F. Laudenbach & V. Poenaru, editors. *Travaux de Thurston sur les surfaces*, second edition, Astérisque 66, Soc. Math. France, Paris (1991)
15. N. Halpern, A proof of the collar lemma. *Bull. London Math. Soc.* **13** (1981), no. 2, 141144.
16. J. H. Hubbard. *Teichmüller theory and applications to geometry, topology, and dynamics. Vol. 1. Teichmüller theory.* Matrix Editions, Ithaca, NY, 2006.
17. Y. Imayoshi & M. Taniguchi. *An introduction to Teichmüller spaces.* Springer-Verlag, Tokyo, 1992. xiv+279 pp.
18. N. V. Ivanov, Nikolai. *Mapping class groups.* Handbook of geometric topology, 523633, North-Holland, Amsterdam, 2002.
19. G. A. Jones & D. Singerman. *Complex functions. An algebraic and geometric viewpoint.* Cambridge University Press, Cambridge, 1987.

20. S. Katok. *Fuchsian groups.* Chicago Lectures in Mathematics. University of Chicago Press, Chicago, IL, 1992.
21. L. Keen, Collars on Riemann surfaces. *Discontinuous groups and Riemann surfaces (Proc. Conf., Univ. Maryland, College Park, Md., 1973)*, pp. 263268. Ann. of Math. Studies, No. 79, Princeton Univ. Press, Princeton, N.J., 1974.
22. S. P. Kerckhoff, The Nielsen realization problem. *Ann. of Math. (2)* **117** (1983), no. 2, 235265.
23. C. J. Leininger, Degeneration of hyperbolic structures on surfaces. Also in this volume.
24. C. M. Series *Hyperbolic Geometry.* Online notes, available from www.maths.warwick.ac.uk/~masbb.
25. W. P. Thurston, On the geometry and dynamics of diffeomorphisms of surfaces, *Bull. Amer. Math. Soc*, **19**(2), 1988.
26. W. P. Thurston. *Three-dimensional geometry and topology. Vol. 1.* Princeton Mathematical Series, 35. Princeton University Press, Princeton, NJ, 1997.

DEGENERATIONS OF HYPERBOLIC STRUCTURES ON SURFACES

Christopher J. Leininger
Department of Mathematics, University of Illinois
Urbana–Champaign, IL, USA
clein@math.uiuc.edu

We give an introduction to some of the tools used in studying the degeneration of hyperbolic structures on surfaces as developed by Thurston, and in particular we describe his construction of a compactification of Teichmüller space.

Keywords: Surface, hyperbolic, lamination, foliation

Mathematics Subject Classification 2000: 57M50, 32G15, 57R30

1. Introduction

These notes are a continuation of lecture notes by J. Aramayona [2] which, beginning with an introduction to the geometry of the hyperbolic plane $\mathbb{H}^2$, introduce the notion of a hyperbolic structure on S, a closed oriented surface of genus $g \geq 2$, and study the space of equivalence classes of all hyperbolic structures, the Teichmüller space of S. In these notes we will provide a somewhat informal introduction to some of the work of W.P. Thurston on degenerations of hyperbolic structures and his compactification of Teichmüller space. After some preliminary discussion of length functions on Teichmüller space (Section 2), we introduce both measured foliations (Section 3) and measured geodesic laminations (Section 4), explain how one can pass back and forth between these two types of objects (Section 5) as well as how to parametrize the space of all such (Section 6), and finally explain how Thurston compactifies Teichmüller space via measured foliations (Section 7). The main references for the material discussed here are [16, 22, 13, 6, 24, 34, 12].

1.1. *Notation, terminology and some conventions*

We denote the Teichmüller space of S by

$$\mathcal{T}(S) = \{[f, X] \mid f : S \to X = \mathbb{H}^2/\Gamma\}$$

where $f : S \to X$ here is an orientation preserving homeomorphism, $\Gamma < \mathrm{Isom}^+(\mathbb{H}^2) = \mathrm{PSL}(2,\mathbb{R})$ is a discrete group, and $[f, X]$ represents the equivalence class via the equivalence relation $(f, X) \sim (h, Y)$ if and only if $h \circ f^{-1} \simeq \psi : X \to Y$, where ψ is an isometry. When there is no confusion, we write X instead of $[f, X]$. We also often think of a point in $\mathcal{T}(S)$ as a hyperbolic metric on S (up to isotopy). From $f : S \to X$ we obtain this hyperbolic metric $\rho = \rho_X$ by pulling back the metric on X by f.

The Teichmüller space naturally injects into the set of discrete faithful representations into $\mathrm{PSL}(2,\mathbb{R})$, modulo conjugation within the full group of isometries

$$\mathcal{T}(S) \to \mathcal{DF}(\pi_1(S), \mathrm{PSL}(2,\mathbb{R}))/conj$$

by the holonomy homomorphism $[f, X] \mapsto f_* : \pi_1(S) \to \Gamma < \mathrm{PSL}(2,\mathbb{R})$. This is a bijection, and we topologize $\mathcal{T}(S)$ by declaring this to be a homeomorphism onto its image. See [22, 33] for more details.

The hyperbolic metric ρ is an example of a *path metric* which means that the distance between any two points is the infimum of lengths of paths between those points. In general, given a path metric on S, if we let $p : \widetilde{S} \to S$ denote the universal covering, then there is a unique path metric on $\widetilde{S}$ for which the covering map is a local isometry. This is defined by declaring the lengths of paths in $\widetilde{S}$ to be the lengths of their images in S and we call this the *induced metric* on $\widetilde{S}$.

Because S is closed, hence compact, it follows that any metric on S is complete. Therefore, any path metric on S is a *geodesic metric*: the distance between two points is the length of some path between the points. In fact, one can check that the induced metric on $\widetilde{S}$ is also a geodesic metric. We refer the reader to [10] for a detailed discussion of path metrics and geodesic metrics.

We make a few remarks about terminology and set some definitions. First, we will use the same name for the metric on S and on $\widetilde{S}$ with the context determining which is intended. Second, we note that there are two uses of the word "geodesic" which we will need, and it is cumbersome to always adorn this word with all the appropriate adjectives. We therefore make the following convention. An isometric embedding of an interval in $\mathbb{R}$ into $\widetilde{S}$ (or its image) is a *geodesic* in $\widetilde{S}$. A *locally* isometric embedding of an interval or circle of some circumference (or its image) is a *geodesic* in S. In particular, geodesic has a different meaning depending on whether we are considering it in S or $\widetilde{S}$. When we require another meaning, we will include the necessary adjectives (e.g. we may refer to a *minimizing geodesic segment* between two points of S to refer to an isometrically embedded arc between the two points).

The set of homotopy classes of (homotopically nontrivial) curves which have a simple representative will be denoted $\mathcal{S} = \mathcal{S}(S)$. The *(geometric) intersection number* $\iota(\alpha, \beta)$ is defined for any pair $\alpha, \beta \in \mathcal{S}(S)$ as the minimal number of intersection points among all simple representatives. If we fix a hyperbolic metric ρ on S, then the geodesic representatives of α and β are in fact simple and realize the geometric intersection number, meaning that they intersect in precisely $\iota(\alpha, \beta)$ points. See [13, 15], for example.

2. Length Functions

Given $\alpha \in \mathcal{S}(S)$, we let

$$\ell_{\cdot}(\alpha) : \mathcal{T}(S) \to \mathbb{R}_+$$

denote the *length function* of α:

$$\ell_{[f,X]}(\alpha) = \ell_X(\alpha) := \text{length}_X(f(\alpha))$$

where $\text{length}_X(f(\alpha))$ is the length of the geodesic representative of the homotopy class of α. Equivalently, this is the ρ_X–length of the geodesic representative of α in S.

Since the length of $f(\alpha)$ in X is equal to the translation length of $f_*(\alpha_0)$, where $\alpha_0 \in \pi_1(S)$ is any element whose conjugacy class represents α, an exercise in hyperbolic geometry shows that

$$\ell_{[f,X]}(\alpha) = 2\cosh^{-1}\left(\frac{|tr(f_*(\alpha_0))|}{2}\right). \tag{2.1}$$

This last equation shows that $\ell_{\cdot}(\alpha)$ is a continuous function on $\mathcal{T}(S)$ for all $\alpha \in \mathcal{S}(S)$.

Our goal in this section is to use the lengths of curves to completely describe the topology of $\mathcal{T}(S)$. More precisely, write

$$\ell_* : \mathcal{T}(S) \to \mathbb{R}^{\mathcal{S}}$$

for the map $X \mapsto \{\ell_X(\alpha)\}_{\alpha \in \mathcal{S}}$.

Theorem 2.1. *The map ℓ_* is a proper embedding. In fact, there exists $\alpha_1, \ldots, \alpha_{9g-9} \in \mathcal{S}$ such that*

$$\mathcal{T}(S) \to \mathbb{R}^{9g-9}$$

given by $X \mapsto \{\ell_X(\alpha_i)\}_{i=1}^{9g-9}$ is a proper embedding.

Before we give the proof, first we recall the Fenchel-Nielsen coordinates as described in Aramayona's Lectures [2]; see also [22, 15, 33, 8] for other descriptions. Let $\mathcal{P} = \{\alpha_1, \ldots, \alpha_{3g-3}\}$ be a pants decomposition with transverse curves $\beta_1, \ldots, \beta_{3g-3}$ (so $\iota(\alpha_i, \beta_i)$ is 1 or 2, and $\iota(\alpha_i, \beta_j) = 0$ for $i \neq j$). The coordinates

$$F = F_{\mathcal{P}} : \mathcal{T}(S) \to \mathbb{R}_+^{3g-3} \times \mathbb{R}^{3g-3}$$

are given by

$$F(X) = (\ell_1(X), \ldots, \ell_{3g-3}(X), t_1(X), \ldots, t_{3g-3}(X))$$

where $\ell_i(X) = \ell_X(\alpha_i)$ and $t_i(X)$ is the twist coordinate of β_i in α_i for X.

To prove the theorem, we need to show that the twist coordinates can be determined from the length functions. Fix all length coordinates ℓ_j and all twist coordinates t_j but one, say t_i. Set $X_{t_i} = F^{-1}(\ell_1, \ldots, \ell_{3g-3}, t_1, \ldots, t_{3g-3})$. We are thinking of X_{t_i} as a function of the single variable t_i.

Lemma 2.2. *For any curve β with $\iota(\alpha_i, \beta) \neq 0$, the map*

$$t_i \mapsto \ell_{X_{t_i}}(\beta)$$

is a proper, strictly convex function.

We postpone the proof of this lemma and sketch the proof of Theorem 2.1.

Proof of Theorem 2.1. Let τ_i denote the Dehn twist in the curve α_i. Since $\iota(\beta_i, \alpha_j) = 0$ if $i \neq j$, we have $\tau_i(\beta_j) = \beta_j$ in this case.

A Dehn twist in α_i applied to a hyperbolic structure changes the ith twist coordinate by subtracting the length ℓ_i of α_i, and therefore

$$\ell_{X_{t_i}}(\tau_i(\beta_i)) = \ell_{\tau_i^{-1} \cdot X_{t_i}}(\beta_i) = \ell_{X_{t_i + \ell_i}}(\beta_i).$$

By Lemma 2.2 $\ell_{X_{t_i}}(\beta_i)$ is strictly convex as a function of t_i, and since $\ell_i > 0$, it follows that the map

$$t_i \mapsto (\ell_{X_{t_i}}(\beta_i), \ell_{X_{t_i + \ell_i}}(\beta_i)) = (\ell_{X_{t_i}}(\beta_i), \ell_{X_{t_i}}(\tau_i(\beta_i)))$$

is injective. Combining this with the fact that F is injective we see that the composition

$$\mathcal{T}(S) \xrightarrow{F} \mathbb{R}_+^{3g-3} \times \mathbb{R}^{3g-3} \to \mathbb{R}^{9g-9}$$

that sends $X \in \mathcal{T}(S)$ to

$$(\ell_X(\alpha_1), \ldots, \ell_X(\alpha_{3g-3}), \ell_X(\beta_1), \ldots, \ell_X(\beta_{3g-3}), \ell_X(\tau_1(\beta_1)), \ldots, \\ \ell_X(\tau_{3g-3}(\beta_{3g-3})))$$

is also injective, as required.

To complete the proof, we need only verify that these lengths determine a proper function. That is, if $X_n \to \infty$ in $\mathcal{T}(S)$, then we must check that the lengths of one of our curves tends to infinity. For this we use the fact that F is a homeomorphism onto $\mathbb{R}_+^{3g-3} \times \mathbb{R}^{3g-3}$. Therefore, up to subsequence, one of the following must happen:

- There exists some i so that $\ell_{X_n}(\alpha_i) = \ell_i(X_n) \to \infty$.
- There exists some i so that $\ell_{X_n}(\alpha_i) \to 0$. In this case we see that α_i has larger and larger collar neighborhoods in X_n (see [19, 23]), and hence $\ell_{X_n}(\beta_i) \to \infty$.
- There exists some i and $0 < c < C < \infty$ so that $c < \ell_{X_n}(\alpha_i) < C$ for all n and $t_i(X_n) \to \pm\infty$. In this case we have $\ell_{X_n}(\beta_i) \to \infty$.

In each case some curve has its length tending to infinity, and this completes the proof. □

To prove the lemma, we take a short detour into a beautiful result of Kerckhoff [24] which gives an expression for the derivative of the length of β (which we will see is increasing so that the function is convex). Since we are considering just one index at a time, let $\alpha = \alpha_i$ and $t = t_i$. Then X_t is the 1–parameter family of hyperbolic structures obtained by *cutting open along* α (that is, deleting α and taking the completion of the associated path metric), twisting length t to the right, and regluing. We consider any curve β with $\iota(\alpha, \beta) = k \neq 0$, and we write $\theta_1(t), \ldots, \theta_k(t)$ for the angles at the k points of intersection, measured counterclockwise from the geodesic representative of α to that of β in X_t. Kerckhoff's result is the following.

Theorem 2.3. *With notation as above,*

$$\frac{d}{dt}(\ell_{X_t}(\beta)) = \sum_{i=1}^{k} \cos(\theta_i(t)).$$

Proof. Fix some $t_0 \in \mathbb{R}$, and we prove the theorem at this t_0. It is useful to change perspective a bit. Imagine the curve β on S with α crossing it some number of times. We assume β is given by its geodesic representative on X_{t_0}. Now cut open along α, twist a distance $s = t - t_0$ and reglue to construct the surface $X_t = X_{t_0+s}$. Because we are not changing the hyperbolic structure in the complement of α, we can view β as a union of geodesic arcs with endpoints on α. We recover the homotopy class by connecting these arcs with arcs of length s running along α. See Figure 1.

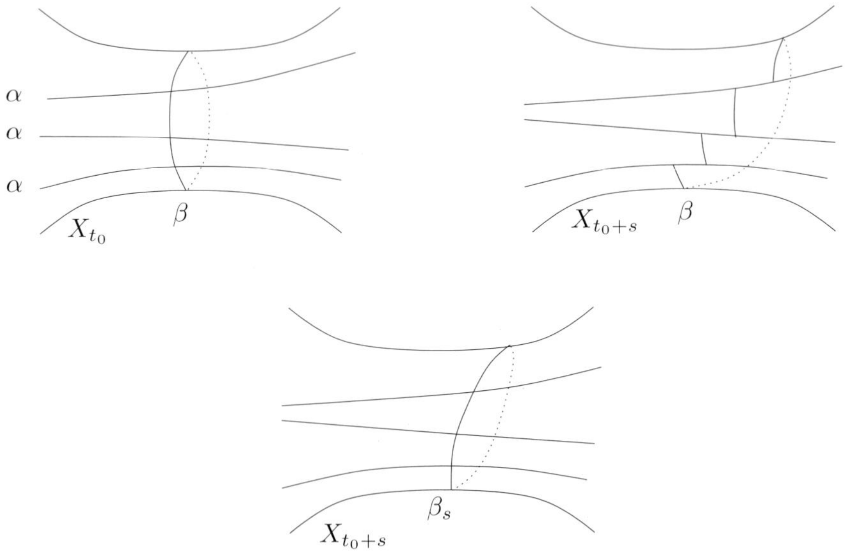

Fig. 1. Upper left: Original hyperbolic structure X_{t_0} with geodesics α and β. Upper right: Hyperbolic structure X_{t_0+s} obtained from X_{t_0} by cutting, shearing to the right a distance s along α, then regluing. Bottom: The geodesic representative β_s of β in X_{t_0+s}.

Inside X_{t_0+s} straighten the result to obtain the geodesic representative for β which we denote β_s. Between consecutive intersection points of α with β_s, we have arcs, $\beta_s^1, \ldots, \beta_s^k$, and we can think of these as arcs whose interiors lie entirely inside the hyperbolic surface X_{t_0+s} with α removed. However, with α removed, all the surfaces X_{t_0+s} are isometric to each other, and hence we can think of β_s^j as a 1-parameter family of geodesic arcs in X_{t_0} with endpoints in α. See Figure 2. When $s = 0$, we are at our initial hyperbolic structure and so $\beta_0^1, \ldots, \beta_0^k$ are the arcs of β cut open along α in X_{t_0}.

Let $0 = a_0 < a_1 < \ldots < a_k$ be such that $a_j - a_{j-1} = \ell_{X_{t_0}}(\beta_0^j)$, and assume we have parametrized the curves as $\beta_s(u)$ so that β_0 has unit speed and $\beta_s(a_j)$ is the jth point of intersection (and $\beta_s(a_0) = \beta_s(a_k)$).

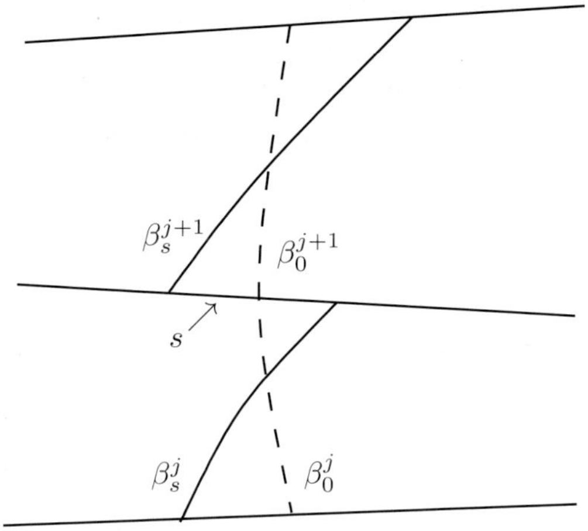

Fig. 2. The geodesic representatives β_s for X_{t_0+s} cut open and viewed back in X_{t_0} giving a 1-parameter family of geodesic arcs β_s^j. Endpoints of the consecutive arcs are distance exactly s apart.

Claim. *For each $j = 1, \ldots, k$ we have*

$$\left.\frac{d}{ds}\right|_{s=0} \ell(\beta_s^j) = \left\langle \beta_0'(a_j), \left.\frac{d}{ds}\right|_{s=0} \beta_s^j(a_j) \right\rangle - \left\langle \beta_0'(a_{j-1}), \left.\frac{d}{ds}\right|_{s=0} \beta_s^j(a_{j-1}) \right\rangle.$$

Proof. This follows from the first variation formula in Riemannian geometry (see e.g. [14]), but can also be seen using the hyperboloid model of the hyperbolic plane inside $(\mathbb{R}^{2,1}, \langle \cdot, \cdot \rangle)$ (see [3]) by appealing to the following two facts:

- $\cosh(d(x,y)) = -\langle x, y \rangle$ for any two points x and y in the hyperboloid.
- For any unit speed parametrization of a geodesic γ in the hyperboloid model and real numbers r, u we have

$$\gamma(r+u) = \gamma(r)\cosh(u) + \gamma'(r)\sinh(u).$$

Lifting to the hyperbolic plane in the hyperboloid model, the first equation implies

$$\cosh(\ell(\beta_s^j)) = -\langle \beta_s(a_{j-1}), \beta_s(a_j) \rangle.$$

Differentiating the left-hand side with respect to s at $s = 0$ yields

$$LHS' = \sinh(\ell(\beta_0^j)) \left.\frac{d}{ds}\right|_{s=0} (\ell(\beta_s^j)),$$

whereas differentiating with respect to s and applying the second fact to the geodesic β_0 twice, the right-hand side becomes

$$\begin{aligned} RHS' &= -\langle \left.\tfrac{d}{ds}\right|_{s=0} \beta_s(a_{j-1}), \beta_0(a_j)\rangle - \langle \beta_0(a_{j-1}), \left.\tfrac{d}{ds}\right|_{s=0} \beta_s(a_j)\rangle \\ &= -\langle \left.\tfrac{d}{ds}\right|_{s=0} \beta_s(a_{j-1}), \beta_0(a_{j-1})\cosh(\ell(\beta_0^j)) + \beta_0'(a_{j-1})\sinh(\ell(\beta_0^j)) \\ &\quad - \langle \beta_0(a_j)\cosh(-\ell(\beta_0^j)) + \beta_0'(a_j)\sinh(-\ell(\beta_0^j)), \left.\tfrac{d}{ds}\right|_{s=0} \beta_s(a_j)\rangle \\ &= -\langle \left.\tfrac{d}{ds}\right|_{s=0} \beta_s(a_{j-1}), \beta_0'(a_{j-1})\sinh(\ell(\beta_0^j))\rangle \\ &\quad + \langle \beta_0'(a_j)\sinh(\ell(\beta_0^j)), \left.\tfrac{d}{ds}\right|_{s=0} \beta_s(a_j)\rangle. \end{aligned}$$

Here we have used the fact that for any point $\mathbf{x}$ in the hyperboloid and any tangent vector $\mathbf{v}$ to the hyperboloid at $\mathbf{x}$ we have $\langle \mathbf{v}, \mathbf{x}\rangle = 0$. Then, factoring out $\sinh(\ell(\beta_0^j))$ from both sides we arrive at the desired equation. □

Next observe that $\beta_s^j(a_j)$ and $\beta_s^{j+1}(a_j)$ are not the same points in X_{t_0}: the former point is exactly a distance s to the right of the latter point along the geodesic α; see Figure 2. Therefore, if we let ξ_j denote the unit tangent vector to α at the jth point of intersection, then it follows that

$$\left.\frac{d}{ds}\right|_{s=0} \beta_s^j(a_j) - \left.\frac{d}{ds}\right|_{s=0} \beta_s^{j+1}(a_j) = \xi_j$$

for $j = 1, \ldots, k-1$ and similarly

$$\left.\frac{d}{ds}\right|_{s=0} \beta_s^k(a_k) - \left.\frac{d}{ds}\right|_{s=0} \beta_s^1(a_0) = \xi_k.$$

Combining the claim with this and summing over j (noting that $\beta'(a_k) = \beta'(a_0)$) we have

$$\left.\frac{d}{ds}\right|_{s=0} \ell_{X_{t_0+s}}(\beta) = \left.\frac{d}{ds}\right|_{s=0} \sum_{j=1}^k \ell(\beta_s^j) = \sum_{j=1}^k \langle \beta'(a_j), \xi_j\rangle = \sum_{j=1}^k \cos(\theta_j)$$

as required. □

Proof of Lemma 2.2. We now simply observe that as we increase t, the angles $\theta_i(t)$ are strictly decreasing; see Figure 3 and [25] for a proof. Since cos is also a strictly decreasing function on the interval $[0, \pi]$, we see that $\frac{d}{dt}\ell_\beta(t)$ is strictly increasing, hence ℓ_β is strictly convex.

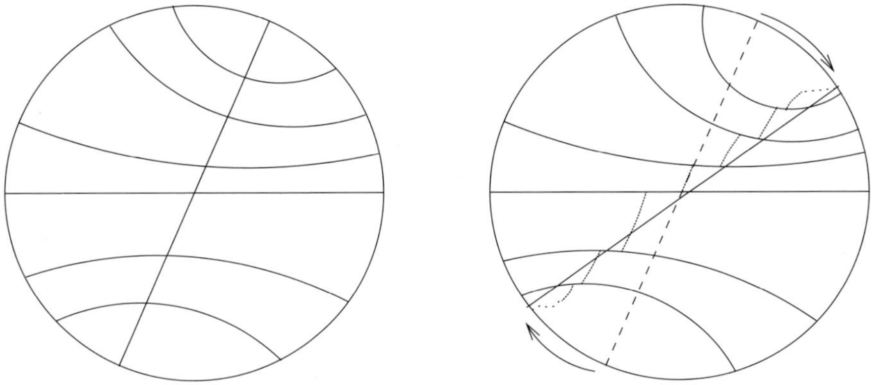

Fig. 3. For $s > 0$, $s \mapsto X_{t_0+s}$ shears everything to the right when viewed in the universal covering.

The fact that $\ell_\beta(t)$ is proper also follows from this picture. As $t \to \infty$ and $t \to -\infty$, we can see that the angle tends to 0 and π, respectively. Therefore, the sum of the cosines are bounded away from 0 as $t \to \pm\infty$, hence the derivatives are bounded away from 0, and $\ell_\beta(t)$ tends to ∞ as $t \to \pm\infty$. □

According to Theorem 2.1, we can describe the topology on $\mathcal{T}(S)$ just in terms of lengths of simple closed curves. Moreover, because the embedding to $\mathbb{R}^\mathcal{S}$ is proper, we can use this theorem to describe how we go to infinity in $\mathcal{T}(S)$: if $X_n \to \infty$ in $\mathcal{T}(S)$, then the length of some curve tends to infinity. In particular, the image of the sequence in $\mathbb{R}^\mathcal{S}$ is tending toward infinity, and we can try to look for a subsequence that becomes asymptotic to some ray—equivalently, we can projectivize and look for a convergent sequence. Thus, we would like to find a compactification of $\mathcal{T}(S)$ by a "projective light cone" just as in the hyperboloid model of hyperbolic space—or after projectivizing, as in the Klein model.

For the latter, we would like to know that the embedding $\mathcal{T}(S) \to \mathbb{R}^\mathcal{S}$ remains injective after projectivizing. Let $\mathbb{P}\ell_* : \mathcal{T}(S) \to \mathbb{PR}^\mathcal{S}$ be the composition of ℓ_* with the projectivization $\mathbb{P} : \mathbb{R}^\mathcal{S} - \{0\} \to \mathbb{PR}^\mathcal{S}$.

Proposition 2.4. *The map $\mathbb{P}\ell_* : \mathcal{T}(S) \to \mathbb{PR}^{\mathcal{S}}$ is also injective.*

Proof. This follows from Theorem 2.1, Equation (2.1), and the fact that for any two matrices $A, B \in \mathrm{SL}(2, \mathbb{R})$,

$$\mathrm{tr}(A)\,\mathrm{tr}(B) = \mathrm{tr}(AB) + \mathrm{tr}(AB^{-1}). \tag{2.2}$$

The point is that if we let α and β be two simple closed curves which intersect transversely in a single point and if we write $f_*(\alpha) = A$, $f_*(\beta) = B$ (up to sign), then letting τ_β denote the Dehn twist in β we have

$$f_*(\tau_\beta(\alpha)) = AB \text{ and } f_*(\tau_\beta^{-1}(\alpha)) = AB^{-1}.$$

So by (2.1) and (2.2) there is a nontrivial relation among the lengths of the four simple closed curves $\alpha, \beta, \tau_\beta(\alpha), \tau_\beta^{-1}(\alpha)$, and one can check that this is not respected by scaling for all choices of α and β. See [16] for details. □

3. Measured Foliations

Here we introduce measured foliations which provide one interpretation for degenerations of hyperbolic structures according to Thurston. An appropriate space of all such objects also embeds into $\mathbb{R}^{\mathcal{S}}$ (Theorem 3.5) disjoint from the image of $\mathcal{T}(S)$ (Proposition 3.6), and this will provide the "light cone" for the image of $\mathcal{T}(S)$ (Theorem 7.1). As our intention is to provide intuition rather than a complete treatment of the subject, we use a fairly restricted definition which allows us to use some geometric arguments. A theorem of Hubbard and Masur [21] proves that any measured foliation is equivalent (in the sense described below) to one by our definition. For a detailed discussion of measured foliations, see [16].

3.1. *Definition by example*

Before giving our definition, we start with a useful example that was described in Aramayona's lectures [2] (see also Thurston's original exposition of this in [32]). Given $\alpha, \beta \in \mathcal{S}(S)$ we represent these homotopy classes by simple closed curves of the same names with exactly $\iota(\alpha, \beta)$ intersection points. Recall that α, β *fill up* S if

$$\iota(\alpha, \gamma) + \iota(\beta, \gamma) > 0$$

for any $\gamma \in \mathcal{S}(S)$. Equivalently, with our choice of representatives, $\alpha \cup \beta$ cut the surface up into topological disks, not one of which is a bigon.

We view $\alpha \cup \beta$ as the 1-skeleton of a cell structure on S with $\iota(\alpha, \beta)$ vertices, each of which is 4-valent. We take the dual cell decomposition and declare each cell to be a Euclidean square. We assume that the squares are glued together by isometries so that the metric on the squares induces a geodesic metric on S by measuring length of a curve as the sum of the lengths of the intersections with each of the Euclidean squares.

The resulting metric is a *Euclidean cone metric*: away from the vertices the metric is locally isometric to the Euclidean plane, while at some of the vertices, there are *cone singularities*, which we now explain. At each vertex v there is some even number, $2k = 2k(v)$, of corners of squares glued together, and because there are no bigons, $2k \geq 4$, so there are at least four corners of squares glued together. If there are exactly four squares, then there is a neighborhood of the vertex which is isometric to a disk in the Euclidean plane. If there are more than four squares, then the metric is singular at v, and v is called a cone point of cone angle $k\pi$; any sufficiently small circle of radius $r > 0$ centered at this point has circumference $k\pi r$. See Figure 4. We can represent α and β by geodesics in this metric which cut through as mid-lines in each square, intersecting each other exactly in the centers.

Because the metric is locally Euclidean, parallel transport (in the complement of the singularities) is locally independent of path. Because S is orientable and because any side of a square is either parallel to α or β (but not both), we see that parallel transport around a loop γ based at x defines

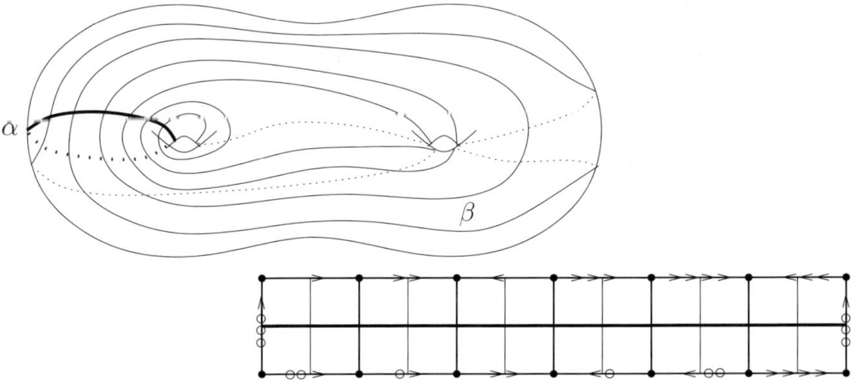

Fig. 4. α and β on S and the Euclidean cone metric obtained from dual cell structure.

Fig. 5. k-prong singularities: $k = 3$ on the left and $k = 4$ on the right.

a linear map on the tangent space $P_\gamma : T_x(S) \to T_x(S)$ which is either the identity I or $-I$. We describe this by saying that the holonomy of the metric lies in $\{\pm I\}$. Therefore, given a line L in the tangent space of a nonsingular point x, we can consider all maximal geodesics in the complement of the singularities *in the direction* L. These geodesics are the leaves of a foliation $\mathcal{F}_0$ on the surface minus the singularities. Adding the singular points back in we obtain a *singular foliation* $\mathcal{F}$ on the entire surface, with a *k-prong singularity* at each cone point with cone angle $k\pi$. See Figure 5.

Away from the cone points we can find locally isometric coordinates (x, y) to $\mathbb{R}^2$ in which the leaves are sent to arcs of vertical lines. The horizontal variation defines a measure $\mu = \mu_\gamma$ on any path $\gamma : [a, b] \to S$:

$$\int_\gamma d\mu = \int_\gamma |dx| = \int_a^b |(x \circ \gamma)'(t)| dt.$$

The pair $(\mathcal{F}, \mu)$ is an example of a singular measured foliation.

We note that μ is really a family of measures, one for every path in S, which satisfies

- If γ_0 and γ_1 are two paths homotopic through a homotopy preserving the intersection with leaves, then $\mu_{\gamma_0} = \mu_{\gamma_1}$, and
- if γ is the concatenation of two paths γ_0 and γ_1, then the restriction of μ_γ to γ_i is μ_{γ_i}, $i = 0, 1$.

As is common, we simply write μ for μ_γ.

We now define a measured (singular) foliation to be a pair $(\mathcal{F}, \mu)$ obtained as above from *some* Euclidean cone metric q on S. More precisely,

q is a Euclidean cone metric (not necessarily built from the construction described above) with cone angles of the form $k\pi$, $k \in \mathbb{Z}_{\geq 3}$ and holonomy in $\{\pm I\}$, the singular foliation $\mathcal{F}$ is obtained from an honest foliation $\mathcal{F}_0$ on S minus the cone points by parallel geodesics in some direction, and μ is the variation orthogonal to the foliation, with respect to the metric q. We let $\widetilde{\mathcal{MF}}(S)$ denote the set of all measured foliation on S.

Remark 3.1. The types of metrics q we are referring to are precisely the metrics induced by *holomorphic quadratic differentials* with respect to some complex structure, and the foliations are the *vertical foliations* for those quadratic differentials; see [17] for a detailed discussion of quadratic differentials.

Given a measured foliation $(\mathcal{F}, \mu)$, we now explain what we mean by a *leaf* of $\mathcal{F}$. Let $\mathcal{F}_0$ be the honest foliation of S in some direction with the cone points removed. In $\mathcal{F}_0$ we know what leaves are: these are maximal, injective locally isometric immersion of (i) $\mathbb{R}$, (ii) a circle of some circumference, (iii) a bounded open interval, or (iv) an open ray in $\mathbb{R}$ (in the given direction). We will call the types (i) and (ii) *nonsingular leaves* of $\mathcal{F}$. Suppose now that $\gamma : (a, b) \to S$ is one of types (iii) or (iv), with $b = \infty$ in the case of type (iv). Observe that because S is complete with respect to q, γ can be extended to $[a, b]$ in the first case and $[a, \infty)$ in the second. We call the images of these paths *separatrices*. A *singular leaf* of $\mathcal{F}$ is an injective locally isometric immersion γ of $\mathbb{R}$ or a circle of some circumference which is a concatenation of separatrices with the following property. After possibly reparametrizing to change the orientation, every time γ passes through a cone point (at the end of one separatrix and the beginning of the next), there is always a cone angle π on the right side of γ. Said differently, recall that all the cone points are k-prong singularities, and so we are requiring that when γ comes in to the singularity, it goes out along the right-most prong. Of course, reversing the orientation we have all left turns. A *leaf* of $\mathcal{F}$ is either a singular leaf or a nonsingular leaf of $\mathcal{F}$.

3.2. *Geometric intersection number I: Measured foliations*

Given a measured foliation $(\mathcal{F}, \mu)$ and $\alpha \in \mathcal{S}$, we define the geometric intersection number of α with μ, or with $(\mathcal{F}, \mu)$, to be

$$\iota(\alpha, \mu) = \iota(\alpha, (\mathcal{F}, \mu)) := \inf_{\alpha_0 \in \alpha} \int_{\alpha_0} d\mu.$$

The next proposition provides the connection with the geometric intersection number between curves.

Proposition 3.2. *Given any $\alpha \in \mathcal{S}$, there exists a measured singular foliation $(\mathcal{F}, \mu)$ such that for all $\gamma \in \mathcal{S}$ we have*

$$\iota(\gamma, \alpha) = \iota(\gamma, \mu).$$

Proof. Let $\beta \in \mathcal{S}$ be any curve (or multicurve) which fills with α, and perform the construction on the pair α, β described at the beginning of Section 3.1 above to produce a Euclidean cone metric. The foliation by geodesics parallel to one of the sides of the squares is a foliation by curves each representing the homotopy class α. Indeed, if we consider the union of the singular leaves we get a graph Γ in S, and the complement of this is an open cylinder C foliated by curves in the homotopy class α; see Figure 4.

Now given any $\gamma \in \mathcal{S}$, any representative $\gamma_0 \in \gamma$ must cross from one end of C to the other at least $\iota(\gamma, \alpha)$ times. Each time it crosses C, there is a contribution of at least 1 to the variation, and hence we have

$$\iota(\gamma, \mu) \geq \iota(\gamma, \alpha).$$

To prove the other inequality, let γ_0 be any representative which intersects one of the leaves α_0 of $\mathcal{F}$ representing α exactly $\iota(\gamma, \alpha)$ times transversely. We can now apply a homotopy to γ_0 "pushing it into Γ", except for the $\iota(\gamma, \alpha)$ arcs which cross directly from one side of C to the other, each giving a contribution of exactly 1 to the total variation. Call this representative γ_1. Since the part of γ_1 that runs around inside Γ does not contribute to the total variation we have

$$\iota(\gamma, \mu) \leq \int_{\gamma_1} d\mu = \iota(\gamma, \alpha)$$

which completes the proof. □

Remark 3.3. We note that the construction depends very much on the choice of curve β, so that $(\mathcal{F}, \mu)$ is not unique.

Analogous to the length function used for Teichmüller space, geometric intersection number for curves and for foliations determines two maps

$$\iota_* : \mathcal{S} \times \mathbb{R}_+ \to \mathbb{R}^{\mathcal{S}} \quad \text{and} \quad \iota_* : \widetilde{\mathcal{MF}(S)} \to \mathbb{R}^{\mathcal{S}}$$

given by

$$\iota_*(t, \beta) = \{t\iota(\alpha, \beta)\}_{\alpha \in \mathcal{S}} \quad \text{and} \quad \iota_*(\mathcal{F}, \mu) = \{\iota(\alpha, \mu)\}_{\alpha \in \mathcal{S}},$$

respectively.

As a consequence of Proposition 3.2 we have the following.

Corollary 3.4.

$$\iota_*(\mathbb{R}_+ \times \mathcal{S}) \subset \iota_*(\widetilde{\mathcal{MF}}) \subset \mathbb{R}^{\mathcal{S}}.$$

Proof. We note that by scaling the metric in the proof of the previous proposition by $t > 0$, we scale the geometric intersection numbers $\iota(\alpha, \mu)$ by t as well. □

It is not too hard to see that the map ι_* on $\mathbb{R}_+ \times \mathcal{S}$ is injective. However, the map ι_* on $\widetilde{\mathcal{MF}}$ is *not* injective. First of all, we can apply an isotopy to any measured foliation. In addition, applying a *Whitehead move* (see Figure 6) will change the isotopy class of singular foliation, but will not change the geometric intersection numbers. We define an equivalence relation on $\widetilde{\mathcal{MF}}(S)$ by declaring two foliations $(\mathcal{F}, \mu)$ and $(\mathcal{F}', \mu')$ to be *measure equivalent* if they differ by isotopy and a sequence of Whitehead moves. We let $\mathcal{MF}(S)$ denote the set of measure equivalence classes of measured foliations.

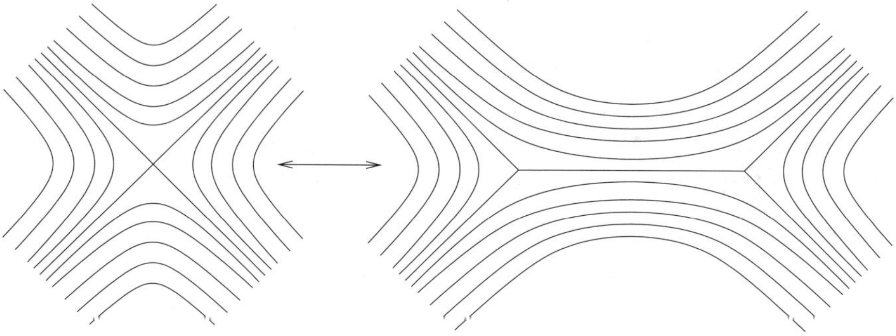

Fig. 6. Whitehead move splitting up/combining singularities.

The following important theorem is beyond the scope of these notes, but it justifies the name "measure equivalent"; see [16].

Theorem 3.5. *Two foliations $(\mathcal{F}, \mu), (\mathcal{F}', \mu') \in \widetilde{\mathcal{MF}}(S)$ are equivalent if and only if $\iota_*(\mu) = \iota_*(\mu')$. Said differently, the equivalence relation is precisely the one for which the map ι_* descends to an injection*

$$\iota_* : \mathcal{MF}(S) \to \mathbb{R}^{\mathcal{S}}.$$

Using this theorem we give $\mathcal{MF}(S)$ the topology making this injection into an embedding. For our purposes we could just define the equivalence relation to be the one on $\widetilde{\mathcal{MF}}(S)$ determined by the fibers of the map ι_*.

Next, we observe that $\mathcal{MF}(S)$ and $\mathcal{T}(S)$ in $\mathbb{R}^{\mathcal{S}}$ are disjoint.

Proposition 3.6. *The image of $\mathcal{MF}(S)$ and $\mathcal{T}(S)$ are disjoint in $\mathbb{R}^{\mathcal{S}}$,*

$$\iota_*(\mathcal{MF}(S)) \cap \ell_*(\mathcal{T}(S)) = \emptyset.$$

Proof. We note that for any $X \in \mathcal{T}(S)$, we have a positive lower bound on $\ell_X(\alpha)$ over all $\alpha \in \mathcal{S}$ (in fact, $\{\ell_X(\alpha)\}_{\alpha \in \mathcal{S}}$ is a closed discrete subset of $\mathbb{R}$ not containing 0). To prove the proposition, it therefore suffices to prove

Claim. *Given $(\mathcal{F}, \mu)$ there exists a sequence of simple closed curves $\alpha_n \subset \mathcal{S}$ so that as $n \to \infty$ we have*

$$\iota(\alpha_n, \mu) \to 0.$$

Proof. If $\mathcal{F}$ has a closed nonsingular leaf, the isotopy class of this leaf, call it α, obviously has $\iota(\alpha, \mu) = 0$, and we let $\alpha_n = \alpha$ for all n. If $\mathcal{F}$ has no closed leaves, then there is a leaf locally isometric to $\mathbb{R}$ in $\mathcal{F}$. This must accumulate at some point x on S. For any n we can construct a curve near x built from these arcs together with short arcs transverse to $\mathcal{F}$ and having variation at most $1/n$ with μ. See [16], as well as [12] for a similar construction. □

This completes the proof of the proposition. □

3.3. *Geodesics in the q-metric*

We now explain one benefit of defining measured foliations with respect to a Euclidean cone metric q. First observe that the geodesics in the q-metric have a nice local description. Namely, in the complement of the singularities, these paths must be straight Euclidean segments (since a local isometry to the Euclidean plane makes them length minimizing paths). Furthermore, whenever a geodesic enters and exits a singularity, it makes and angle $\geq \pi$ on both sides (if the angle were less than π on one side, you can easily find a nearby path which is shorter). An important point for us is that the geodesics in $\widetilde{S}$ with the induced metric are unique.

Proposition 3.7. *For any Euclidean cone metric q on S with cone angles $\geq 2\pi$, the induced metric q on $\widetilde{S}$ has the property that any two points are connected by a unique (distance minimizing) geodesic.*

This is a consequence of the fact that the q-metric on $\widetilde{S}$ is CAT(0), or *nonpositively curved* in the sense of Alexandrov together with the Cartan-Hadamard Theorem; see [1] or [10]. Instead of defining the CAT(0) condition, we sketch an elementary proof of this proposition using the Gauss-Bonnet Theorem.

Proof. Suppose we have two geodesics between a pair of points z and w. By passing to sub-arcs of these two geodesics we can assume that the interiors of the geodesics are disjoint. So, these two geodesics form a bigon as in Figure 7, and we can double this bigon over its boundary to produce a sphere with a Euclidean cone metric. While the cone angles may not all be greater than 2π, one can check that the only place where this can fail is at the two endpoints z and w. Now, taking a sufficiently high degree branched cover of the sphere over these two points, we again get a sphere with a Euclidean cone metric, and all cone angles are now greater than 2π. The Gauss-Bonnet Theorem shows that this is impossible.

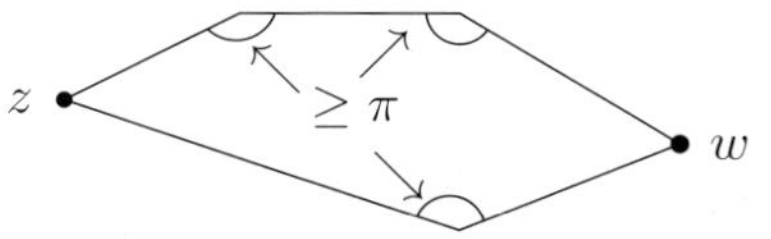

Fig. 7. An (impossible) pair of distinct geodesics connecting two points in the Euclidean cone metric on $\widetilde{S}$. □

Remark 3.8. In fact, this proof (together with the existence of geodesics) also shows that the two properties of geodesics preceding the statement of the proposition are in fact both necessary and sufficient to guarantee that a path is a geodesic. This also follows from the fact that the metric is CAT(0) and so local geodesics are global geodesics; see [10].

For a Riemannian metric a pair of compact geodesic segments which are not part of a larger single geodesic must intersect in only finitely many points. However, because the q-metric is singular, two geodesics γ_1 and γ_2 in $\widetilde{S}$ can run together for some part of their length without being contained in a larger geodesic. If γ_1 and γ_2 are disjoint for a while, then come together (possibly only at a point), then become disjoint again, then there are two possibilities: either γ_1 exits γ_2 on the opposite side it started on, and we say that γ_1 *crosses* γ_2, or it exits on the same side.

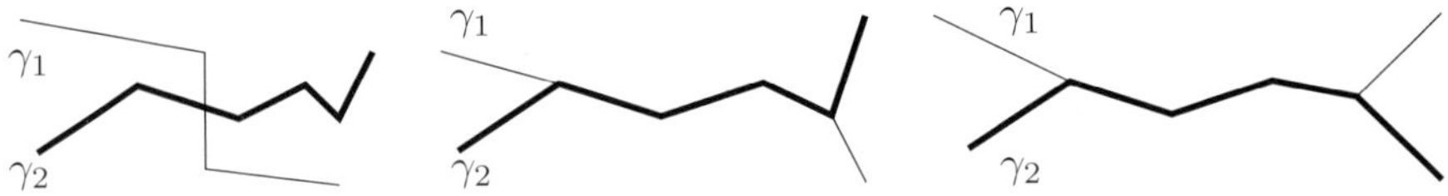

Fig. 8. Two geodesics γ_1, γ_2 in $\widetilde{S}$. The left and center figures show γ_1 crossing γ_2, whereas γ_1 and γ_2 do not cross in the right figure.

Returning to the foliation $\mathcal{F}$, we also observe that this can be pulled back to the universal covering to obtain a foliation $\widetilde{\mathcal{F}}$. The leaves of $\mathcal{F}$, hence also $\widetilde{\mathcal{F}}$, satisfy the local conditions described above for being geodesic and so are indeed geodesics. It follows that any such leaf is proper in the universal covering. We will describe an even stronger property that q-geodesics in $\widetilde{S}$ enjoy in Section 5, but for now, we use properness to sketch a proof of the following; see [17].

Proposition 3.9. *Let $(\mathcal{F}, \mu)$ be a measured foliation on S defined in terms of some Euclidean cone metric q. Given a homotopy class of curve α on S, let $\alpha*$ denote the q-geodesic representative. Then*

$$\iota(\alpha, \mu) = \int_{\alpha*} d\mu.$$

That is, the geodesic representative realizes the geometric intersection number.

Proof. For any representative α_0 of α, $\int_{\alpha_0} d\mu$ measures "how many leaves" α_0 crosses. Of course, α_0 will generally cross infinitely many leaves, but the transverse measure provides a way to make this precise. Pulling this picture back to the universal covering, we can choose lifts $\widetilde{\alpha}*$ a biinfinite geodesic and $\widetilde{\alpha}_0$ a biinfinite path in $\widetilde{S}$, each invariant under an element $\delta \in \pi_1(S)$, hence remaining a bounded distance from each other. Observe that since $\widetilde{\alpha}*$ is a geodesic, it will not cross any leaf of $\widetilde{\mathcal{F}}$ more than once: if $\widetilde{\alpha}*$ crossed a leaf γ of $\widetilde{\mathcal{F}}$, left it, then came back to cross it again, we would have two distinct geodesics between a pair of points (the arc of $\widetilde{\alpha}*$ and the arc of γ), which is impossible. Furthermore, if $\widetilde{\alpha}*$ crosses some leaf γ of $\widetilde{\mathcal{F}}$, then its two ends must diverge from γ (this is another fact about CAT(0) spaces), and since $\widetilde{\alpha}_0$ remains a bounded distance from $\widetilde{\alpha}*$, we see that $\widetilde{\alpha}_0$ must also cross γ. So, α_0 crosses at least as many leaves as $\alpha*$ and hence

$$\int_{\alpha*} d\mu \leq \int_{\alpha_0} d\mu$$

as required. □

4. Measured Laminations

An alternative to measured singular foliations that serve equally well in many situation (and sometimes better) are measured geodesic laminations. To define these, we first fix a hyperbolic metric ρ on S. By a *complete simple geodesic* on S (with respect to ρ) we mean the image of an injective geodesic defined on either a circle or the entire real line. A *geodesic lamination* on S is a closed set which is a union of pairwise disjoint, complete simple geodesics

$$\mathcal{L} = \bigcup_{x \in \mathcal{L}} \ell_x.$$

Here each ℓ_x is a complete simple geodesic called a *leaf* of $\mathcal{L}$. Appealing to the Poincaré-Hopf Index Theorem, we see that the $\mathcal{L}$ cannot be the entire surface, and in fact, it is nowhere dense and has zero Lebesgue measure. From this one sees that it can only be expressed as a disjoint union of complete simple geodesics in one way. In particular, the closed set $\mathcal{L}$ determines the decomposition into a disjoint union of simple geodesics. See [13, 34, 12, 7] for more on geodesic laminations.

A *transverse measure* on $\mathcal{L}$ can be defined as in the case of foliations as a measure on paths transverse to $\mathcal{L}$, so that the support of the measure is precisely the intersection of the path with $\mathcal{L}$. Furthermore we require that these are invariant under homotopy preserving transverse intersection with the leaves of $\mathcal{L}$ and behave appropriately under concatenation (compare this with the properties of the transverse measure on a foliation described in Section 3.1). However, there is a more concrete way to describe a transverse measure which we now explain; see [34, 6, 30] for a more on what follows.

As usual write ρ for the pull back of ρ to $\widetilde{S}$, which makes $\widetilde{S}$ isometric to the hyperbolic plane $\mathbb{H}^2$. If we use the Poincaré disk model of $\mathbb{H}^2$, then the isometry $\widetilde{S} \to \mathbb{H}^2$ provides a compactification of S by a *circle at infinity* which we denote S^1_∞ so that the isometry extends to a homeomorphism $\widetilde{S} \cup S^1_\infty \to \overline{\mathbb{H}^2}$. Geodesics in $\widetilde{S}$ determine a pair of points on S^1_∞, the limit points of the two ends of the geodesic, and any pair of distinct points determines a geodesic in this way. This allows us to identify the set of all (unoriented biinfinite) ρ-geodesics in $\widetilde{S}$ with the set of unordered pairs of distinct points in S^1_∞. We denote this $G(\widetilde{S})$ so that

$$G(\widetilde{S}) = \big(S^1_\infty \times S^1_\infty - \Delta\big)/\sim \tag{4.1}$$

where Δ is the diagonal and $(x, y) \sim (y, x)$. This also topologizes the set of geodesics in a natural way. For example, a sequence of geodesics γ_n

converges to γ if and only if they can be parametrized to converge uniformly on compact sets; or alternatively, if and only γ_n converges to γ in the *Chabauty topology* (see [12, 3]) on the set of closed subsets of $\widetilde{S}$. We leave the verification of these claims to the reader.

Remark 4.1. At first glance, the notation $G(\widetilde{S})$ and S^1_∞ seem to be imprecise since the metric ρ was instrumental in defining these, yet it does not appear in the notation. In fact, if ρ and $\hat{\rho}$ are any two hyperbolic metrics on S and S^1_∞ and $\hat{S}^1_\infty$ are the two circle used to compactify $\widetilde{S}$, then in fact the identity map $\widetilde{S} \to \widetilde{S}$ extends to a homeomorphism between the two compactifications; see e.g. [13]. So, abusing notation we do not mention the metric ρ in the notation.

The action of $\pi_1(S)$ on $\widetilde{S}$ by covering transformations extends to $\widetilde{S} \cup S^1_\infty$. This action determines an action of $\pi_1(S)$ on $G(\widetilde{S})$ (either from the fact that the action is by isometries, or by appealing to Equation (4.1)). A geodesic lamination $\mathcal{L}$ on S gives rise to a geodesic lamination $\widetilde{\mathcal{L}} = p^{-1}(\mathcal{L})$ on $\widetilde{S}$ which is invariant under the action of $\pi_1(S)$. The lamination $\widetilde{\mathcal{L}}$ determines a closed $\pi_1(S)$-invariant subset of $G(\widetilde{S})$, which we also call $\widetilde{\mathcal{L}}$ and we define a *transverse measure* on $\mathcal{L}$ to be a $\pi_1(S)$-invariant Radon measure on $G(\widetilde{S})$ whose support is precisely this set $\widetilde{\mathcal{L}}$ (recall that a Radon measure is a Borel measure which measures compact sets finitely). Since the support of μ is required to be $\widetilde{\mathcal{L}}$, we note that μ actually determines $\widetilde{\mathcal{L}}$ and hence also $\mathcal{L}$. As such, we sometimes refer to μ as a measured lamination, without specifically mentioning the underlying lamination $\mathcal{L}$. We refer to $\mathcal{L}$ as the *supporting lamination of* μ.

Suppose μ is a transverse measure in this sense. To see how this is related to the more geometric notion of a transverse measure, observe that given any path transverse to $\mathcal{L}$ we can lift this to a path in $\widetilde{S}$ transverse to $\widetilde{\mathcal{L}}$, and any two such lifts differ by an element of $\pi_1(S)$. For any sufficiently small arc of the path (so that the arc intersects any geodesic of $\widetilde{L}$ at most once), we look at the set of geodesics in $\widetilde{\mathcal{L}}$ which intersect this arc, and define the measure of this arc to the measure of this set of geodesics. Invariance under $\pi_1(S)$ implies that we obtain a well-defined measure on the original path in S supported on the intersection of the path with $\mathcal{L}$.

We now give the basic example of a measured geodesic lamination. If α is a simple closed geodesic on S, then this is a geodesic lamination with one leaf, and the preimage $p^{-1}(\alpha) = \widetilde{\alpha} \subset \widetilde{S}$ defines a closed *discrete* set of geodesics on which $\pi_1(S)$ acts transitively. Up to scaling there is exactly one $\pi_1(S)$-invariant Radon measure on $G(\widetilde{S})$ supported on $\widetilde{\alpha}$, namely the counting measure, and we denote this μ_α. This is the measure which assigns to a subset of geodesics $E \subset G(\widetilde{S})$ the number of geodesics in $\widetilde{\alpha}$ which lie in E. The associated measure on a path in S counts the number of times α crosses the path.

Not every geodesic lamination admits a transverse measure since we have required the measure to have support exactly equal to the lamination. For example, if $(\mathcal{L}, \mu)$ is a measured geodesic lamination, then any isolated leaf of $\mathcal{L}$—a leaf with a neighborhood in S intersecting $\mathcal{L}$ only in that leaf—must be a simple closed geodesic. In particular, a lamination with an infinite isolated leaf, for example a "spiraling leaf" (see e.g. [7]), does not admit a transverse measure. Given a measured geodesic lamination $(\mathcal{L}, \mu)$, let $\mathcal{L}_0 \subset \mathcal{L}$ be the union of the simple closed geodesics. We observe that $\mathcal{L}_0$ is necessarily a finite set of geodesics (indeed, it contains at most $3g-3$ simple closed geodesics), and the measure μ decomposes as $\mu_0 + \mu_1$ with μ_0 supported on $\mathcal{L}_0$ and μ_1 supported on the lamination $\mathcal{L}_1 = \mathcal{L} - \mathcal{L}_0$. Since $\mathcal{L}_1$ has no closed leaves, the transverse measure μ_1 has no atoms in $G(\widetilde{S})$—that is, no one-point subset of $G(\widetilde{S})$ has positive measure. Equivalently, the measure on any arc transverse to $\mathcal{L}_1$ has no atoms.

We let $\mathcal{ML}(S)$ denote the set of all measured geodesic laminations on S, which we can think of as a subset of the space of all measures on $G(\widetilde{S})$. This latter space is given the weak* topology, and $\mathcal{ML}(S)$ is given the subspace topology. From the construction of a measured lamination μ_α from a simple closed curve α, we obtain an injection $\mathbb{R}_+ \times \mathcal{S}(S) \to \mathcal{ML}(S)$, by $(t, \alpha) \mapsto t\mu_\alpha$. We note that $\mathcal{ML}(S)$ depends on the initial hyperbolic structure on S. However, as in Remark 4.1 we see that for any two hyperbolic metrics, the spaces so constructed are homeomorphic by a homeomorphism extending the inclusion of $\mathbb{R}_+ \times \mathcal{S}(S)$.

4.1. *Geometric intersection number II: Measured laminations*

Given $\alpha \in \mathcal{S}(S)$ and $(\mathcal{L}, \mu) \in \mathcal{ML}(S)$, there is a geometric intersection number

$$\iota(\alpha, \mu) = \iota(\alpha, (\mathcal{L}, \mu)) = \inf_{\alpha_0 \in \alpha} \int_{\alpha_0 \in \alpha} d\mu_{\alpha_0} .$$

The infimum is realized by the geodesic representative of α (which we denote by α). This is similar to the situation for measured foliations described in Section 3.3, and is perhaps easier to visualize since we have concrete models for $(\widetilde{S}, \rho) \cong \mathbb{H}^2$. As before, the point is that any other representative of

α must cross the leaves of $\mathcal{L}$ at least as much as the geodesic representative, and hence the total variation is at least as large.

As with measured foliations, the geometric intersection number determines a map

$$\iota_* : \mathcal{ML}(S) \to \mathbb{R}^{\mathcal{S}}$$

by $\iota_*(\mathcal{L}, \mu) = \{\iota(\alpha, \mu)\}_{\alpha \in \mathcal{S}}$.

Though we will not use this below, we note that the geometric intersection number has the following nice extension due to Thurston. Recall that μ_α is the measured lamination determined by $\alpha \in \mathcal{S}(S)$.

Theorem 4.2. *There is a continuous, symmetric, bilinear function*

$$\iota : \mathcal{ML}(S) \times \mathcal{ML}(S) \to \mathbb{R}$$

such that for all $\alpha, \beta \in \mathcal{S}(S)$ *and* $t, s \in \mathbb{R}_+$

$$\iota(t\mu_\alpha, s\mu_\beta) = ts\iota(\alpha, \beta).$$

Bilinearity means that for all t, s, λ, μ, ν we have $\iota(t\mu, \lambda) = t\iota(\mu, \lambda) = \iota(\mu, t\lambda)$ and $\iota(\mu + \nu, \lambda) = \iota(\mu, \lambda) + \iota(\nu, \lambda)$, when $\mu + \nu$ is again a measured lamination. In fact, $\mathcal{ML}(S)$ is a subspace of infinite codimension in the space of all $\pi_1(S)$–invariant Radon measures on $G(\widetilde{S})$, called the space of *geodesic currents*, and Bonahon proved that ι has a continuous, symmetric, bilinear extension to this space as well; see [5, 6] for this and more on currents. One can describe $\iota(\mu, \lambda)$ as the total mass of $\mu \times \lambda$ in an appropriate sense. For example, we can think of $\mu \times \lambda$ as defining a measure on S supported on the transverse intersections of the supporting laminations of μ and λ, and then $\iota(\mu, \lambda)$ is the total mass of this measure.

5. Measured Foliations and Measured Laminations

We have seen that $\mathbb{R}_+ \times \mathcal{S}(S)$ can be naturally injected into both $\mathcal{MF}(S)$ as well as $\mathcal{ML}(S)$. The next theorem shows that these spaces are in some sense the same.

Theorem 5.1. *There is an* $\mathbb{R}_+$*-invariant homeomorphism* $\mathrm{Str} : \mathcal{MF}(\mathrm{S}) \to \mathcal{ML}(\mathrm{S})$ *which is the identity on the image of* $\mathbb{R}_+ \times \mathcal{S}(S)$ *in each.*

One proof of this goes through an intermediate set of combinatorial objects called *train tracks*. See [30] and [11] for more on this.

We explain the idea for Theorem 5.1 here without mention of train tracks, leaving most of the details to the reader; see also [13, 26] for more on this approach.

To begin, recall that if (X, q) and (Y, ρ) are metric spaces, then given non-negative numbers K and C, a map $f : X \to Y$ is called a (K, C)-*quasi-isometry* if

$$\frac{1}{K}(\rho(x,y) - C) \leq q(x,y) \leq K\rho(x,y) + C$$

for all $x \in X$ and $y \in Y$, and if there exists $A > 0$ so that the A-neighborhood of $f(X)$ is the entire range, $Y = N_A(f(X))$.

We will need the following.

Proposition 5.2. *Given any q and ρ as above, there exists $K \geq 1$ so that the identity $(\widetilde{S}, q) \to (\widetilde{S}, \rho)$ is a $(K, 1)$-quasi-isometry.*

Proof. This is really a special case of a much more general result, often called the Švarc-Milnor Lemma; see e.g. [10]. The basic idea is as follows. Since S is compact and since ρ and q determine the same topology, there exists $0 < \epsilon \leq \epsilon' \leq 1$ so that

$$B^\rho_\epsilon(x) \subset B^q_{\epsilon'}(x) \quad \text{and} \quad B^q_\epsilon(x) \subset B^\rho_{\epsilon'}(x) \tag{5.1}$$

for all $x \in \widetilde{S}$.

Now connect any two points $z, w \in \widetilde{S}$ by a ρ-geodesic and subdivide it into at most $1 + (\rho(z, w)/\epsilon)$ intervals of ρ-length at most ϵ. It follows that the q-distance between consecutive endpoints of the subdivided intervals is at most ϵ', and hence

$$q(z,w) \leq \epsilon' \left(1 + \frac{\rho(z,w)}{\epsilon}\right) \leq 1 + \frac{\epsilon' \rho(z,w)}{\epsilon}.$$

Setting $K = \epsilon'/\epsilon$ this proves one inequality and a symmetric argument proves the other. □

From this it follows that any geodesic $\gamma : \mathbb{R} \to \widetilde{S}$ in the q-metric is also a *quasi-geodesic* in the ρ-metric. That is, there exists constants $K, C > 0$ so that

$$\frac{1}{K}(|t - s| - C) \leq \rho(\gamma(t), \gamma(s)) \leq K|t - s| + C.$$

Since γ is a q-geodesic, we have $|t - s| = q(\gamma(s), \gamma(t))$, and so this is immediate from Proposition 5.2 for some K and $C = 1$. The importance of this lies in its application to the next fact.

Proposition 5.3. *There exists $A > 0$ so that for any q-geodesic γ in $\widetilde{S}$, there is ρ-geodesic γ' so that*

$$\gamma \subset N^{\rho}_{A}(\gamma') \text{ and } \gamma' \subset N^{\rho}_{A}(\gamma).$$

If γ is compact (respectively, half-infinite), then γ' can be chosen to have the same endpoints (respectively, endpoint). With these choices γ' is unique. If γ is biinfinite, then γ' is unique.

The proposition is true for any quasi-geodesic where A depends only on the quasi-geodesic constants; see [10] for a proof of existence. The key fact that is used is that in the hyperbolic plane, closest point projection to a geodesic decreases arc-length by a factor which is exponential in the distance to the geodesic. The uniqueness statements follow from uniqueness of geodesics between any pair of distinct points in $\overline{\mathbb{H}^2}$. Given a q-geodesic γ, we call the corresponding ρ-geodesic the *(ρ-)straightening of* γ.

We now construct a map $\mathrm{Str} : \mathcal{MF}(\mathrm{S}) \to \mathcal{ML}(\mathrm{S})$, obtained by straightening all the leaves of a foliation to leaves of a lamination. Let $(\mathcal{F}, \mu)$ be a measured foliation defined via a Euclidean cone metric q, and let $\widetilde{\mathcal{F}}$ be the foliation pulled back to $\widetilde{S}$. For each leaf ℓ_q of $\widetilde{\mathcal{F}}$, let ℓ_ρ denote the ρ-straightening. From this we see that ℓ_q of $\widetilde{\mathcal{F}}$ has two endpoints on the circle at infinity S^1_∞ (this strengthens the properness of ℓ_q mentioned in Section 3.3).

Next, observe that no two leaves of $\widetilde{\mathcal{F}}$ cross each other—they may however run together for a finite or infinite amount of time, but if and when they depart, they do so on the same side they started on: this is a consequence of our definition of leaves, specifically, the fact that they always "turn right"; see the end of Section 3.1. Therefore the endpoints on S^1_∞ of any two leaves do not *link* each other, meaning that given two pairs of points so that all four points are distinct, then the first pair lies in a single component of the complement of the other pair. From this it follows that the ρ-straightenings of distinct leaves are also either disjoint, or equal. Furthermore, if a sequence of these ρ-straightening ℓ_{ρ_n} converge to a geodesic ℓ_ρ, one can check that (up to passing to a subsequence) the corresponding leaves ℓ_{q_n} also converge to a leaf ℓ_q which has ℓ_ρ as its ρ-straightening. Therefore $\widetilde{\mathcal{L}}$, the union of all ρ-straightenings of leaves of $\widetilde{\mathcal{F}}$, is a closed set of disjoint biinfinite geodesics, hence a geodesic lamination; see Figure 9.

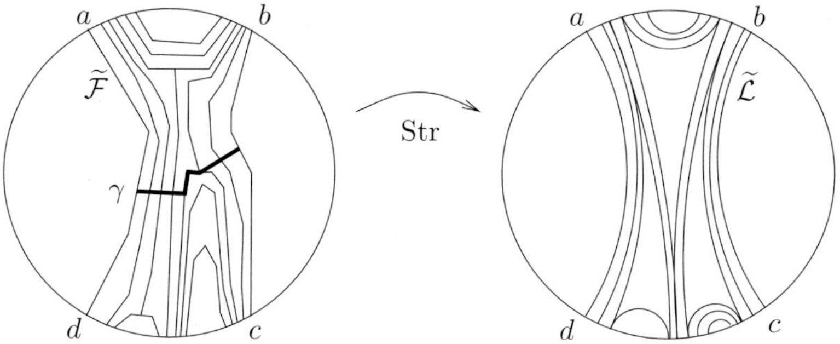

Fig. 9. The left shows part of $\widetilde{\mathcal{F}}$ and the right shows part of the straightening $\widetilde{\mathcal{L}}$. The arc γ crossing the leaves of $\widetilde{\mathcal{F}}$ is shown thickened.

Next observe that because $\widetilde{\mathcal{F}}$ is invariant by $\pi_1(S)$, so is $\widetilde{\mathcal{L}}$, and hence $\widetilde{\mathcal{L}}$ descends to a geodesic lamination $\mathcal{L}$ on S. We wish to impose a transverse measure λ on $\mathcal{L}$. For this, we let $[a,b] \times [c,d]$ denote any rectangle in $G(\widetilde{S})$. Let $\Omega([a,b] \times [c,d])$ denote the (possibly empty) set of leaves of $\widetilde{\mathcal{F}}$ which have one endpoint in $[a,b]$ and the other in $[c,d]$. Because no two leaves of $\widetilde{\mathcal{F}}$ cross, $\Omega([a,b] \times [c,d])$ has a linear order, and we let ℓ_{q_0} and ℓ_{q_1} denote the two extreme leaves. Let γ be a q-geodesic arc between ℓ_{q_0} and ℓ_{q_1}, and observe that γ crosses each leaf in $\Omega([a,b] \times [c,d])$ exactly once (though it may run along one or more of these leaves); see Figure 9. Define

$$\lambda([a,b] \times [c,d]) = \mu(\gamma).$$

It is not hard to check that λ is a measure on $G(\widetilde{S})$ with support $\widetilde{\mathcal{L}}$, and hence $(\mathcal{L}, \lambda)$ is a measured lamination. We define $\mathrm{Str}(\mathcal{F}, \mu) = (\mathcal{L}, \lambda)$.

Generically each leaf of $\mathcal{L}$ is the straightening of exactly one leaf of $\mathcal{F}$. The situation where this is not the case is when S contains an annulus foliated by parallel simple closed curves (for example, in the construction of a foliation from a simple closed curve). In this case, all the parallel leaves are straightened to a single simple closed curve.

We can also construct an inverse to Str. For this, we first observe that the straightening procedure has the effect of "inserting space" between all the leaves of a foliation to produce the lamination (this is a little bit like the operation of *Denjoying* a foliation, see [Example 4.14, [11]], in which more leaves are added to a foliation, except here we don't add the leaves). To reverse this, we want to take a lamination and "collapse" all the space between the leaves. This is problematic when the lamination $\mathcal{L}$ has simple closed curve components, as might be expected given the discussion of the previous paragraph. For this reason, we first modify our measured geodesic laminations as follows.

Given $(\mathcal{L}, \lambda)$, let $\mathcal{L}_0 = \beta_1 \cup \cdots \cup \beta_k$ be the union of the simple closed geodesics in $\mathcal{L}$, so that β_i is assigned some t_i-times the transverse counting measure by λ for each $i = 1, \ldots, k$. Replace each β_i with the ϵ-neighborhood of β_i for some very small ϵ, and foliate each neighborhood by curves parallel to the core curve, β_i. Finally, give each foliated annulus a transverse measure (with no atoms) so that the total measure of a transverse arc from one boundary component to the other is precisely t_i. Let $\mathcal{L}'$ denote the resulting "partial foliation/lamination" and continue to denote the transverse measure by λ.

Remark 5.4. The point here is that the transverse measure assigned to any arc transverse to $\mathcal{L}$ will have atoms at each point of intersection with one of the β_i. The procedure just described removes all the atoms.

We are now ready to begin describing the collapsing procedure. We make some choices to carry this out, which is not at all surprising: many different foliations will straighten to the same measured geodesic lamination (although all such measured foliations are equivalent to each other). Let $\mathcal{P} = \alpha_1, \ldots, \alpha_{3g-3}$ denote any pants decomposition of S by geodesics for which $\mathcal{L}$ transversely intersects each α_i. Let $P_1, \ldots, P_{2g-2}$ be the pants in S with geodesic boundary determined by $\mathcal{P}$. An exercise shows that any essential arc in a pair of pants is isotopic to one of the six shown in Figure 10; see for example [30]. By choosing the $\epsilon > 0$ used in the construction of $\mathcal{L}'$ from $\mathcal{L}$ sufficiently small, we can assume that each leaf of $\mathcal{L}'$ is transverse to $\mathcal{P}$, and that each arc of intersection of a leaf of $\mathcal{L}'$ with each P_j is isotopic to one of these six.

For each $j = 1, \ldots, 2g-2$ consider the intersection of $\mathcal{L}'$ with P_j. In P_j there are two or three rectangles which contain all arcs of intersection of leaves of $\mathcal{L}'$ with P_j (or briefly, arcs of $\mathcal{L}' \cap P_j$), and up to homeomorphism, these rectangles must be as shown in one of the two pictures of Figure 11. We can choose these rectangles to be bounded on two sides by arcs of $\mathcal{L}' \cap P_j$ and on the other two sides by arcs in the boundary of P_j . Furthermore, all arcs of $\mathcal{L}' \cap P_j$ run between the latter two sides.

Now we describe the collapsing procedure in each pair of pants. Fix one such $P = P_j$ and assume that we are in situation on the left in Figure 11, say, and that each rectangle contains a nonempty set of arcs of $\mathcal{L}' \cap R$. We will explain how to modify this construction in the general case. Let R denote any one of the three rectangles in P.

We arbitrarily pick Euclidean coordinates on R providing a homeomorphism $G_R : R \to [0,1] \times [0,1]$ so that each arc of $\mathcal{L}' \cap R$ is of the form

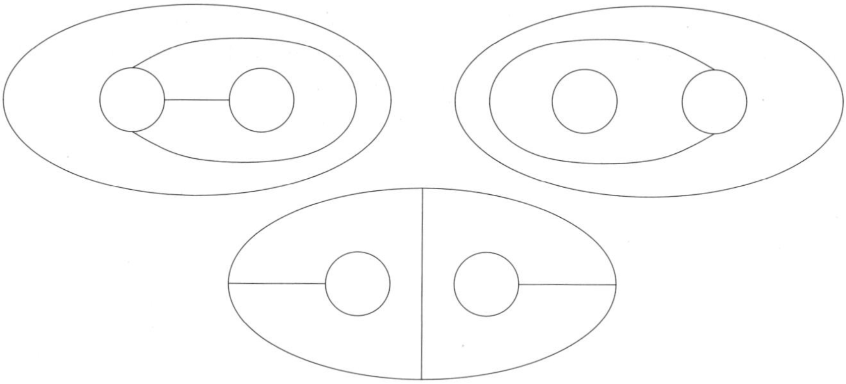

Fig. 10. The six isotopy classes of arcs in a pair of pants.

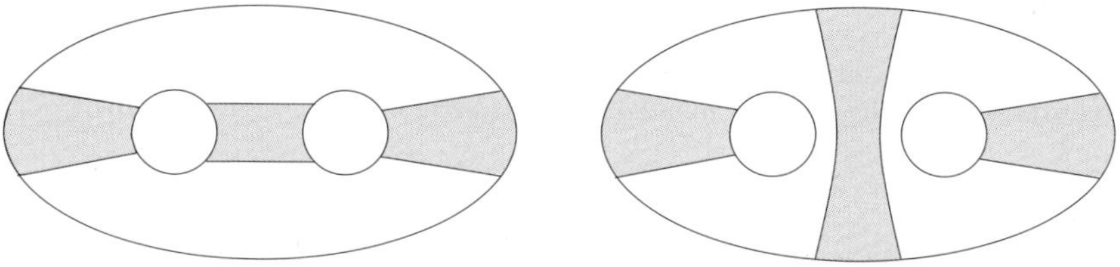

Fig. 11. All arcs of intersection of $\mathcal{L}' \cap P_j$ must be contained in the shaded rectangles.

$[0,1] \times \{y\}$, for some $y \in [0,1]$. The transverse measure λ determines a measure (which we also denote λ) on the second factor $[0,1]$. If the total variation of $[0,1]$ with respect to this measure is r, we define a map

$$F_\lambda : [0,1] \times [0,1] \to [0,1] \times [0,r]$$

by

$$F_\lambda(x,y) = \left(x, \int_0^y d\lambda\right).$$

Because λ has no atoms, it follows that this defines a continuous, surjective mapping. Observe that every arc $[0,1] \times \{y\} \subset [0,1] \times [0,r]$ is the F_λ-image of an arc of $\mathcal{L}' \cap R$. Composing we obtain a map $H = F_\lambda \circ G_R : R \to [0,1] \times [0,r]$.

We do this for each of the three rectangles producing maps

$$H_i : R_i \to [0,1] \times [0,r_i]$$

for $i = 1, 2, 3$. The three Euclidean rectangles $\{[0,1] \times [0, r_i]\}_{i=1}^3$ can be glued together along two tripods *by isometries* as in Figure 12 and the result is a pair of pants $\hat{P}$ with a Euclidean cone metric and a measured singular foliation $(\mathcal{F}_P, \mu_P)$ by geodesics. It is straightforward to construct a quotient map $H : P \to \hat{P}$ which is precisely H_i on each R_i. This map H sends each arc of $\mathcal{L}' \cap P$ to a leaf of the $\mathcal{F}_P$.

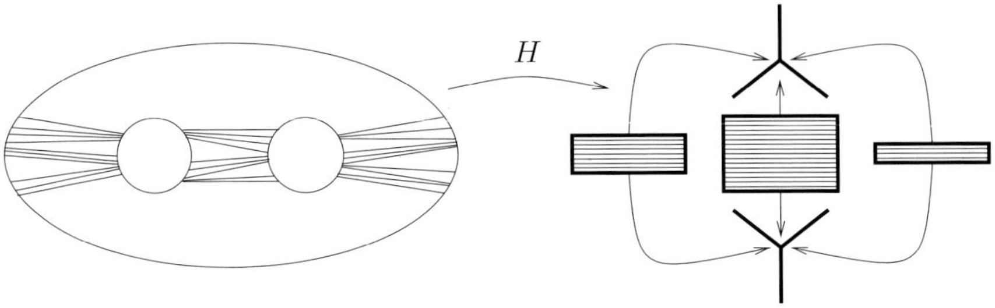

Fig. 12. The collapsing map H on a pair of pants. The image of the pants is a pair of pants expressed as the union of foliated Euclidean rectangles.

There are two other possibilities: there may be no arcs of $\mathcal{L}' \cap P$ in one of the rectangles or the rectangles may be as in Figure 11 on the right. For each of these situations we can construct a quotient map $H : P \to \hat{P}$ in a similar fashion with minor modifications. For example, in the case that all arcs are contained in just two rectangles, we glue these rectangles together along a single 4-prong graph (which can be seen as the degenerate case of the one we considered in more detail).

Because the measure λ was used to define the Euclidean metric on the arcs which met the boundary of the pants, it is straightforward to check that the resulting pants $\hat{P}_1, \ldots, \hat{P}_{2g-1}$ can be glued together by isometries to produce a surface $\hat{S}$ with a Euclidean cone metric and singular measured foliation $(\mathcal{F}, \mu)$. Moreover, this gluing can be done so that the maps $H : P_j \to \hat{P}_j$ are the restrictions of a single map

$$H : S \to \hat{S}$$

which sends leaves of $\mathcal{L}'$ to leaves of $\mathcal{F}$.

The map H is a homotopy equivalence, and as such is homotopic to a homeomorphism $\hat{H} : S \to \hat{S}$: this is straightforward from the construction here, but also follows from the Dehn-Nielsen-Baer Theorem; see [15]. Now we pull back the Euclidean cone metric and foliation to one on S by $\hat{H}$ and we continue to denote it $(\mathcal{F}, \mu)$. Then one can check that $\mathrm{Str}(\mathcal{F}, \mu) = (\mathcal{L}, \lambda)$ so that at least this is a one-sided inverse. To prove that straightening, then collapsing gives an equivalent measured foliation is beyond the scope of these notes since we have not developed any of the tools for studying Whitehead equivalence; see [16] for a detailed treatment of Whitehead equivalence.

6. Dehn-Thurston Coordinates

From now on we allow ourselves the freedom of passing back and forth between elements of $\mathcal{ML}(S)$ and $\mathcal{MF}(S)$ via Theorem 5.1. We now refer to an element of either space simply as μ when this is convenient. We will either explicitly state which we are referring to, or else let the context determine the meaning.

Our next goal is to describe the topology of the space $\mathcal{ML}(S)$ by constructing coordinates. This is a version of the Dehn-Thurston coordinates, using as much hyperbolic geometry as possible to simplify the discussion. See [30] for a more thorough discussion of Dehn-Thurston coordinates.

As with Fenchel-Nielsen coordinates, we begin with a pants decomposition $\mathcal{P} = \{\alpha_1, \ldots, \alpha_{3g-g}\}$ for S. We choose a hyperbolic metric ρ on S so that the lengths of the α_i are all small and equal. The Collar Lemma [23, 19] implies that by taking the α_i sufficiently small, we can find very wide collar neighborhoods of each of the α_i, so that the boundaries of these collars have length 1/2, say. Note that this is not the geodesic length, but the actual length of the boundary curve. Now let $C_1, \ldots, C_{3g-3}$ be these collars and let $P_1, \ldots, P_{2g-2}$ be the pants which are the complements of these collars. We emphasize that the P_i are not taken to have geodesic boundary as is usually the case, but instead the pants share their boundaries with the collars.

We know that any complete simple geodesic ℓ on S will intersect any P_i in an arc which is isotopic to one of the 6 pictured in Figure 10. By choosing the length of the α_i sufficiently small, we can assume that if two such geodesics intersect the pants in isotopic arcs, then these arcs are very close to each other. More precisely, we observe the following.

Lemma 6.1. *Given $\epsilon > 0$ there is a $\delta > 0$ with the following property. Suppose we choose each α_i to have length $\delta_0 < \delta$ with the boundary of each P_j of length 1/2. Then if ℓ_1 and ℓ_2 are two complete simple geodesics in S which intersect some P_j in isotopic arcs γ_1, γ_2, then γ_1 and γ_2 are within ϵ of each other.*

Proof. For each small $\delta > 0$ look at a hyperbolic metric ρ_δ where the α_i all have length δ (the twisting parameter for the metrics is irrelevant when δ is small). Now suppose ℓ is a simple geodesic intersecting one of the pants P in an arc γ. Suppose the two boundary components of P are $\partial_0 P$ and $\partial_1 P$ and the corresponding geodesics in the pants decomposition are α_0 and α_1, respectively. All of this can be pulled back to the universal covering to obtain geodesics $\widetilde{\alpha}_0, \widetilde{\alpha}_1, \widetilde{\ell}$ and arcs $\widetilde{\partial_0 P}, \widetilde{\partial_1 P}$ as shown in Figure 13 in the disk model. These arcs $\widetilde{\partial_i P}$ are equidistant lines to $\widetilde{\alpha}_i$, for $i = 0, 1$.

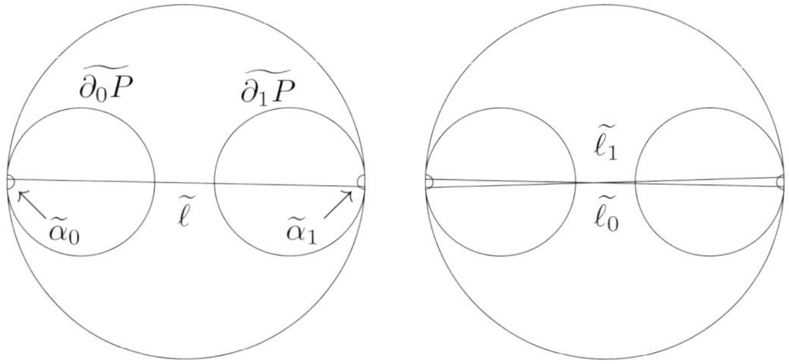

Fig. 13. Left: The lifted picture. Right: Any two geodesics $\widetilde{\ell}_0, \widetilde{\ell}_1$ which intersect $\widetilde{\alpha}_0$ and $\widetilde{\alpha}_1$ must be close between $\widetilde{\partial_0 P}$ and $\widetilde{\partial_1 P}$.

Because we have chosen ∂P to have all components of length $1/2$, the distance between $\widetilde{\partial_0 P}$ and $\widetilde{\partial_1 P}$ is bounded above, independent of δ. It is now clear that if δ is chosen very small, the Euclidean diameter of the $\widetilde{\alpha}_i$ will be very small, and any two geodesics $\widetilde{\ell_0}, \widetilde{\ell_1}$ which intersect both $\widetilde{\alpha_0}, \widetilde{\alpha_1}$ will be very close to each other between $\widetilde{\partial_0 P}$ and $\widetilde{\partial_1 P}$. $\square$

From this we can assume that the α_i have been chosen so small that every time a complete simple geodesic in S enters (or exits) any pants P_j through a component of ∂P_j it does so through one of four very small "windows" in that boundary component—which window it goes through depends on which boundary component it exits through. See Figure 14.

For each collar C_j, we choose a geodesic arc c_j from one boundary component to the other, with endpoints *outside* the windows. If $(\mathcal{L}, \mu)$ is any measured geodesic lamination, observe that $\mathcal{L}$ will intersect each of these arcs c_j transversely (possibly not at all), and so determines a number $t_j(\mu)$ for each j:

$$t_j(\mu) = \int_{c_j} d\mu.$$

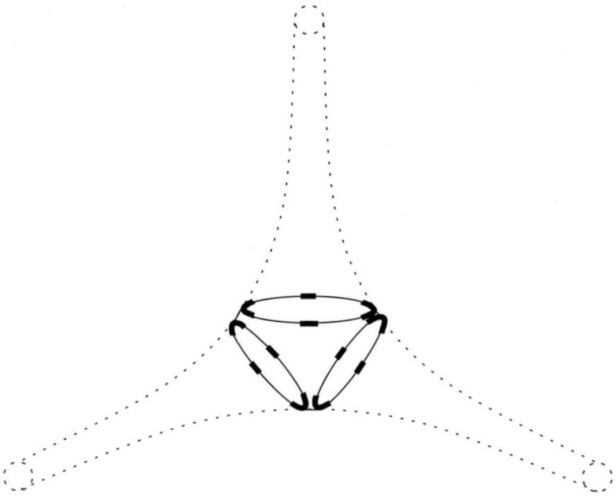

Fig. 14. Pants P_j and part of the adjacent collars (dotted). The dark lines are the windows through which all simple geodesics intersecting P_j must cross ∂P_j.

This measures how much $(\mathcal{L}, \mu)$ crosses c_j, and relative to $\iota(\alpha_i, \mu)$, this measures "how much $(\mathcal{L}, \mu)$ twists" inside of C_j. Moreover, if no α_i is a component of $\mathcal{L}$, then the orientation on the surface allows us to count this with a sign: as the arcs run from one side of C_j to the other, they either all cross left to right across c_j, or right to left. Call the former positive and the latter negative, and hence $t_j(\mu)$ takes values in $\mathbb{R}$, instead of just $\mathbb{R}_{\geq 0}$.

We also have the intersection number of any μ with α_i. Write this as

$$s_i(\mu) = i(\mu, \alpha_i)$$

for each $i = 1, \ldots, 3g-3$.

Let $\mathcal{ML}_{\mathcal{P}}(S)$ denote the set of measured geodesic laminations $(\mathcal{L}, \mu)$ for which no α_i is contained in $\mathcal{L}$. We have a map

$$\Phi_{\mathcal{P}} : \mathcal{ML}_{\mathcal{P}}(S) \to (\mathbb{R}_{\geq 0} \times \mathbb{R})^{3g-3}$$

given by

$$\Phi_{\mathcal{P}}(\mu) = \{(s_i(\mu), t_i(\mu))\}_{i=1}^{3g-3}.$$

We can extend $\Phi_{\mathcal{P}}$ to all of $\mathcal{ML}(S)$ as follows. First observe that if α_i is a component of $\mathcal{L}$, then $(\mathcal{L}, \mu)$ has

$$s_i(\mu) = i(\mu, \alpha_i) = 0.$$

We can still define $t_i(\mu)$ so that

$$|t_i(\mu)| = \int_{c_i} d\mu$$

but now $\mathcal{L} \cap C_i = \alpha_i$ so there is no obvious choice of sign for $t_i(\mu)$. Therefore, we simply give it *both signs*, and so to obtain a well-defined map, we must then modify the range. Namely, in each factor

$$\mathbb{R}_{\geq 0} \times \mathbb{R}$$

we identify $(0, t) \sim (0, -t)$. Note that the result is the same as the cone on $\mathbb{RP}^1$ which we denote $C\mathbb{RP}^1$, and this is homeomorphic to the plane: $C\mathbb{RP}^1 \cong \mathbb{R}^2$. We have therefore defined a map

$$\Phi_{\mathcal{P}} : \mathcal{ML}(S) \to (C\mathbb{RP}^1)^{3g-3} \cong \mathbb{R}^{6g-6}.$$

Theorem 6.2. *The map $\Phi_{\mathcal{P}}$ is a homeomorphism onto the complement of* 0 *(= the cone point), and furthermore $\Phi_{\mathcal{P}}(t\mu) = t\Phi_{\mathcal{P}}(\mu)$ for all $t > 0$, $\mu \in \mathcal{ML}(S)$. Consequently, $\mathbb{P}\mathcal{ML}(S) \cong S^{6g-7}$.*

We will just explain the idea, rather than carrying out the proof. It is not too hard to see that $\Phi_{\mathcal{P}}$ is continuous: the coordinates can all be described in terms of integrals of compactly supported continuous functions on $G(\widetilde{S})$. To prove that $\Phi_{\mathcal{P}}$ is a bijection, we construct an inverse by building a "partial measured foliation" from foliated rectangles glued together according to the data specified by the coordinates $\{(s_i, t_i)\}_{i=1}^{3g-3}$, and then "straightening" this to a measured geodesic lamination (compare with Section 5). Finally, the map is proper by homogeneity, and hence a homeomorphism.

From this it is not too difficult to see that the map

$$\iota_* : \mathcal{ML}(S) \to \mathbb{R}^{\mathcal{S}}$$

defined in Section 4.1 is injective. This is similar to the proof of Theorem 2.1. Indeed, the same $9g - 9$ curves

$$\alpha_1, \ldots, \alpha_{3g-3}, \beta_1, \ldots, \beta_{3g-3}, \tau_1(\beta_1), \ldots, \tau_{3g-3}(\beta_{3g-3})$$

suffice for essentially the same reason.

As a consequence we have the following.

Corollary 6.3. *The subset of $\mathcal{ML}(S)$ consisting of $(\mathcal{L}, \mu)$, where $\mathcal{L}$ is a finite union of simple closed geodesics is dense in $\mathcal{ML}(S)$.*

Proof. If $\Phi_{\mathcal{P}}(\mathcal{L}, \mu)$ has all integer coordinates, then $\mathcal{L}$ is a finite union of simple closed geodesics. The set of positive real multiples of integral points in $(C\mathbb{RP}^1)^{3g-3}$ is dense, hence the preimage of this set by $\Phi_{\mathcal{P}}$ is dense in $\mathcal{ML}(S)$. □

With more work, one can see that $\mathbb{R}_+ \times \mathcal{S}(S)$ is dense in $\mathcal{ML}(S)$; see [16, 30].

7. Degenerations of Hyperbolic Structures

In this final section we explain Thurston's Theorem which provides a compactification of Teichmüller space. We primarily follow the discussion in [16] in what follows.

Recall that by Proposition 2.4, the map

$$\mathbb{P}\ell_* : \mathcal{T}(S) \to \mathbb{PR}^{\mathcal{S}}$$

is injective. Similarly, Theorem 3.5 implies

$$\mathbb{P}\iota_* : \mathbb{P}\mathcal{MF}(S) \to \mathbb{PR}^{\mathcal{S}}$$

is injective. We identify $\mathcal{T}(S)$ and $\mathbb{P}\mathcal{MF}(S)$ with their images in $\mathbb{PR}^{\mathcal{S}}$. According to Proposition 3.6 these are disjoint in $\mathbb{PR}^{\mathcal{S}}$. We let $\overline{\mathcal{T}}(S)$ denote the closure of $\mathcal{T}(S)$ in $\mathbb{PR}^{\mathcal{S}}$.

Theorem 7.1. *The closure of* $\mathcal{T}(S)$ *in* $\mathbb{R}^{\mathcal{S}}$ *is given by*

$$\overline{\mathcal{T}}(S) = \mathcal{T}(S) \cup \mathbb{P}\mathcal{MF}(S)$$

and $(\overline{\mathcal{T}}(S), \mathbb{P}\mathcal{MF}(S)) \cong (\overline{B}^{6g-6}, S^{6g-7})$.

Here $\overline{B}^{6g-6}$ is the closed ball in $\mathbb{R}^{6g-6}$ with boundary the sphere S^{6g-7}. By Theorem 6.2 and Theorem 5.1 we already know that $\mathbb{P}\mathcal{MF}(S) \cong S^{6g-7}$.

According to this theorem, any sequence of hyperbolic structures X_n which exits every compact set has a subsequence (which we also denote X_n) with the following property. There exists $\mu \in \mathcal{ML}(S)$ and a sequence of numbers t_n so that for every $\alpha \in \mathcal{S}(S)$ we have

$$t_n \ell_{X_n}(\alpha) \to \iota(\mu, \alpha)$$

as $n \to \infty$. Alternatively we see that for any *two* curves α, β so that $\iota(\mu, \alpha) \neq 0$ we have

$$\frac{\ell_{X_n}(\beta)}{\ell_{X_n}(\alpha)} \to \frac{\iota(\mu, \beta)}{\iota(\mu, \alpha)}$$

as $n \to \infty$. Thus, the theorem says that, up to scaling, hyperbolic lengths are approximated by geometric intersection numbers with laminations/foliations.

7.1. *The fundamental lemma*

Thurston's proof of the theorem is based on making the last sentence of the previous section quantitative. Specifically, given a point of Teichmüller space X we wish to construct a measured foliation $(\mathcal{F}_X, \mu_X)$ so that lengths in X are well-approximated by intersection numbers with μ_X. To describe this construction, we let $\mathcal{P} = \{\alpha_1, \ldots, \alpha_{3g-g}\}$ denote a pants decomposition of S. The similarity in the Dehn-Thurston coordinates on $\mathcal{ML}(S) = \mathcal{MF}(S)$ and Fenchel-Nielsen coordinates on $\mathcal{T}(S)$ provides the inspiration for the following construction.

Fix attention on one of the pairs of pants P and write $\alpha_1, \alpha_2, \alpha_3$ for the boundary curves (which may or may not all be distinct in S). For each $i = 1, 2, 3$ let $\ell_i = \ell_X(\alpha_i)$. Then we say that the triple (ℓ_1, ℓ_2, ℓ_3) *satisfies the triangle inequality* if for each $i = 1, 2, 3$ we have

$$\ell_i \leq \ell_{i+1} + \ell_{i+2}$$

with indices taken $mod\, 3$. If this is true with strict inequality for each i we say that the triple *satisfies the strict triangle inequality.* If this does not hold for all i then there is a unique i so that with indices take $mod\, 3$

$$\ell_i > \ell_{i+1} + \ell_{i+2}.$$

Call this α_i the *offending boundary component.*

In each of these three cases we first construct a partial measured foliation. Suppose for example that (ℓ_1, ℓ_2, ℓ_3) satisfies the strict triangle inequality. In this case, we let $\gamma_{12}, \gamma_{13}, \gamma_{23}$ denote the embedded hyperbolic geodesic arcs in P so that γ_{ij} is orthogonal to α_i and α_j, for each $i \neq j$. Let $r_{ij} = (\ell_i + \ell_j - \ell_k)/4$, where $k \notin \{i, j\}$, and consider the neighborhoods

$$N_{ij} = N_{r_{ij}}(\gamma_{ij}).$$

Observe that each of these neighborhoods is a "rectangle" in P with two boundary components in ∂P. Any two of the rectangles meet in exactly two points, both of which lie in the boundary. Specifically, $N_{ij} \cap N_{ik}$ is precisely two points in α_i. Foliate N_{ij} by arcs equidistant to γ_{ij}. See Figure 15.

When (ℓ_1, ℓ_2, ℓ_3) does not satisfy the triangle inequality, with offending boundary component α_1, say, we can carry out a similar construction. In this case, we again consider the geodesic arcs γ_{12}, γ_{13} and neighborhoods

$$N_{1j} = N_{r_{1j}}(\gamma_{1j})$$

for $j = 2, 3$ where $r_{1j} = \ell_j/2$ here. Because $\ell_1 > \ell_2 + \ell_3$, we see that $\alpha_1 - (N_{12} \cup N_{13})$ is a pair of arcs, call them δ_1, δ_2, each with length $(\ell_1 -$

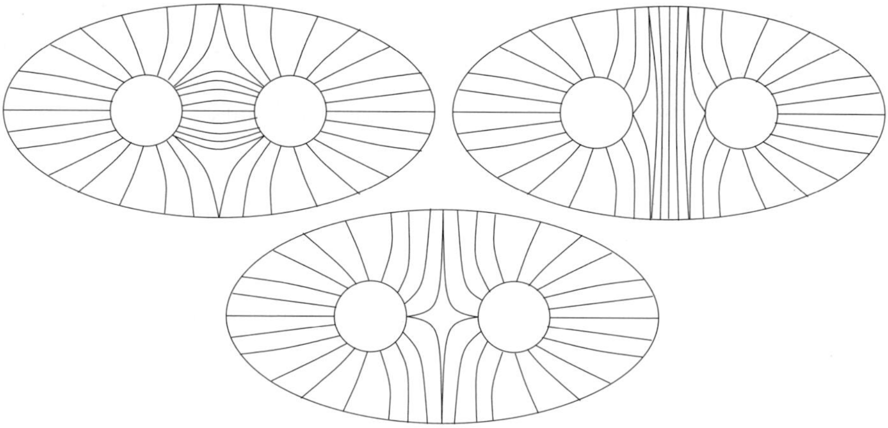

Fig. 15. The three types of partial measured foliations seen in a pair of pants. Upper left: Strict triangle inequality. Upper right: No triangle inequality (with "outer boundary component" the offending component). Bottom: Non-strict triangle inequality

$\ell_2 - \ell_3)/2$. We let N_1 be the union of arcs running between δ_1 and δ_2 which are equidistant to γ_1, the embedded geodesic arc which connects α_1 to itself orthogonally. We foliate N_1 by these arcs. See Figure 15. We caution the reader that γ_1 may not lie N_1. The case of nonstrict triangle inequality is obviously the transition between the two, and if $\ell_1 = \ell_2 + \ell_3$, say, then we can carry out a similar construction using only the two geodesic arcs γ_{12} and γ_{13} and neighborhoods N_{12} and N_{13}; see Figure 15.

We give this foliation a transverse measure which is just the distance between leaves. More precisely, given a tangent vector v let $\pi(v)$ denote the projection onto the orthogonal complement of the tangent space to the leaves (and $\pi(v) = 0$ if v is not based on a leaf), and define the transverse measure of a curve $\gamma : [0,1] \to S$ by

$$\int_0^1 |\pi(\gamma'(t))| dt$$

where $|v|$ is the norm of the vector v using the hyperbolic metric ρ_X.

This produces a partial measured foliation in each pair of pants P_i for which the leaves meet the boundary orthogonally. From this and the definition of the transverse measure, it is straightforward to check that this is the restriction to each P_i of a partial measured foliation on the entire surface. Denote this $(\widehat{\mathcal{F}}_X, \widehat{\mu}_X)$.

This is a *partial* measured foliation because it is only defined on the complement of a collection of triangles and/or quadrilaterals (the latter occurring in the case of nonstrict triangle inequality). The triangles can be collapsed down to tripods (and 4-prong graphs for quadrilaterals) to produce a measured foliation $(\mathcal{F}_X, \mu_X)$ following the same idea as in the construction in the proof of Theorem 5.1. Here we build a Euclidean cone metric q_X from Euclidean rectangles, foliations on the rectangles as in the proof of Theorem 5.1, and a collapsing map, all done one pair of pants at a time. For example, in the case that (ℓ_1, ℓ_2, ℓ_3) satisfies the triangle inequality, we map N_{ij} to the Euclidean rectangle $[0,1] \times [0, 2r_{ij}]$ with horizontal foliation (recall that $2r_{ij}$ is the hyperbolic width of N_{ij} described above) so that the transverse measures are preserved. By an abuse of notation, we think of the collapsing map as a map from S to itself, $f : S \to S$. This is a homeomorphism on the interiors of the rectangles, takes leaves of $\widehat{\mathcal{F}}_X$ to leaves of $\mathcal{F}_X$, and preserves transverse measures.

The key ingredient in the proof of Theorem 7.1 is the following "Fundamental Lemma" of [16]. Given $\epsilon > 0$, let $V(\mathcal{P}, \epsilon)$ denote the set of hyperbolic structures in which each curve α_i of $\mathcal{P}$ has length at least ϵ

$$V(\mathcal{P}, \epsilon) = \{X \in \mathcal{T}(S) \,|\, \ell_X(\alpha_i) \geq \epsilon \text{ for all } \alpha_i \in \mathcal{P}\}.$$

Lemma 7.2. *Given ϵ and any curve $\gamma \in \mathcal{S}$ there exists $C \geq 0$ so that for all $X \in V(\mathcal{P}, \epsilon)$ we have*

$$\iota(\gamma, \mu_X) \leq \ell_X(\gamma) \leq \iota(\gamma, \mu_X) + C$$

where $(\mathcal{F}_X, \mu_X)$ is the measured foliation constructed above.

Observe that for the pants curves α_i, we can take $C = 0$ by construction of μ_X.

Proof. The first inequality is easy, and does not require $X \in V(\mathcal{P}, \epsilon)$. Let $f : S \to S$ be the collapsing map as described above. Let γ denote the geodesic representative with respect to X. If $\pi(\gamma'(t))$ is as defined above, then we have

$$\iota(\gamma, \mu_X) \leq \int_{f(\gamma)} d\mu_X = \int_0^1 |\pi(\gamma'(t))| dt \leq \int_0^1 |\gamma'(t)| dt = \ell_X(\gamma)$$

as required.

The second inequality is a bit more subtle, but we sketch the idea. The first fact that we need is the following.

Claim. *For all $\epsilon > 0$, there exists $K > 0$ with the following property. If $X \in V(\mathcal{P}, \epsilon)$ and P is one of the pants determined by $\mathcal{P}$, then the length of the leaves of the partial measured foliation in P are bounded above by K.*

Proof. Let $\alpha_1, \alpha_2, \alpha_3$ be the boundary geodesics of P so that $\ell_1, \ell_2, \ell_3 \geq \epsilon$. The proof is slightly different in each of the three cases, but the idea is the same. For concreteness, we consider the case that (ℓ_1, ℓ_2, ℓ_3) satisfies the triangle inequality and we use the notation as above. Without loss of generality, suppose $\ell_1 \geq \ell_2 \geq \ell_3 \geq \epsilon$. Observe first that the length of any leaf in a rectangle N_{ij} is bounded above by the lengths of the boundary leaves. Furthermore, if we bound the length of the boundary leaves for two of the rectangles, then it is not hard to see that the third is also bounded.

We will bound the lengths of the boundary leaves of N_{12} and N_{13}. First recall that the transverse boundary arc of N_{ij} has length

$$2r_{ij} = \frac{\ell_i + \ell_j - \ell_k}{2}$$

where i, j, k are all distinct. From the triangle inequality, and our assumptions on the lengths, it follows that this is at least $\epsilon/2$ for N_{12} and N_{13}. Fix one of these rectangles N and suppose that the boundary length is L. A calculation in hyperbolic geometry tells us that the length of an arc of the boundary of an R-neighborhood of a geodesic running between two geodesics orthogonal to the original geodesic is $\cosh(R)$ times the length of the geodesic segment between the orthogonals; see Figure 16. From this, a very crude lower bound for the area of N is $\epsilon L/(2\cosh(\epsilon/2))$ (obtained by mapping via a 1-Lipschitz map to a Euclidean rectangle). However, N is contained in P which has area 2π, and hence

$$L \leq \frac{4\pi\cosh(\epsilon/2)}{\epsilon}.$$

This is the required bound in this case. The other two cases are similar. □

Recall that the foliation $(\mathcal{F}_X, \mu_X)$ was described in terms of a Euclidean cone metric q_X, built by making each of the rectangles in the pants into Euclidean rectangles and collapsing out the complementary regions. A simple but useful observation about the metric q_X on each pair of pants P is that it can be obtained as the union of three Euclidean cylinders, each of height $1/2$ (and circumference equal to the X-length of the given curve). The three annuli are glued together along a spine for P as shown in Figure 17 in the case that (ℓ_1, ℓ_2, ℓ_3) satisfies the triangle inequality. To see this, consider the foliation orthogonal to $\mathcal{F}_X$ (compare Figure 12).

Let γ be any simple closed curve on S, and let γ_0 to be a q_X-geodesic representative of γ. Every time γ_0 enters and exits a pair of pants P, the description of geodesics from Section 3.3 tells us that it does so in a relatively simple way: first it crosses one of the annuli, traverses at most one

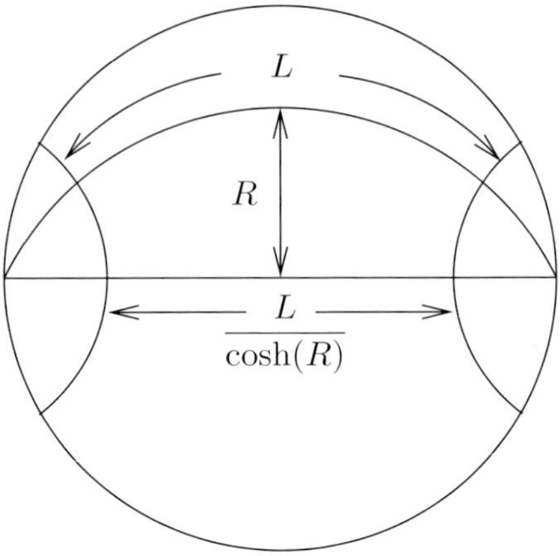

Fig. 16. Lengths of equidistant curves increase exponentially.

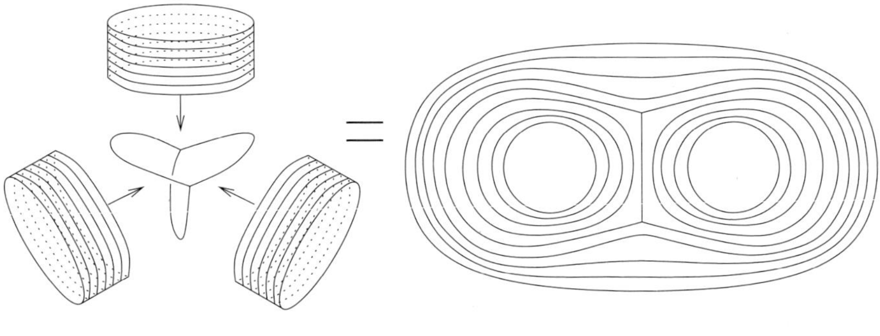

Fig. 17. Pants built from Euclidean rectangles can also be obtained by gluing Euclidean cylinders. The union of the separatrices forms the spine.

of the arcs of the spine, then it crosses another annulus and exits P. Observe that when γ_0 crosses an annulus, it generally does this "diagonally". We replace each such diagonal subpath of γ_0 with a path having the same total μ_X-variation that first traverses around the boundary of the pants, then follows a leaf of $\mathcal{F}_X$ to the spine. Similarly, if γ_0 traverses an edge of the spine, we can modify this, also keeping the μ_X-variation fixed, so that it runs along a leaf of $\mathcal{F}_X$ to the boundary of the pants, runs along the boundary of the pants, then back along a leaf of $\mathcal{F}_X$. Let γ_1 denote the resulting representative of the homotopy class γ. We have

$$\int_{\gamma_1} d\mu_X = \int_{\gamma_0} d\mu_X = \iota(\gamma, \mu_X).$$

On the other hand, γ_1 determines a representative γ_2 which does essentially the same thing in the hyperbolic metric: it traverses the boundary of the pants in the same fashion as γ_1 and traverses leaves of the partial foliation $\widehat{\mathcal{F}}_X$ whenever γ_1 does.

Now we observe that for a fixed curve γ, the number of times γ_1, and hence γ_2, traverses any arc of a leaf is uniformly bounded by some number $D > 0$ (depending only on the number of times γ crosses the pants curves). The hyperbolic length of γ_2 is the sum of the lengths of the arcs contained in the pants curves plus the lengths of the arcs of leaves traversed. By the claim above the hyperbolic length of each arc of a leaf traversed is at most K, and hence

$$\ell_X(\gamma) \leq \ell_X(\gamma_2) \leq \int_{\gamma_1} d\mu_X + KD = \iota(\gamma, \mu_X) + KD.$$

Setting $C = KD$ completes the proof. □

From this we have the following corollary. To avoid confusion, we do not identify $\mathcal{T}(S)$ and $\mathbb{P}\mathcal{MF}(S)$ with their images in $\mathbb{PR}^{\mathcal{S}}$, but refer to their respective embeddings $\mathbb{P}\ell_*$ and $\mathbb{P}\iota_*$ explicitly.

Corollary 7.3. *Let $X_n \in V(\mathcal{P}, \epsilon)$ with $X_n \to \infty$ in $\mathcal{T}(S)$. Then*

$$\{\mathbb{P}\ell_*(X_n)\}_{n=1}^{\infty} \text{ and } \{\mathbb{P}\iota_*(\mu_{X_n})\}_{n=1}^{\infty}$$

either both converge or both diverge. Furthermore, if they both converge, then they have the same limit.

Proof. From Theorem 2.1 we know that there is some curve γ so that $\ell_{X_n}(\gamma) \to \infty$ as $n \to \infty$. Suppose now that $\mathbb{P}\ell_*(X_n)$ converges to $\mathbb{P}(\{x_\alpha\}_{\alpha\in\mathcal{S}})$. Let $\{t_n\}_{n=1}^{\infty}$ be such that $t_n\ell_*(X_n)$ converges to $\{x_\alpha\}_{\alpha\in\mathcal{S}}$. By definition then we have

$$\lim_{n\to\infty} t_n \ell_{X_n}(\alpha) = x_\alpha$$

for all $\alpha \in \mathcal{S}(S)$. Applying this to $\alpha = \gamma$ we see that $t_n \to 0$.

For each $\alpha \in \mathcal{S}$, let $C(\alpha)$ be the constant from Lemma 7.2. Then $t_nC(\alpha) \to 0$ and so by the Lemma 7.2 and the squeeze theorem we have

$$\lim_{n\to\infty} t_n \iota(\mu_{X_n}, \alpha) = x_\alpha$$

as required.

The argument is easily reversed to prove that if $\{\mathbb{P}\iota_*(\mu_{X_n})\}$ converges, then $\{\mathbb{P}\ell_*(X_n)\}$ does as well and the limits are the same. $\square$

We are now ready for the proof of Theorem 7.1.

Proof of Theorem 7.1. Given any sequence $X_n \to \infty$ in $\mathcal{T}(S)$, we can find a pants decomposition $\mathcal{P} = \{\alpha_1, \ldots, \alpha_{3g-3}\}$ an $\epsilon > 0$ and a subsequence (also called $\{X_n\}$) so that $\{X_n\} \subset V(\mathcal{P}, \epsilon)$. To find such a pants decomposition, start with any pants decomposition and consider one of the curves. If there is a subsequence where the length of that curve is bounded below, pick that subsequence and move on to the next curve in the pants decomposition. If there is no such subsequence, then the length of that curve is tending to zero, and we can replace it with a curve that transversely intersects it. This has length tending to infinity, and we can then move on to the next curve. We continue this until we arrive at the desired $\mathcal{P}$ and subsequence $\{X_n\}$.

Since $\mathbb{P}\mathcal{MF}(S) \cong S^{6g-7}$ is compact, we can pass to a subsequence so that μ_{X_n} converges to some $\mu \in \mathbb{P}\mathcal{MF}(S)$. From Corollary 7.3 we see that X_n also converges to μ. Therefore $\overline{\mathcal{T}}(S) \subset \mathcal{T}(S) \cup \mathbb{P}\mathcal{MF}(S)$.

To see that we get equality, we observe that by using Fenchel-Nielsen coordinates and Dehn-Thurston coordinates, the map $X \mapsto \mu_X$ maps $V(\mathcal{P}, \epsilon)$ onto the subset $\mathcal{MF}_{(\mathcal{P},\epsilon)}(S) \subset \mathcal{MF}(S)$ consisting of those μ with $\iota(\mu, \alpha_i) \geq \epsilon$ for all $\alpha_i \in \mathcal{P}$. In particular, given any $\mu \in \mathcal{MF}(S)$, there is a pants decomposition $\mathcal{P}$ and $\epsilon > 0$ so that $n\mu$ is in $\mathcal{MF}_{(\mathcal{P},\epsilon)}(S)$ for all n. That is, $n\mu = \mu_{X_n}$ for some $X_n \in V(\mathcal{P}, \epsilon)$. The sequence $n\mu$ is constant projectively, but clearly $X_n \to \infty$ in $\mathcal{T}(S)$—the lengths of all pants curves tend to ∞. By Corollary 7.3 we have $X_n \to \mu$ and so

$$\mathcal{T}(S) \cup \mathbb{P}\mathcal{MF}(S) \subset \overline{\mathcal{T}}(S)$$

and hence these sets are equal.

Let $\mathbb{P}\mathcal{MF}_{\mathcal{P}}(S)$ be the set of measured foliations with $\iota(\alpha_i, \mu) \neq 0$ for all $\alpha_i \in \mathcal{P}$. Clearly, $\mathbb{P}\mathcal{MF}_{\mathcal{P}}(S)$ is the image of $\mathcal{MF}_{(\mathcal{P},\epsilon)}(S)$ for any $\epsilon > 0$ under the projectivization $\mathbb{P}$. To see that $\overline{\mathcal{T}}(S)$ is a manifold with boundary equal to $\mathbb{P}\mathcal{MF}(S)$, we set

$$U_{\mathcal{P}} = \mathcal{T}(S) \cup \mathbb{P}\mathcal{MF}_{\mathcal{P}}(S) \subset \mathbb{P}\mathbb{R}^{\mathcal{S}}.$$

Being an open subset of $\mathbb{P}\mathcal{MF}(S) \cong S^{6g-7}$, we see that $\mathbb{P}\mathcal{MF}_{\mathcal{P}}(S)$ is homeomorphic to an open subset $W \subset \mathbb{R}^{6g-7}$. We define

$$\Psi_{\mathcal{P}} : U_{\mathcal{P}} \to \mathbb{P}\mathcal{MF}_{\mathcal{P}}(S) \times (0, \infty]$$

by

$$\Psi_{\mathcal{P}}(X) = (\mathbb{P}(\mu_X), \min_{\alpha_i \in \mathcal{P}}(\ell_X(\alpha_i)))$$

for $X \in \mathcal{T}(S)$, and for $\mathbb{P}(\mu) \in \mathbb{P}\mathcal{MF}_{\mathcal{P}}(S)$ we set $\Psi_{\mathcal{P}}(\mathbb{P}(\mu)) = (\mathbb{P}(\mu), \infty)$. One can check that this is a homeomorphism. Then since

$$\mathbb{P}\mathcal{MF}_{\mathcal{P}}(S) \times (0, \infty] \cong W \times [0, \infty)$$

and since $\overline{\mathcal{T}}(S)$ is covered by such sets, we see that $\overline{\mathcal{T}}(S)$ is a manifold with boundary $\mathbb{P}\mathcal{MF}(S) \cong S^{6g-7}$ (the second countability and Hausdorff conditions follow from the embedding into $\mathbb{P}\mathbb{R}^{\mathcal{S}}$)

That $\overline{\mathcal{T}}(S) \cong \overline{B}^{6g-6}$ now follows from this, the fact that $\mathcal{T}(S) \cong \mathbb{R}^{6g-6}$, and purely topological considerations; see [16] for more details. □

7.2. *Final comments on Theorem 7.1*

7.2.1. *Application to the mapping class group*

The utility of the Thurston compactification $\overline{\mathcal{T}}(S)$ comes from the fact that this is a mapping class group invariant compactification: the action of the mapping class group $\mathrm{Mod}(S)$ on $\mathcal{T}(S)$ (see Aramayona's Lectures [2]) extends to an action on $\overline{\mathcal{T}}(S)$. This is immediate from the naturality of the construction. Indeed, $\mathrm{Mod}(S)$ acts on $\mathcal{S}$, and hence on $\mathbb{P}\mathbb{R}^{\mathcal{S}}$ by acting on the coordinates

$$\phi \cdot \{x_\alpha\}_{\alpha \in \mathcal{S}} = \{x_{\phi \cdot \alpha}\}_{\alpha \in \mathcal{S}}.$$

It is straightforward to check that the inclusion $\mathbb{P}\ell_*$ is equivariant with respect to the action of $\mathrm{Mod}(S)$ (as is $\mathbb{P}\iota_*$). Therefore, as we have identified $\mathcal{T}(S)$ with its image, we see that the induced action on $\mathcal{T}(S)$ as a subspace of $\mathbb{P}\mathbb{R}^{\mathcal{S}}$ agrees with the usual action, and hence the action extends to the closure $\overline{\mathcal{T}}(S)$ as the restriction of the action on all of $\mathbb{P}\mathbb{R}^{\mathcal{S}}$. This is a fundamental ingredient in Thurston's proof of his classification Theorem for mapping classes stated in Aramayona's lectures, [2]. Indeed, using the Brouwer Fixed Point Theorem, the action on $\overline{\mathcal{T}}(S)$ of any element of $\mathrm{Mod}(S)$ has a fixed point, and a careful analysis of this fixed point provides all the necessary information needed to classify the elements of $\mathrm{Mod}(S)$. See [16] for Thurston's original proof, or [13, 4, 20, 15, 9] for other proofs.

7.2.2. *Other approaches*

There are now several alternative constructions of the Thurston compactification of $\mathcal{T}(S)$ of Theorem 7.1. One goes by way of actions on $\mathbb{R}$-trees, and is due to Skora [31] (see also [18, 29, 28]), building on the work of Morgan and Shalen [29, 27]. Another proof is due to Bonahon [6], obtained by embedding $\mathcal{T}(S)$ into the space of projective geodesic currents mentioned in Section 4.1, and then showing that the closure is precisely the image of $\mathcal{T}(S)$ together with $\mathbb{P}\mathcal{ML}(S)$. It is worth mentioning however that these constructions do not give the full strength of Theorem 7.1. More precisely, these provide a construction of the compactification, but not a description of $\overline{\mathcal{T}}(S)$ as a closed ball.

Acknowledgements

I would like to thank Javier Aramayona, David Dumas, Richard Kent, Dan Margalit and Caroline Series for helpful discussions throughout the writing of these notes, as well as Ser-Wei Fu, Pradthana Jaipong and the referee for many useful comments. I would especially like to thank Ser Peow Tan and the IMS at the National University of Singapore for their hospitality during the summer school on *Geometry, Topology and Dynamics of Character Varieties* in July 2010.

References

1. S. B. Alexander and R. L. Bishop. The Hadamard-Cartan theorem in locally convex metric spaces. *Enseign. Math.* (2) 36 (1990), no. 3-4, 309–320.
2. J. Aramayona. *Hyperbolic structures on surfaces.* This review volume.
3. R. Benedetti & C. Petronio. *Lectures on hyperbolic geometry.* Universitext, Springer-Verlag, Berlin, 1992. xiv+330 pp.
4. L. Bers. An extremal problem for quasiconformal mappings and a theorem by Thurston. *Acta Math.* 141 (1978), no. 1-2, 73–98.
5. F. Bonahon. Bouts des variétés hyperboliques de dimension 3. *Ann. of Math. (2)*, 124(1):71–158, 1986.
6. F. Bonahon. The geometry of Teichmüller space via geodesic currents. *Invent. Math.*, 92(1):139–162, 1988.
7. F. Bonahon. Geodesic laminations on surfaces. In *Laminations and foliations in dynamics, geometry and topology (Stony Brook, NY, 1998)*, volume 269 of *Contemp. Math.*, pages 1–37. Amer. Math. Soc., Providence, RI, 2001.
8. M. Bestvina, K. Bromberg, K. Fujiwara, J. Souto. Shearing coordinates and convexity of length functions. Preprint, 2009.
9. M. Bestvina and M. Handel. Train-tracks for surface homeomorphisms. *Topology* 34 (1995), no. 1, 109–140.
10. M. R. Bridson & A. Haefliger. *Metric spaces of non-positive curvature.* Gdrundlehren der Mathematischen Wissenschaften [Fundamental Principles of Mathematical Sciences], 319. Springer-Verlag, Berlin.

11. D. Calegari. *Foliations and the geometry of 3-manifolds.* Oxford Mathematical Monographs. Oxford University Press, Oxford, 2007. xiv+363 pp.
12. R. D. Canary, D. B. A. Epstein & P. L. Green. *Notes on notes of Thurston With a new foreword by Canary.* London Math. Soc. Lecture Note Ser., 328, Fundamentals of hyperbolic geometry: selected expositions, 1–115, Cambridge Univ. Press, Cambridge, 2006.
13. A. J. Casson & S. A. Bleiler. *Automorphisms of surfaces after Nielsen and Thurston.* London Mathematical Society Student Texts, 9. Cambridge University Press, Cambridge, 1988. iv+105 pp.
14. M. P. do Carmo. *Riemannian geometry.* Mathematics: Theory & Applications. Birkhäuser Boston, Inc., Boston, MA, 1992.
15. B. Farb & D. Margalit. *A primer on mapping class groups.* Princeton University Press, to appear.
16. A. Fathi, F. Laudenbach, and V. Poenaru. *Travaux de Thurston sur les surfaces*, volume 66 of *Astérisque.* Société Mathématique de France, Paris, 1979. Séminaire Orsay, With an English summary.
17. F. P. Gardiner. *Teichmüller theory and quadratic differentials.* Pure and Applied Mathematics (New York). John Wiley & Sons Inc., New York, 1987. A Wiley-Interscience Publication.
18. H. Gillet & P. B. Shalen. Dendrology of groups in low **Q**-ranks. *J. Differential Geom.* 32 (1990), no. 3, 605–712.
19. N. Halpern, A proof of the collar lemma. *Bull. London Math. Soc.* **13** (1981), no. 2, 141–144.
20. M. Handel & W. P. Thurston, New proofs of some results of Nielsen, *Adv. in Math.* 56 (1985), no. 2, 173–191.
21. J. Hubbard & H. Masur. Quadratic differentials and foliations. *Acta Math.*, 142(3-4):221–274, 1979.
22. Y. Imayoshi & M. Taniguchi. *An introduction to Teichmüller spaces.* Springer-Verlag, Tokyo, 1992. xiv+279 pp.
23. L. Keen, Collars on Riemann surfaces. *Discontinuous groups and Riemann surfaces (Proc. Conf., Univ. Maryland, College Park, Md., 1973)*, pp. 263–268. Ann. of Math. Studies, No. 79, Princeton Univ. Press, Princeton, N.J., 1974.
24. S. P. Kerckhoff. The Nielsen realization problem. *Ann. of Math. (2)* **117** (1983), no. 2, 235–265.
25. S. P. Kerckhoff. Earthquakes are analytic. *Comment. Math. Helv.* **60** (1985), no. 1, 17–30.
26. G. Levitt. Foliations and laminations on hyperbolic surfaces. *Topology*, 22(2):119–135, 1983.
27. J. W. Morgan. Group Actions on Trees and the Compacti
cation of the Spaces of Classes of SO(n; 1)-representations. *Topology*, 25 (1986) 1–34.
28. J. W. Morgan and J.-P. Otal. Relative growth rates of closed geodesics on a surface under varying hyperbolic structures. *Comment. Math. Helv.* 68 (1993), no. 2, 171–208.

29. J. W. Morgan and P. B. Shalen. Degenerations of Hyperbolic Structures, I: Valuations, Trees and Surfaces. *Ann. of Math. (2)* 120 (1984), no. 3, 401–476.
30. R. C. Penner and J. L. Harer. *Combinatorics of train tracks*, volume 125 of *Annals of Mathematics Studies*. Princeton University Press, Princeton, NJ, 1992.
31. R. K. Skora. Splittings of surfaces. *J. Amer. Math. Soc.* 9 (1996), no. 2, 605–616.
32. W. P. Thurston. *On the geometry and dynamics of diffeomorphisms of surfaces*, Bull. Am. Math. Soc., New Ser. **19** (1988), 417–431.
33. W. P. Thurston. *Three-dimensional geometry and topology. Vol. 1.* Princeton Mathematical Series, 35. Princeton University Press, Princeton, NJ, 1997.
34. W. P. Thurston. *The Geometry and Topology of Three-Manifolds*, Princeton University course notes, available at www.msri.org/publications/books/gt3m/ (1980).

PING-PONG LEMMAS WITH APPLICATIONS TO GEOMETRY AND TOPOLOGY

Thomas Koberda

Department of Mathematics
Harvard University
Cambridge, MA 02138, USA
koberda@math.harvard.edu

We study the actions of certain groups in geometric and topological contexts. Of particular interest are free groups, various free products and right-angled Artin groups. We discuss various tools which can be used to show that a particular action is faithful. We illustrate applications of these principles in the context of hyperbolic geometry and mapping class groups of surfaces.

Keywords: Ping-pong lemmas, mapping class groups, word problem

Mathematics Subject Classification 2000: 57M60, 20F36, 20E06

1. Introduction

Let G be a group and let X be a set equipped with an action of G. By this we mean a homomorphism $\alpha : G \to \mathcal{S}(X)$, the group of permutations of X. One of the first basic questions one can ask about an action is whether it is faithful, which is to say whether or not α is an injective map. If G is a free group for example, and $g \in G$ is 10^{100} letters long, it may be impossible to verify by hand whether or not g acts nontrivially on X via α. This article concerns itself with various more efficient methods for showing that an action is faithful. These tools are collectively called **ping-pong lemmas**, and they find their origins in the work of Klein in [16].

Let us begin by illustrating the inner workings of ping-pong lemmas with a basic example. Consider the complex plane $\mathbb{C}$ and let $r, \tau \in \mathbb{C}$, where $r \in \mathbb{R}$ and $\Im(\tau) \neq 0$. Both r and τ act by homeomorphisms on $\mathbb{C}$ by

$$z \mapsto z + x,$$

$x \in \{r, \tau\}$. Since translation by r and τ commute with each other, we obtain a homomorphism $\mathbb{Z}^2 \to \mathrm{Homeo}(\mathbb{C})$. It is obvious that this homomorphism is injective. We shall give a proof using a ping-pong lemma for abelian groups which will exhibit many of the salient features of general ping-pong lemmas.

Proposition 1.1. *Let the vectors v and w generate $\mathbb{Z}^2$. The homomorphism $\mathbb{Z}^2 \to \mathrm{Homeo}(\mathbb{C})$ given by $v \mapsto \{z \mapsto z + r\}$ and $w \mapsto \{z \mapsto z + \tau\}$ is injective.*

Proof. Recall that $\mathbb{C}$ naturally sits inside of $\widehat{\mathbb{C}}$, the Riemann sphere. A system of neighborhoods of $\{\infty\}$ is given by complements of compact sets of $\mathbb{C}$. Let $1 \neq g \in \mathbb{Z}^2$ and let O be the origin in $\mathbb{C}$. We claim there is a neighborhood N of $\{\infty\}$ in $\widehat{\mathbb{C}}$ which is independent of g and does not contain O such that $g(O) \in N$. It will follow that the image of g in $\mathrm{Homeo}(\mathbb{C})$ is nontrivial, so that the action is faithful as claimed.

Consider the complement U_1 of a closed infinite rectangular neighborhood of the imaginary axis of width r (so that the real coordinates of the boundaries are $\pm r/2$), and consider also the complement U_2 of a closed rectangular neighborhood of the real axis of width $\Im(\tau)$ (so that the imaginary coordinates of the boundaries are $\pm\Im(\tau)/2$). Let $N = U_1 \cup U_2$.

To see that this N works as claimed, write $1 \neq g \in \mathbb{Z}^2$ as a reduced word of the form $v^n w^m$. If $m = 0$ then the absolute value of the real coordinate of $g(O)$ is at least r. If $m \neq 0$ then the absolute value of the imaginary coordinate of $g(O)$ is at least $|\Im(\tau)|$. It follows that $g(O) \in N$. $\square$

A principle which allows us to formulate, prove and use a ping-pong lemma for a group G is the existence of certain distinguished subsets

$$\{A_1, \ldots, A_n\}$$

of the phase space X and a basepoint

$$x_0 \in X \setminus \bigcup_{i=1}^{n} A_i$$

such that whenever $g \in G$ is nontrivial,

$$g(x_0) \in \bigcup_{i=1}^{n} A_i.$$

In this paper, the "distinguished subsets" are usually small disjoint neighborhoods (in some topology) of fixed points of certain generators of G. In the case of Proposition 1.1, we had $A_1 = N$ and $x_0 = O$. The distinguished subset N is a neighborhood of $\{\infty\}$ in $\widehat{\mathbb{C}}$, the fixed point of the $\mathbb{Z}^2$-action.

The general philosophy of ping-pong lemmas will allow us formulate and prove them for various classes of groups, most importantly:

(1) Finitely generated free groups F_n.
(2) More generally, finitely generated free products.
(3) Right-angled Artin groups.

Recall that **free groups** are given by a set of generators and no relations. **Free products** are given by the generating sets of a collection of groups $\{G_i\}$ with no relations other than the ones which hold in the factor groups. **Right-angled Artin groups** are given by a finite graph Γ. We declare the generators to be the vertices of Γ. We declare two generators to commute if they are connected by an edge, and we impose no further relations.

We will be able to prove or sketch proofs of an assortment of results from the ping-pong lemmas for these groups, such as:

Theorem 1.2. *Let ϕ, ψ be two non-elliptic isometries of hyperbolic space $\mathbf{H}^n$ which do not fix any common point on the boundary. Then there exists an N such that $\langle \phi^N, \psi^M \rangle \cong F_2$.*

Theorem 1.2 was one of the motivations for studying ping-pong in the first place in [16].

Theorem 1.3. *Let G be a finite free product of finitely generated torsion-free abelian groups. Then there exists an n such that G embeds discretely into the isometry group of $\mathbf{H}^n$. Groups of this form are the only right-angled Artin groups which embed discretely into the isometry groups of hyperbolic spaces.*

Theorem 1.3 can be found in the author's paper [18], but it was probably known before. See [12] for a discussion of a similar ilk.

A **mapping class group** of a manifold M is defined to be the group of orientation-preserving homeomorphisms of M modulo isotopy (homotopy through homeomorphisms). The mapping class group of M is often written $\mathrm{Mod}(M)$. Considering $\mathrm{Homeo}(M)$ as a topological group, we can write

$$\mathrm{Mod}(M) = \pi_0(\mathrm{Homeo}(M)).$$

The elements of a mapping class group are called mapping classes.

Theorem 1.4. *Let ϕ and ψ be two mapping classes of some surface Σ. There exists an N such that $\langle \phi^N, \psi^N \rangle \in \{\{1\}, \mathbb{Z}, \mathbb{Z}^2, F_2\}$.*

Theorem 1.4 has been known at least since the time of McCarthy and Papadopoulos in [20].

One important type of mapping classes is a so-called Dehn twist. To specify a Dehn twist, we first need an essential (homotopically nontrivial) simple closed curve c on the surface Σ. The twist T_c about c is the homeomorphism of Σ given by cutting Σ open along c and regluing with a full twist. We say that two isotopy representatives of c and d are in minimal position if they intersect a minimal number of times over all isotopy representatives of c and d.

Theorem 1.5. *Let T_c and T_d be Dehn twists about non-isotopic simple closed curves c and d. Then:*

(1) T_c and T_d commute if and only if c and d have isotopy representatives which are disjoint.
(2) $\langle T_c, T_d\rangle \cong F_2$ if and only if c and d intersect at least twice when in minimal position. Also, $\langle T_c^{n_1}, T_d^{n_2}\rangle \cong F_2$ whenever $|n_1|, |n_2| > 1$.
(3) T_c and T_d satisfy the relation $T_cT_dT_c = T_dT_cT_d$ whenever c and d intersect exactly once in minimal position.

These are quite well-known relations which hold in mapping class groups. See Farb and Margalit's book [9] for more information.

Finally, we will be able to give some indication of the proof of the following special case of a result due to the author:

Theorem 1.6. *Let $\{\psi_1, \ldots, \psi_k\}$ be Dehn twists about essential simple closed curves on a surface Σ. There exists an N such that $\{\psi_1^N, \ldots, \psi_k^N\}$ generate a right-angled Artin group.*

The full proof can be found in [18]. M. Kapovich has also proposed a proof in [14]. Clay, Leininger and Mangahas studied similar questions using a tool called the complex of curves in [6].

2. Free Groups and Free Products

The first version of the ping-pong lemma we give here is for free groups. The statement of the ping-pong lemma had originally been formulated by Klein in [16] for the study of discrete groups automorphisms of the Riemann sphere (which coincides with the modern study of so-called Kleinian groups).

Theorem 2.1. *Let F_n be a free group on n generators $\{g_1, \ldots, g_n\}$ and let X be a set equipped with an F_n-action. Suppose there exist subsets*

$X_1, \ldots, X_n$ *of* X *and a point*

$$x_0 \in X \setminus \bigcup_{i=1}^{n} X_i$$

such that for each nonzero k, $g_i^k(x_0) \in X_i$ *and* $g_i^k(X_j) \subset X_i$ *for every* $j \neq i$. *Then the* F_n*-action on* X *is faithful.*

Proof. Let $w \in F_n$ be a reduced word. We proceed by induction on the length of w. The base case of length one words is addressed by the hypotheses of the theorem. Write $w = g_i^k w'$, where k is maximal in absolute value. Note that the length of w' is strictly less than that of w. Since w is reduced and k is maximal, we have that the left-most letter of w' is neither g_i nor g_i^{-1}. It follows by the inductive hypothesis that $w'(x_0) \in X_j$ for some $j \neq i$. But since $g_i^k(X_j) \subset X_i$, we have that $w(x_0) \in X_i$. In particular, $w(x_0) \neq x_0$ so that w does not act by the identity on X. □

This is the most basic nontrivial example of a ping-pong lemma. Note that from the proof it is apparent that the ping-pong lemma is almost a tautology. Indeed, let X be the set of reduced words in F_n, equipped with an action of F_n by left multiplication. Let X_i be the set of reduced words in F_n whose left-most letter is $g_i^{\pm 1}$. If x_0 is the identity, it becomes clear that the hypotheses of the ping-pong lemma are satisfied.

We would often like to think of ping-pong lemmas in a geometric context. As such, the previous example can be made more geometric using the Cayley graph of F_n as X, and the X_i are certain subsets of the vertices of the Cayley graph.

For our first application of the ping-pong lemma, we have the following:

Proposition 2.2. *The matrices*

$$\begin{pmatrix} 1 & 2 \\ 0 & 1 \end{pmatrix}, \begin{pmatrix} 1 & 0 \\ 2 & 1 \end{pmatrix}$$

generate a free subgroup of $SL_2(\mathbb{Z})$.

Proof. The group $\Gamma(2)$ generated by these matrices acts naturally on $\mathbb{R}^2$ by linear maps, and we let $X = \mathbb{R}^2$. Let $x_0 = (1, 1)$, $X_1 = \{(x, y) \mid |y| > |x|\}$ and $X_2 = \{(x, y) \mid |x| > |y|\}$. It is straightforward to apply Theorem 2.1. □

The interested reader is encouraged to show that the subgroup $\Gamma(2)$ generated by the matrices in the previous proposition has finite index in

$SL_2(\mathbb{Z})$. Furthermore, the reader might try and find a relation which holds between the matrices

$$\begin{pmatrix} 1 & 1 \\ 0 & 1 \end{pmatrix}, \begin{pmatrix} 1 & 0 \\ 1 & 1 \end{pmatrix}.$$

We now state a more general version of Theorem 2.1 which characterizes free products. The reader is also directed to P. de la Harpe's book [8]:

Theorem 2.3. *Let G be a group and let X be a set equipped with a G-action. Suppose G is generated by subgroups Γ_1 and Γ_2 with $|\Gamma_1| \geq 3$ and $|\Gamma_2| \geq 2$. Suppose furthermore that X_1 and X_2 are two subsets of X with X_2 not included in X_1. Finally, suppose that for each nonidentity $\gamma_i \in \Gamma_i$, $\gamma_1(X_2) \subset X_1$ and $\gamma_2(X_1) \subset X_2$. Then $G \cong \Gamma_1 * \Gamma_2$.*

Note that the conclusions of Theorem 2.3 imply that the action of G on X is faithful, as will be apparent from the proof.

Proof of Theorem 2.3. Let $w \in \Gamma_1 * \Gamma_2$, and assume that w can be written as $w = a_1 b_1 a_2 b_2 \cdots a_k$ with each $a_i \in \Gamma_1$ and each $b_i \in \Gamma_2$. We claim that w does not act by the identity on X. Indeed, $w(X_2) \subset X_1$. Since X_2 is not contained in X_1, w does not act by the identity.

If $w = b_1 a_1 \cdots b_k$ then we simply conjugate w by a nonidentity element of Γ_1 to show that the conjugate of w is nontrivial. If $w = a_1 b_1 \cdots a_k b_k$ then we conjugate by an element of Γ_1 not equal to a_1 or a_1^{-1}, which is possible since Γ_1 has at least three elements. We argue similarly when $w = b_1 \cdots a_k$, replacing w by an appropriate conjugate. □

The reader will notice in the sequel that there is some overlap between ping-pong for free products and for right-angled Artin groups, especially within the discussion of right-angled Artin groups of isometries of hyperbolic spaces.

We will not be discussing linear groups at length in the present article, but we will remark that the ping-pong lemma was of fundamental importance in the proof of the following theorem of Tits, usually cited as the "Tits Alternative" (see [23]):

Theorem 2.4. *Let $\Gamma < GL_n(\mathbb{C})$ be a finitely generated subgroup. Then Γ either contains a solvable group with finite index (i.e. Γ is virtually solvable) or contains a nonabelian free group.*

Tits' theorem is so well-known that a common question which is now asked about classes of groups is whether they "satisfy a Tits alternative", which is to say whether or not every finitely generated subgroup either contains a nonabelian free group or is virtually solvable (or virtually nilpotent or virtually abelian).

3. Hyperbolic Geometry

Let $\mathbf{H}^n$ denote n-dimensional hyperbolic space. Recall that this is the unique simply connected, complete Riemannian n-manifold with constant sectional curvatures equal to -1. For more details and background on hyperbolic geometry, the reader is directed to Benedetti and Petronio's book [1] and Katok's book [15].

The hyperboloid model of hyperbolic geometry furnishes hyperbolic n-space as a component of the sphere of radius -1 inside of the indefinite inner product space $\mathbb{R}^{n,1}$. The "sphere" of radius -1 is given by points whose coordinates $(x_1, \ldots, x_{n+1})$ satisfy $x_1^2 + \cdots + x_n^2 - x_{n+1}^2 = -1$. This sphere has two components. The orthogonal group of linear maps of $\mathbb{R}^{n,1}$ which preserve the inner product

$$\langle \mathbf{x} \mid \mathbf{y} \rangle = x_1 y_1 + \cdots + x_n y_n - x_{n+1} y_{n+1}$$

is written $O(n, 1)$. The index 2 subgroup $SO(n, 1)$ of $O(n, 1)$ acts by isometries on $\mathbf{H}^n$. It turns out that this group has finite index in the full group of isometries of $\mathbf{H}^n$. Usually one considers a further finite index subgroup $SO^+(n, 1)$ of orientation-preserving isometries.

Hyperbolic n-space has a natural boundary, the sphere at infinity. It is homeomorphic to S^{n-1}. The sphere at infinity coincides with the Gromov boundary of hyperbolic space.

Isometries of hyperbolic space come in three flavors: **elliptic**, **parabolic** and **loxodromic**. An isometry ϕ falls into one of these three categories if it has respectively zero, one or two fixed points on the sphere at infinity. Occasionally loxodromic isometries are called **hyperbolic**. If ϕ is elliptic then it fixes at least one point in $\mathbf{H}^n$ and is conjugate to an orthogonal homeomorphism of S^{n-1}. If ϕ is parabolic then it fixes a unique point on S^{n-1}. Conjugating this point to be $\{\infty\}$, ϕ acts on $\mathbb{R}^{n-1}$ (which is viewed as the complement of $\{\infty\}$ in the sphere) by a fixed-point free Euclidean isometry. Such isometries have directions in which they act by translations. Thus, infinity is an attracting fixed point in the sense that if $K \subset \mathbb{R}^{n-1}$ is any compact set and N is any neighborhood of infinity, then there is an M_K such that for all $n \geq M_K$, we have $\phi^n(K) \subset N$. If ϕ is loxodromic then it stabilizes a unique hyperbolic geodesic, and in particular its endpoints at infinity. Along this geodesic, ϕ acts by a translation. In general, ϕ acts on $\mathbf{H}^n \cup S^{n-1}$ by so-called north-south dynamics. One of the fixed points at infinity is a repelling fixed point, which we view as the north pole, and the other is an attracting fixed point, which we view as the south pole. Outside of a neighborhood of the north and south poles, ϕ moves each point a definite distance toward the south pole. Observe that if an isometry has finite order then it must be elliptic. It is a good exercise for the reader to show that, independently of the commentary in this paragraph, a prime order homeomorphism of $\mathbf{H}^n$ fixes a point in the interior of $\mathbf{H}^n$.

We now consider discrete subgroups of $SO^+(n,1)$, which is to say subgroups which are discrete with respect to the Lie group topology on $SO^+(n,1)$. Since point stabilizers in $\mathbf{H}^n$ are compact, it follows that the orbit of a point in $\mathbf{H}^n$ under a discrete subgroup is a discrete subset of $\mathbf{H}^n$. It is a fact that if Γ is a discrete subgroup of $SO^+(n,1)$, there is a finite index subgroup $\Gamma' < \Gamma$ such that $\mathbf{H}^n/\Gamma'$ is a manifold with fundamental group Γ' (called a hyperbolic n-manifold). Observe that this implies that Γ' contains no elliptic elements. Indeed, the fundamental group of a hyperbolic manifold cannot contain torsion by an elementary argument using cohomological dimension (see [3], for instance). On the other hand, an infinite order elliptic element cannot be contained in any discrete subgroup of $SO^+(n,1)$.

If a discrete subgroup Γ of $SO^+(n,1)$ has at least one global fixed point in the interior or boundary of hyperbolic space, we call Γ **elementary**. One can show that an elementary discrete subgroup of $SO^+(n,1)$ is either finite or contains an infinite cyclic group with finite index. In other words, a discrete elementary subgroup Γ is **virtually cyclic**. Otherwise, Γ is called **nonelementary**. Most subgroups of isometries of hyperbolic space which we will be considering will be nonelementary.

We are now ready to prove some of the theorems mentioned in the introduction:

Proof of Theorem 1.2. Let g and h be the two candidate isometries of $\mathbf{H}^n$. Suppose first that g and h are both loxodromic with fixed points x, y and w, z respectively. Choose small neighborhoods of x, y, w, z which are disjoint. Because of north-south dynamics, there are sufficiently large powers of g and h so that the complements of the small neighborhoods of x and y (respectively w and z) are sent into the small neighborhoods of w and z (respectively x and y) under the powers of h (respectively g). The theorem follows in this case by choosing a basepoint outside of the union of the neighborhoods of the four fixed points and applying Theorem 2.1.

Suppose that g is parabolic. We think of the fixed point as $\{\infty\}$, so that g acts on $S^{n-1} \backslash \{\infty\} \cong \mathbb{R}^{n-1}$ by a Euclidean isometry. Fix a small neighborhood N_h of the fixed point or fixed points of h. Choosing a small enough neighborhood N_g of $\{\infty\}$ and a sufficiently large power of g, we may arrange for N_h to lie outside of N_g and for sufficiently large powers of g to send N_h into N_g. Replacing h by a sufficiently large power as well, the claim follows by Theorem 2.1. □

We remark that Theorem 1.2 does not hold if we do not allow ourselves to pass to sufficiently high powers of g and h. The reader may have already observed that the matrices

$$\begin{pmatrix} 1 & 1 \\ 0 & 1 \end{pmatrix}, \begin{pmatrix} 1 & 0 \\ 1 & 1 \end{pmatrix}$$

satisfy a relation which persists in the quotient $SL_2(\mathbb{R}) \to PSL_2(\mathbb{R})$, the latter of which is identified with the group of orientation-preserving isometries of $\mathbf{H}^2$. Yet, those matrices are parabolic and do not share a fixed point at infinity.

Furthermore, the hypothesis that the two isometries of $\mathbf{H}^n$ do not share any fixed points at infinity is essential, since the stabilizer of a point at infinity is a nonabelian solvable group.

We are now in a position to discuss free products within isometry groups of hyperbolic spaces. We will save the second half of Theorem 1.3 until the section on right-angled Artin groups.

Proof of Theorem 1.3, Part 1. Let $G = \mathbb{Z}^{n_1} * \cdots * \mathbb{Z}^{n_k}$. Suppose that n_1 is the maximal rank of a free factor. We will exhibit G as a discrete group of isometries of $\mathbf{H}^{n_1+1}$. To do this, choose k distinct points on S^n, and let $P_1, \ldots, P_k$ be the parabolic subgroups stabilizing these points. In each P_i, choose n_i linearly independent Euclidean translations. Choose small disjoint neighborhoods $N_1, \ldots, N_k$ of each fixed point on the sphere and a basepoint x_0 contained outside of the k neighborhoods of the fixed points. Choose sufficiently large powers of all the Euclidean translations so that if such an isometry $p \in P_i$, $p(N_j) \subset N_i$ for each $j \neq i$ and $p(x_0) \in N_i$. One can use Theorem 2.3 or a variation on Theorem 2.1 to show that the resulting isometries give a copy of G inside of the isometry group of $\mathbf{H}^{n_1+1}$. We will finish the proof of the theorem in the next section. □

4. Right-Angled Artin Groups

In this section we will formulate and prove a ping-pong lemma for right-angled Artin groups with a view towards applications in hyperbolic geometry and mapping class groups. We first give a precise definition of right-angled Artin groups. Let Δ be a finite graph with vertex set V and edge set $E \subset V \times V$. The right-angled Artin group on Δ, written $G(\Delta)$, has a presentation

$$\langle V \mid [v_i, v_j] \text{ if and only if } (v_i, v_j) \in E\rangle.$$

We tacitly identify the vertices of Δ with generators of $G(\Delta)$ and occasionally refer to the vertices as vertex generators of $G(\Delta)$. A general reference for right-angled Artin groups is Charney's article [5]. In the literature, right-angled Artin groups are often called **graph groups**.

A ping-pong lemma for a right-angled Artin groups is going to be a combination of Theorem 2.1 and Proposition 1.1. The statement can be found originally in a preprint of Crisp and Farb in [7] (see also [18] and [14]):

Theorem 4.1. *Let $G = G(\Delta)$ and let X be a set with a G-action. Suppose the following hold:*

(1) There exist subsets $X_i \subset X$ for each vertex v_i of Δ whose union is properly contained in X.
(2) For each nonzero $k \in \mathbb{Z}$ and $(v_i, v_j) \notin E$, $v_i^k(X_j) \subset X_i$.
(3) For each nonzero $k \in \mathbb{Z}$ and $(v_i, v_j) \in E$, $v_i^k(X_j) \subset X_j$.
(4) There exists

$$x_0 \in X \setminus \bigcup_{i \in V} X_i$$

such that for each nonzero $k \in \mathbb{Z}$, $v_i^k(x_0) \in X_i$.

Then the G-action on X is faithful.

First Proof of Theorem 4.1. As usual, we show that the action is faithful by induction on the length of words in $G = G(\Delta)$, using V as a generating set. The base case is resolved by the conditions on x_0. For the inductive step, write $w = v_i^k w'$, where k is maximal in absolute value and w' has length strictly less than the length of w. By the inductive hypothesis, $w'(x_0) \in X_j$ for some j. If (v_i, v_j) does not form an edge of Δ then $g_i^k(w'(x_0)) \in X_i$. Otherwise, (v_i, v_j) forms an edge of Δ, so that $v_i^k(w'(x_0)) \in X_j$. In particular, $w(x_0) \neq x_0$. □

Note that when Δ is a completely disconnected graph then Theorem 4.1 restricts to Theorem 2.1. It is interesting to note what Theorem 4.1 says in the case when Δ is a complete graph, so that $G(\Delta)$ is a free abelian group. It says that there are subsets of X, one for each vertex, which are disjoint from the basepoint and which are each preserved by the vertices of Δ. When we consider n linearly independent Euclidean translations acting on $\mathbb{R}^n$, such subsets can be taken to be certain neighborhoods of infinity which do not contain the origin, for instance. Note that this is the argument which we used in the proof of Proposition 1.1.

We now give a proof of Theorem 4.1 which is due to M. Kapovich in [14]. The argument gives a slightly different perspective on right-angled Artin groups and ping-pong:

Second Proof of Theorem 4.1. Write $w \in G(\Delta)$ as a reduced word in the vertices of Δ. We call a word **central** if it is a product of vertex generators which commute with each other (in other words, the word is a product of vertices which sit in a complete subgraph of Δ). We say that $w \in G(\Delta)$ is in **central form** if it is written as a product $w = w_n \cdots w_1$ of central words which is maximal, in the sense that the last letter of w_j does not commute with the last letter of w_{j-1} (where we read from left to right).

Now let x_0 be the basepoint and let $w = w_n \cdots w_1$ be a nontrivial element of $G(\Delta)$, written in central form. We claim that $w(x_0)$ is contained in the X_i corresponding to the last letter of w_n. We proceed by induction on n.

If $n = 1$ then conditions 3 and 4 of the assumptions on G and X gives the claim. Now we consider $w_n(w_{n-1} \cdots w_1(x_0))$. By induction, $w_{n-1} \cdots w_1(x_0)$ is contained in the X_i corresponding to the last letter v_i of w_{n-1}. The last letter v_j of w_n does not commute with v_i, so conditions 2 and 3 imply that $w_n(X_i) \subset X_j$, so that $w(x_0) \in X_j$. In particular, $w(x_0) \neq x_0$. □

Much like in Theorem 2.1, given a right-angled Artin group G, one can use the action of G on its Cayley graph (with respect to the generators given by the vertices of Δ) to produce a set X with a G-action which satisfies the hypotheses of Theorem 4.1.

Note the following easy corollary of Theorem 4.1:

Corollary 4.2. *Let $v_1, \ldots, v_k$ be vertices of a graph Δ, viewed as generators of $G(\Delta)$, and let $n_1, \ldots, n_k$ be nonzero integers. Then*

$$\langle v_1^{n_1}, \ldots, v_k^{n_k} \rangle \cong G(\Delta).$$

We can use Corollary 4.2 to give a proof of the second part of Theorem 1.3. Before giving the proof, we leave as an exercise to the reader to show that if g and h are two loxodromic isometries of $\mathbf{H}^n$ which both stabilize $\{\infty\}$ but do not have the same fixed set at infinity, then g and h do not generate a discrete group of isometries of $\mathbf{H}^n$.

Proof of Theorem 1.3, Part 2. Let $G(\Delta)$ be a right-angled Artin group which is discretely embedded in the group of isometries of some hyperbolic space $\mathbf{H}^n$. We have noted that by discreteness and the fact that $G(\Delta)$ has no torsion, the image of $G(\Delta)$ under the embedding cannot contain any elliptic elements. It follows in particular that the vertices of Δ are all sent to either parabolic or loxodromic isometries, which have associated to them their fixed points at infinity. We can group the vertices according to their fixed points. If any two loxodromic isometries stabilize a point at infinity then discreteness implies that they must generate a virtually cyclic group of isometries which stabilizes a unique geodesic in $\mathbf{H}^n$. This results from the fact that any discrete subgroup of the stabilizer of a point in the sphere at infinity is virtually abelian. One can similarly show that one cannot have a discrete group of isometries which contains a loxodromic isometry and a parabolic isometry stabilizing one of the loxodromic isometry's fixed points. It follows that if v_i and v_j are two vertex generators of $G(\Delta)$ then their fixed point sets at infinity are either disjoint or equal.

If we replace the vertex generators by any nonzero powers, the isomorphism type of the group of isometries does not change. Choosing as our collection $\{X_i\}$ small disjoint neighborhoods of the fixed points of the vertex generators at infinity, we can apply either Theorem 4.1 or Theorem 2.3 to show that after replacing the generators of $G(\Delta)$ by large powers, ping-pong implies that $G(\Delta)$ splits as a free product. Precisely, we get one cyclic free factor for each geodesic stabilized by a vertex generator of $G(\Delta)$, and we get a torsion-free abelian free factor coming from each parabolic fixed point of a vertex generator. The rank of stabilizer of this parabolic fixed point is the rank of the free factor. The second conclusion of the theorem follows. □

5. Mapping Class Groups

In this section, we apply the machinery above to the somewhat more complicated context of mapping class groups. Recall that the mapping class group of a manifold M is defined to be the orientation-preserving homeomorphism group of M, modulo isotopy. It is a deep theorem of Dehn that when Σ is a surface of genus g and n punctures, the mapping class group $\mathrm{Mod}(\Sigma)$ is finitely generated by certain homeomorphisms known as Dehn twists (see Farb and Margalit's book [9] for more background and a proof of this fact). Other general references on mapping class groups and tools for studying them are [4], [10], [11] and [21].

We will sketch some of the relevant features of mapping class group theory here without proofs. Proofs can be found in [4], [9], [10] and the references in those books. The first important result about mapping class groups which we shall require is the classification of mapping classes, due originally to J. Nielsen and further developed by W. Thurston. A mapping class is called **finite order** if it has finite order as a mapping class. It is a deep theorem that there exists a representative of ψ which is a finite order homeomorphism of Σ. A mapping class is called **reducible** if there is some nonempty collection of isotopy classes of essential simple closed curves which is preserved by it. A mapping class ψ is called **pseudo-Anosov** if it has infinite order and for any isotopy class of essential simple closed curves c, we have that $\psi^k(c)$ and c are not isotopic for each nonzero k.

An important feature of pseudo-Anosov homeomorphisms is their dynamics on a certain sphere, called the space of **projective measured laminations**. Each pseudo-Anosov homeomorphism has a unique pair of fixed points on this sphere, and each pseudo-Anosov homeomorphism has north-south dynamics with respect to these two fixed points. Furthermore, if ψ and ϕ are both pseudo-Anosov and fix a common projective measured lamination, then ψ and ϕ share a common power.

Corollary 5.1. *If ψ and ϕ are pseudo-Anosov homeomorphisms then there exists an $N > 0$ such that $\langle \psi^N, \phi^N \rangle$ is either cyclic or free.*

A mapping class ψ is called **pure** if there is a finite collection $\mathcal{C}$ of pairwise nonisotopic simple closed curves which are stabilized componentwise by ψ, and on each component of the complement of $\mathcal{C}$, ψ restricts to either a pseudo-Anosov homeomorphism or the identity. The mapping class group contains a finite index subgroup which consists entirely of pure mapping classes (see [9] or [11]). It is somewhat easier to show that each mapping class has a power which is pure:

Lemma 5.2. *Let ψ be any mapping class. Then ψ has a power which is pure.*

Proof. If ψ has finite order then some power of ψ is the identity which is pure. If ψ is pseudo-Anosov then ψ is already pure. If ψ is reducible then we have that $\mathcal{C}$ is a nonempty essential multicurve preserved by ψ. By passing to a power of ψ, we may assume $\mathcal{C}$ is component-wise preserved. Cutting Σ open along $\mathcal{C}$, we apply the Nielsen-Thurston classification to see that ψ restricts to either a finite order, reducible or pseudo-Anosov homeomorphism on each component of the complement of $\mathcal{C}$. Since cutting open along an essential simple closed curve increases the Euler characteristic of the components of the complement of the curve, we obtain the conclusion of the lemma by induction on the Euler characteristic. □

An example of a reducible pure mapping class is a Dehn twist about an essential simple closed curve. Dehn twists do not have the favorable dynamics enjoyed by pseudo-Anosov homeomorphisms on the space of projective measured laminations. Fortunately, the explicit nature of Dehn twists allows for an easy description of the action of a twist on the homology of a surface.

Let c be a simple closed curve with homology class $[c]$ If $[d]$ is another homology class, then the twist about c acts on $[d]$ by a transvection

$$[d] \mapsto [d] \pm i([c],[d])[c].$$

Here, $i([c],[d])$ denotes the algebraic intersection number of $[c]$ and $[d]$, and the sign of the transvection depends on whether one does a left-handed or right-handed twist about $[c]$. If $[c]$ is a nonzero homology class, it follows that a twist about c has infinite order.

If the homology class of c is trivial, it is not obvious that the twist about c should be a nontrivial mapping class. To see that it is, one can lift the twist about c to a finite cover where it becomes a simultaneous twist about a multicurve. If the chosen cover is abelian, has degree at least three, and provided that c lifts to the cover, then each lift of c represents a nontrivial homology class, and one can find homologically nontrivial curves with nontrivial algebraic intersection with the total lift of c (see [17], for instance). It follows that the twist about any essential simple closed curve is a nontrivial infinite order mapping class.

Suppose c and d are essential simple closed curves which intersect exactly once. These assumptions imply that the homology classes of c and d are nontrivial and linearly independent. It is easy to see that there is a tubular neighborhood of $c \cup d$ which is homeomorphic to a torus with one boundary component in which c and d form the longitude and meridian. One can then check that on the homology of this torus (which injects into the homology of Σ), the twists about c and d act by the matrices

$$\begin{pmatrix} 1 & 1 \\ 0 & 1 \end{pmatrix}, \begin{pmatrix} 1 & 0 \\ 1 & 1 \end{pmatrix}$$

respectively, up to a sign.

Before sketching some of the ideas in the proof of Theorem 1.5, we need one more algebraic fact about free groups. A group G is called **Hopfian** if any surjection $G \to G$ is injective. G is called **residually finite** if every nontrivial element of G survives in a finite quotient of G. If G is finitely generated, residual finiteness is equivalent to the statement that the intersection of all finite index subgroups of G is trivial. The reader should compare the statement and proof with similar results in [8] and [19]:

Lemma 5.3. *A finitely generated free group F_n is Hopfian.*

Proof. This fact will follow from the residual finiteness of F_n. To see that residual finiteness implies Hopficity, suppose $\phi : F_n \to F_n$ is a surjection with nontrivial kernel K. Since F_n is finitely generated, there are only finitely many subgroups of a given index k. Pulling back these subgroups along ϕ gives us a finite collection of finite index subgroups of F_n which contain K. Since the number of these subgroups is finite, every subgroup of index k contains K. It follows that K is contained in the intersection of all finite index subgroups of F_n, which is to say K is trivial by residual finiteness.

To see the residual finiteness of F_n, let X be a rose with n petals so that $\pi_1(X) \cong F_n$. Endow X with a graph metric. We have a map $F_n \to (\mathbb{Z}/m\mathbb{Z})^n$ for any integer $m > 1$ given by reducing homology modulo m. This finite quotient of F_n gives rise to a finite cover X^1 of X. We pull back the graph metric to this cover. We repeat this construction of taking the homology modulo m, producing a cover and pulling back the graph metric. We thus obtain a tower of finite covers

$$\cdots \to X^3 \to X^2 \to X^1 \to X.$$

Note that in any graph, a simple loop (one which visits each vertex at most once) is always homologically nontrivial modulo any integer greater than one. It follows by induction that no loop of length less than k lifts to a closed loop on the kth level of the tower of covers. In particular for each k, there is a finite index subgroup of F_n which does not contain any words of length less than k. □

Sketch of a Proof of Theorem 1.5. The discussion in the previous paragraph shows that two twists about disjoint nonisotopic essential simple closed curves generate a free abelian subgroup of the mapping class group. Suppose c and d intersect exactly once. We have seen that there is a summand of the homology of Σ where T_c and T_d act by the matrices

$$\begin{pmatrix} 1 & 1 \\ 0 & 1 \end{pmatrix}, \begin{pmatrix} 1 & 0 \\ 1 & 1 \end{pmatrix}$$

respectively, up to a sign, and by the identity on a complement to this summand. One can check that these matrices satisfy a braiding relation, namely $ABA = BAB$. It is not obvious but true that T_c and T_d also satisfy this relation and that this is the only relation satisfied by these twists. This fact can be seen by studying relations in the braid group and lifting simple braids to Dehn twists in certain branched covers. For more detail, the reader should consult [9].

Note that if we replace the matrices

$$\begin{pmatrix} 1 & 1 \\ 0 & 1 \end{pmatrix}, \begin{pmatrix} 1 & 0 \\ 1 & 1 \end{pmatrix}$$

by nonzero powers, they generate a free subgroup of $SL_2(\mathbb{Z})$. It follows that T_c^N and T_d^M generate a free group for any $N, M > 1$ by the Hopfian property of free groups.

The case where the geometric intersection number of c and d is greater than one is thoroughly resolved in [9], and we shall not treat this case here for the sake of brevity. □

Sketch of a Proof of Theorem 1.4. Let ψ and ϕ be two mapping classes. By passing to powers of ψ and ϕ, we may assume that ϕ and ψ are both pure. To both ψ and ϕ we can associate multicurves $\mathcal{C}_1$ and $\mathcal{C}_2$ such that after cutting Σ open along these multicurves, ψ and ϕ restrict to pseudo-Anosov homeomorphisms or the identity on each component. It is possible that ψ and ϕ restrict to a twist about certain components of these multicurves. We assume that the multicurves $\mathcal{C}_1$ and $\mathcal{C}_2$ have no unnecessary components, so that they do not cut Σ open along surfaces where the mapping classes restrict to the identity. In standard terminology, we call these multicurves **canonical reduction systems** for ψ and ϕ (see [2], [11] and [18], for instance).

Thus, ψ and ϕ have canonical supports. Clearly if the supports of ψ and ϕ are disjoint then ψ and ϕ commute. If two components of their support overlap essentially then the restrictions of ψ and ϕ to those components might share a common power. If this happens for every pair of overlapping components of the supports of ϕ and ψ then the two mapping classes evidently commute. Otherwise, ψ and ϕ admit two overlapping pseudo-Anosov components or two intersecting curves about which ψ and ϕ twist. Using standard theory of measured laminations as in [21] and Theorem 2.1, one can see that the restrictions of sufficiently large powers of ψ and ϕ to these two components or curves generate a free subgroup of the mapping class group. By the Hopfian property of free groups, it follows that for any sufficiently large N, $\langle \psi^N, \phi^N \rangle \cong F_2$. □

A further analysis of the action of mapping classes on the space of projective measured laminations and the application of ping-pong furnish us with a proof of Theorem 1.6. The details of the proof are beyond the scope of this article.

Most questions which are posed for mapping class groups are also posed for outer automorphisms of free groups. An example of similar results proved for outer automorphisms of free groups can be found in [13], where the authors show that after passing to powers, any two outer automorphisms of a free group usually generate a free subgroup of the outer automorphism group.

6. Effective Ping-Pong and the Word Problem

We make a few final remarks about effective actions, ping-pong and the word problem. For general references on computability theory as it is used here, we refer the reader to Soare's book [22]. We wish to gain a perspective on the following:

Question 6.1. Which groups can be expected to have a ping-pong lemma?

If $G = \langle F \mid R \rangle$ is a finitely presented group, one can formulate the word problem for G (cf. Lyndon and Schupp's book [19]). Namely, one would like to have an algorithm which decides whether or not a word w in the free group on F represents the identity in G.

We will call an action of G on a set X effective if there is a Turing machine which computes the action. In particular, the Turing machine takes as an input a word w in the generators of G and a point $x \in X$ and returns the image of x under w. If G acts effectively on a set X and the action satisfies a ping-pong lemma, the conclusion of the lemma provides nothing more than a certificate of nontriviality of an element of G.

With these observations, it is very easy to see how one might solve the word problem for finitely generated free groups and right-angled Artin groups. For free products, the situation is slightly more subtle:

Proposition 6.2. *Let $\Gamma_1, \ldots, \Gamma_k$ be finitely presented groups. Then the free product*

$$\Gamma = \coprod_{i=1}^{k} \Gamma_i$$

has a solvable word problem if and only if each factor has a solvable word problem.

Proof. The "only if" direction is clear: if Γ_i has an unsolvable word problem then we cannot determine if a word written in a given finite generating set for Γ_i (viewed as part of a finite generating set for Γ) represents the identity. Conversely, a general element of Γ can be written as a product $w = w_1 \cdots w_n$, where each w_j is a non-identity element of some Γ_i, and w_j and w_{j+1} are in different factor groups for each j. In this sense, w is "locally reduced". A word w in the generators of $\Gamma_1, \ldots, \Gamma_k$ represents a nonidentity element of Γ if and only if it can be written as a nonempty product $w = w_1 \cdots w_n$ as above. This can be algorithmically checked if the word problems in the $\{\Gamma_i\}$ are solvable. □

One might ask what a general ping-pong lemma might look, and what the content of the lemma might have to do with the algebraic properties of the group. The following metatheorem is in line with the content of this article:

Proposed Theorem 6.3. *Let G be a countable group and let X be a set equipped with a computable G-action. Suppose there is a (possibly infinite) collection of subsets $\{X_i\}$ of X which do not cover all of X, a computable partition of the elements of G into subsets $\{\mathcal{G}_i\}$ in bijection with the $\{X_i\}$, and a basepoint x_0 in the complement of their union such that:*

(1) For each i and $g \in \mathcal{G}_i$, $g(x_0) \in X_i$.
(2) For each i, $j \neq i$, and $g \in \mathcal{G}_i$, g^k either maps each X_j into itself, into X_i, or to x_0 according to a computable rule.

Then the action of G on X is faithful.

The phrases "computable partition" and "computable rule" are rather vague, but for a ping-pong lemma to be useful, the partition and the rule should be effectively computable.

Suppose that G is finitely generated and that G satisfies a ping-pong lemma. Let w be a word in the generators of G. One follows the prescriptions of the ping-pong lemma to find the image of x_0 under w and thus one can verify whether or not w represents the identity in G. Conversely if G has a solvable word problem, one can let X be the set of elements of G with the $\{X_i\}$ being the nonidentity elements and x_0 being the identity. If w is a word that represents a nonidentity element of G then

$$w(x_0) \in \bigcup_i X_i.$$

If w represents the identity then $w(x_0) = x_0$.

In some sense, the class of finitely presented groups which admit ping-pong lemmas should be the ones with solvable word problems. The usefulness of a ping-pong lemma for a particular group will depend a lot on the algebraic nature of the group, apart from the solvability of the word problem.

Acknowledgements

The author thanks B. Goldman, C. Series and S. P. Tan for organizing a workshop on the Dynamics, Geometry and Topology of Character Varieties at the Institute for Mathematical Sciences in Singapore during the summer of 2010. He thanks the organizers for inviting him to contribute this article. The author also thanks J. Forehand and the referee for carefully reading the manuscript and for their numerous helpful comments.

References

1. Ricardo Benedetti and Carlo Petronio. *Lectures on hyperbolic geometry.* Universitext, Springer, 1992.
2. Joan S. Birman and Alex Lubotzky and John McCarthy. Abelian and solvable subgroups of the mapping class groups. *Duke Math. J.* 50, no. 4, 1107–1120, 1983.
3. Kenneth S. Brown. *Cohomology of groups.* Graduate Texts in Mathematics, no. 87, Springer, New York, 1982.
4. Andrew J. Casson and Steven A. Bleiler. *Automorphisms of surfaces after Nielsen and Thurston.* London Mathematical Society Student Texts, 9, 1988.
5. Ruth Charney. An introduction to right-angled Artin groups. *Geom. Dedicata*, 125, 141–158, 2007.

6. Matt Clay, Chris Leininger and Johanna Mangahas. The geometry of right angled Artin subgroups of mapping class groups. Preprint.
7. John Crisp and Benson Farb. The prevalence of surface groups in mapping class groups. Preprint.
8. Pierre de la Harpe. *Topics in Geometric Group Theory. University of Chicago Press*, 2000.
9. Benson Farb and Dan Margalit. *A Primer on the mapping class group*. Online book, to be published by *Princeton University Press*.
10. A. Fathi and F. Laudenbach and V. Poénaru. *Travaux de Thurston sur les Surfaces*. Soc. Math. de France, Paris, Astérisque 66–67, 1979.
11. Nikolai Ivanov. Subgroups of Teichmüller modular groups. *Translations of Mathematical Monographs*, 115, 1992.
12. Tadeusz Januszkiewicz and Jacek Świątkowski. Hyperbolic Coxeter groups of large dimension. *Comment. Math. Helv.* 78, 555–583, 2003.
13. Ilya Kapovich and Matin Lustig. Ping-pong and outer space. *J. Topol. Anal.* 2, no. 2, 173–201, 2010.
14. Michael Kapovich. RAAGs in Ham. Preprint.
15. Svetlana Katok. *Fuchsian groups*. Chicago Lectures in Mathematics. *University of Chicago Press*, Chicago, IL, 1992.
16. Felix Klein. Neue Beiträge zur Riemann'schen Functionentheorie. *Math. Ann.* 21 (1883), no. 2, 141–218.
17. Thomas Koberda. Asymptotic linearity of the mapping class group and a homological version of the Nielsen–Thurston classification. To appear in *Geom. Dedicata*.
18. Thomas Koberda. Right-angled Artin groups and a generalized isomorphism problem for finitely generated subgroups of mapping class groups. Preprint.
19. Roger C. Lyndon and Paul E. Schupp. *Combinatorial group theory*. Springer, New York, 1977.
20. John McCarthy and Athanase Papadopoulos. Dynamics on Thurston's sphere of projective measured foliations. *Comment. Math. Helv.* 64, no. 1, 133–166, 1989.
21. R.C. Penner and J.L. Harer. *Combinatorics of train tracks*. Annals of Mathematics Studies, no. 125, *Princeton University Press*, 1992.
22. Robert I. Soare. *Computability Theory and Applications: The Art of Classical Computability*. To be published by *Springer-Verlag*.
23. Jacques Tits. Free subgroups in linear groups. *J. Algebra*, 20, 250–270, 1972.

CREATING SOFTWARE FOR VISUALIZING KLEINIAN GROUPS

Yasushi Yamashita

Department of Information and Computer Sciences
Nara Women's University
Kita-uoya-nishi-machi, Nara, 630-8506, Japan
yamasita@ics.nara-wu.ac.jp

We consider a simple class of Kleinian groups called once punctured torus groups. In this note, we will show how to create a computer program from scratch that can visualize fundamental sets and limit sets of the groups. No knowledge of computer programming is assumed.

Keywords: Punctured torus group, limit set, Ford domain

Mathematics Subject Classification 2000: 30F40

1. Introduction

This note is based on six lectures given by the author at the summer school "Geometry, Topology and Dynamics of Character Varieties" at the Institute for Mathematical Sciences, National University of Singapore in the summer of 2010. The aim of the lectures was to show how to create a computer program from scratch that can visualize fundamental sets and limit sets of punctured torus groups. The main audience was graduate students and no knowledge of computer programming was assumed.

In low dimensional topology and Kleinian group theory, computation and experiment become very important. There are many amazing software, and using them is not only helpful for study but fun. The author recommends [6] for this. In these lectures, we create our own computer program, starting with how to install the programming language *Python*. After introducing the basic grammar and rules of this programming language, we will make a program that can visualize the deformation of hyperbolic structures on a punctured torus and the limit set. This is a very simple version of M. Wada's software OPTi [13]. The size of our program is about 300 lines long. It is a simple and small program, but the user can deform the group by moving the mouse and save the picture in PostScript format.

The author hopes that this note helps the reader become familiar with the computer programing.

The paper is organized as follows. In §2 we begin by showing how to install the Python language system into the computer. Then, the programing language is introduced informally using many examples. In §3 we give background material for punctured torus groups and describe the algorithm which will be used in our program. In §4 we explain our program *OptPy* in detail. In §5 we list the source code of our program. In §6 we describe how to install and use it.

2. Python

This section gives a brief informal introduction to Python programming language. We refer to [11] for an in-depth introduction. It is a general purpose computer programming language. It is relatively easy to learn, and runs on Windows, Mac OS X, Linux. Also, it is free to use. The official website of Python is `http://www.python.org/`.

Python is already used in several pieces of software for hyperbolic geometry and related fields. The user interface to SnapPea was written by M. Culler and N. Dunfield in Python [4]. Its command-line interface is essentially Python's interactive shell that we will use later. See `http://www.math.uic.edu/~t3m/` for other Python libraries by the same authors. *Regina* by B. Burton is another example [3]. It is software for normal surface theory and has the ability to write and run scripts in Python so that it can perform repetitive tasks over large sets of data, such as SnepPea's census.

As mentioned in the introduction, we will begin by installing Python.

2.1. *Python setup*

This part gives information on the setup of the Python environment on different platforms.

Mac OS X or Linux/Unix Python comes pre-installed on most Linux distributions and Mac OS X. Start Python interactive shell with the command `python` in your terminal. Python will display version information and a prompt. See Figure 1. The prompt for Python interactive shell is `>>>` (and `...` for multi-line construct). After this prompt, you can type your Python program. In Figure 1, it is `1+2` and the result is `3`.

You can use your favorite editor to write a longer program. For example, emacs and vim support Python. To run the program, save the list into a

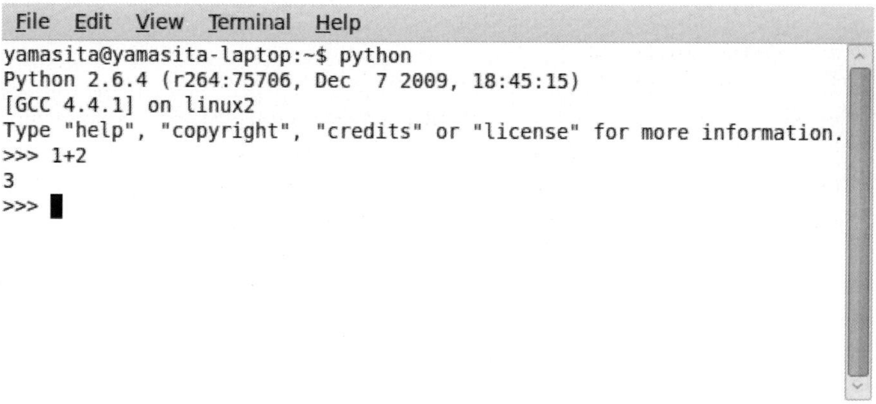

Fig. 1. Python on Linux terminal.

file called, say, “sample.py”, and type

```
% python sample.py
```

in the terminal. Then you can see the result of the computation.

Windows You can download Python from the official web site

```
http://www.python.org/download/.
```

Choose `Python 2.X.X Windows Installer` in this page. (As of this note writing, the version number is `2.7.1`. From now on, we write 2.7.1 instead of 2.X.X.) A file named `python-2.7.1.msi` will be saved. Then double click this file. The setup program will start and Python will be installed in the Windows system.

Next, we need an editor. In the lecture at the summer school, we used PyScripter [12]. This is an integrated development environment — a package consisting of editor, compiler and debugger. You can download it from `http://code.google.com/p/pyscripter/`. See Figure 2 for a screen shot. The upper right window is the editor. At the bottom window, you can run your Python program.

To test a short program, the interactive shell *IDLE*, which comes with the above Python 2.7.1 Windows Installer, is useful. You can start IDLE by `Start` ⟶ `Python 2.7` ⟶ `IDLE (Python GUI)`. See Figure 3. The prompt for Python interactive shell is `>>>` (and `...` for multi-line construct). After this prompt, you can type your Python program. In the Fig-

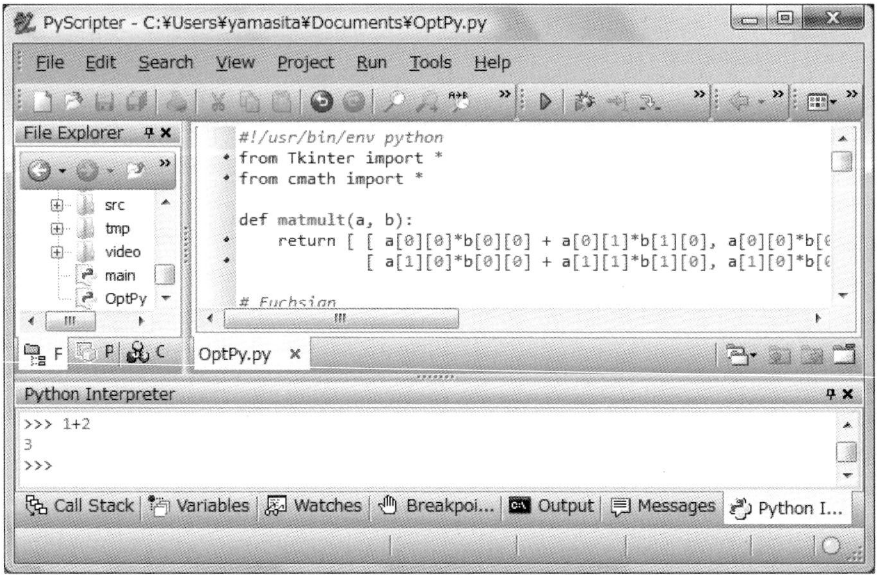

Fig. 2. PyScripter (IDE for Python).

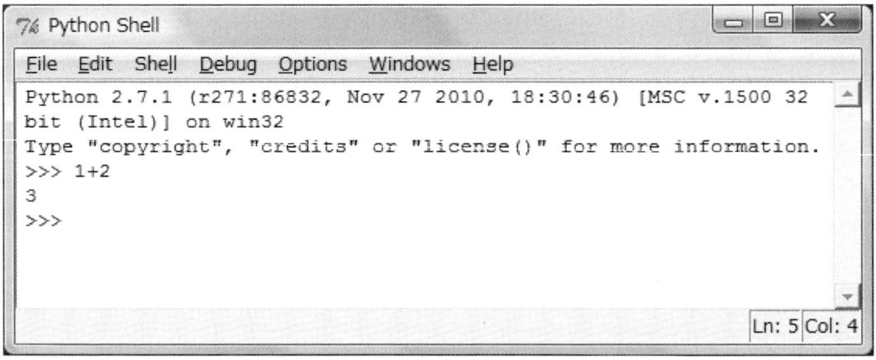

Fig. 3. IDLE (Python interactive shell).

ure 3, it is `1+2`, and the result is `3`. Otherwise, you can use the bottom window of PyScripter.

Remark 2.1. In the summer school, we used the library *SciPy*. But, since some Mac OS X users have problems installing SciPy, we will use only the standard library which comes with python-2.7.1 and not SciPy.

2.2. *Basic data types and operations*

We describe the first steps of Python programming by showing the examples. The shaded boxes are what you see in the interactive shell. (Recall that `>>>` and `...` are the prompt.)

First of all, you can use it as a calculator.

```
>>> 1 + 2
3
>>> 3.4 * 5.6 - 7.8
11.239999999999998
```

Remark 2.2. The second result looks strange. It must be 11.24. This is called *representation error*. The input and output were in decimal (base 10) representation, but computers use binary (base 2) representation. The conversion between them causes this type of error.

In Python, the power 2^{10} is written as `2 ** 10`. The complex number $1.2 + 3.4\sqrt{-1}$ is written as `1.2+3.4j`. Note that `j` (not `i`) is used for $\sqrt{-1}$.

```
>>> 2 ** 10
1024
>>> (1.2 + 3.4j) * (5.6 + 7.8j)
(-19.800000000000001+28.399999999999999j)
```

You can assign a value to a variable. To get the real and imaginary part, type `.real` and `.imag` after the name of the variable. The function `abs` returns the absolute value of the number.

```
>>> a = 12.3 - 45.6j
>>> a
(12.300000000000001-45.600000000000001j)
>>> a.real
12.300000000000001
>>> a.imag
-45.600000000000001
>>> abs(a)
47.229757568719322
```

For later purpose, we want to use *cmath module.* Cmath module consists of many functions for complex numbers. To use it, type `from cmath import *`. The function `sqrt` returns the square root of the parameter and it is defined in cmath module. (The example below is continued from the previous shaded box.)

```
>>> from cmath import *
>>> sqrt(a)
(5.4557198227511341-4.179100236218277j)
```

To express a character or a string, use single-quotations.

```
>>> c = 'a'
>>> c
'a'
>>> s = 'abc'
>>> s
'abc'
```

Python contains so called *compound data types.* They are used to group together other values. Our first example is *tuple.* A tuple consists of a number of values separated by commas. Each component object can be accessed by offset (index).

```
>>> point = (300, 400)   # a tuple
>>> point[0]
300
>>> point[1]
400
```

A list is a series of objects in square brackets, separated by commas. The usage is almost similar to tuples.

```
>>> a = [2, 3, 5, 7, 11, 13]
>>> a[0]
2
>>> a[1] = 4
>>> a
[2, 4, 5, 7, 11, 13]
```

For the difference between tuple and list, see [11].

2.3. *Control flow*

There are three control flow statements in Python - `if`, `for` and `while`.

If statement The `if` statement is used for conditional execution. The syntax is as follows:

```
if <condition 1>:
    statements A
elif <condition 2>:
    statements B
elif <condition 3>:
    statements C
else:
    statements D
```

There can be zero or more `elif` parts, and the `else` part is optional. The body (statements) must be indented. If `condition 1` is satisfied, `statements A` is executed, and so on.

```
>>> x = 3
>>> if x > 0:
...     print "x is positive"
...     print x
... elif x == 0:
...     print "x is zero"
... else:
...     print "x is negative"
...     print x
...
x is positive
3
>>> tr = 2+3j
>>> if tr == 2 or tr == -2:
...     print "parabolic"
... elif tr.imag == 0 and (-2 < tr.real and tr.real < 2):
...     print "elliptic"
... else:
...     print "loxodromic"
...
loxodromic
```

The standard comparison operators are < (less than), > (greater than), == (equal to), <= (less than or equal to), >= (greater than or equal to) and != (not equal to). Note that = is used for assignment and == for equality.

For statements The `for` statement is used for sequential looping. The `for` statement iterates over the items of a list, in the order that they appear in the list. The syntax is as follows:

```
for <variable(s)> in <list>:
    statements
```

Each item of `list` is assigned to the `variable(s)` in turn, and then `statements` is executed.

```
>>> for x in [0, 1, 2, 3]:
...     print x
...
0
1
2
3
>>> for i, j, k in [(0,1,2), (1,2,0), (2,0,1)]:
...     print "i=", i
...     print "j=", j
...     print "k=", k
...
i= 0
j= 1
k= 2
i= 1
j= 2
k= 0
i= 2
j= 0
k= 1
```

The built-in function `range()` generates a list of integers.

```
>>> range(-3, 3)
[-3, -2, -1, 0, 1, 2]
>>> for i in range(-3, 3):
```

```
...     print i
...
-3
-2
-1
0
1
2
```

While statements The `while` statement is used for conditional looping. The syntax is as follows:

```
while <condition>:
    statements
```

The while statement repeatedly executes the block of statements as long as the condition is true.

```
>>> x = -3
>>> while x < 3:
...     print x
...     x = x+1
...
-3
-2
-1
0
1
2
```

If you want to interrupt a for or a while loop, you can use the statement `break`.

```
>>> x = -3
>>> while x < 3:
...     print x
...     if x == 0:
...         break
...     x = x+1
...
```

```
-3
-2
-1
0
```

2.4. *Function*

We have seen several built-in functions, like `abs()`, `sqrt()`. To create our own function, write as follows:

```
def <function name> ( <parameters> ):
    statement 1
    statement 2
    ...
    return <expression>
```

The definition of the function starts with the key word `def`. The parameters passed to the function must be written within the parentheses. Note that the first line ends with a colon (:). The body (`statement 1, statement 2, ...`) of the function must be indented. The line `return <expression>` is optional. As an example, let us define a Möbius function $T(z) = (2z + 3)/(4z + 5)$. The name of the function is `T` and the name of the parameter is `z`. To call this function with parameter 7, type `T(7.0)`, like built-in functions.

```
>>> def T(z):
...     w = (2*z+3)/(4*z+5)
...     return w
...
>>> T(7.0)
0.51515151515151514
```

The next example decides whether the input is prime number or not.

```
>>> def isprime(n):
...     i = 2
...     while i*i <= n:
```

```
...         if n % i == 0:
...             return False
...         i = i+1
...     return True
...
```

Here are more examples of functions.

```
>>> def norm(z):
...     return z.real*z.real + z.imag*z.imag
...
>>> def nextstx(st, x, ac = 1):
...     st = st + x[0]/(x[1]*x[2])
...     x = (x[1]*x[2] - x[0], x[2], x[1])
...     return st, x
...
>>> a,b = nextstx(2, (3,4,5), 6)
>>> a
2
>>> b
(17, 5, 4)
```

The second function `nextstx` returns two values (`st, x`). When it is called like `a,b = nextstx(2, (3,4,5), 6)`, the value of `st` in this function is assigned to `a` and the value of `x` is assigned to `b`. This function has the *default parameter value* for the third parameter `ac`. If you type `nextstx(2, (3,4,5))` and the third argument is omitted, `ac` becomes `1`.

2.5. *Global variable and local variable*

In this subsection, we consider the variable's visibility within a program. In Python, there are two kinds of variables — *global variables* and *local variables.* Global variables are accessible inside and outside of functions. Local variables are only accessible inside the function. If a variable is assigned a value outside the functions, it is a global variable. If a variable is assigned a value in a function, it is a local variable.

```
>>> x = 10         # this "x" is global
>>> def f():
...     print x  # get the value of the global variable "x"
...
>>> f()
10
>>> def g():
...     x = 20   # this "x" is local
...     print x
...
>>> g()
20
>>> x            # this "x" is global
10
```

The variable `x` in the function `g()` above is local, because its value is set in this function, even though the global variable with the same name `x` already exists. To assign a global variable a value from within a function, `global` statement is needed.

```
>>> x = 10         # this "x" is global
>>> def f():
...     global x # global statement
...     x = 20   # this "x" is global
...     print x
...
>>> x
10
>>> f()
20
>>> x    # the value of the global variable "x" is changed.
20
```

2.6. *GUI*

To open a window and draw lines and circles on it, we use *Canvas* in *Tkinter module*. In the following example, we

(1) declare that we will use 'Tkinter module',

(2) create a canvas with width 400 pixel and height 250 pixel, and assigned it to a variable named `c`.
(3) draw a line segment from (100, 50) to (300, 200) on `c`, and
(4) draw an oval in a square with upper left = (200, 30) and lower right = (300, 130).

```
>>> from Tkinter import *
>>> c = Canvas(width=400, height=250)
>>> c.create_line(100, 50, 300, 200)
>>> c.create_oval(200, 30, 300, 130)
>>> c.focus_set()
>>> c.pack()
>>> c.mainloop()
```

See Figure 4 for the result of running this example. Note that, in the co-

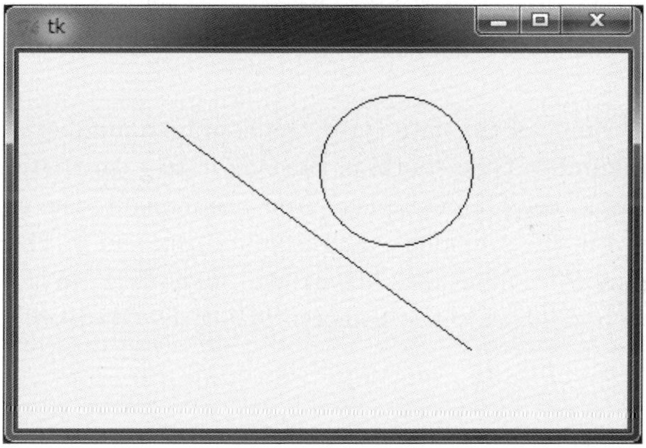

Fig. 4. TkInter.

ordinate system of `Canvas`, $(0, 0)$ is the upper left corner. The one unit in this coordinate system is pixel on the screen.

Next, we want to capture the mouse move in our canvas. In the example below, when the (left) mouse button (`Button-1`) is clicked, the function `handler_b1` is called with one parameter which describes this event.

We first define the function `handler_b1`. This is a usual function and takes one parameter named `event` as input and prints `event.x` and event.y. Here, print is a built-in function. Connecting *mouse press event* and handler_b1 is done by the line c.bind("<Button-1>", handler_b1).

```
>>> def handler_b1(event):
...     print "mouse is at (", event.x, ",", event.y, ")"
...     return
...
>>> c = Canvas(width=800, height=600)
>>> c.bind("<Button-1>", handler_b1)
'44804936handler_b1'
>>> c.focus_set()
>>> c.pack()
>>> c.mainloop()
mouse is at ( 166 , 127 )
mouse is at ( 109 , 138 )
```

After c.mainloop() is called, the Python system takes care of this canvas, waiting mouse event. When the left mouse button is pressed, the Python system calls handler_b1 with one parameter putting x and y coordinate of the mouse in it.

Remark 2.3. Some of the data types can contain a number of data. Recall that complex number type contains real and imag, so that you can write like z.imag and z.imag if a complex value is assigned to the variable z. The event type used in the previous example contains x and y. The Canvas type contains even more — functions, like bind(), create_line(), etc. They are called *classes.* and provide the features of *Object Oriented Programming.* See [11].

2.7. *An example*

The following is a sample python program using functions (create_line, paint, handler_button), GUI (canvas), global variables (p1, p2, canvas), and an if statement.

```
#!/usr/bin/env python
from cmath import *
from Tkinter import *
p1 = (333, 300)
p2 = (466, 300)
canvas = Canvas(width=800, height=600)

```

```
def create_line(z1, z2, color="black"):
    x1 = z1.real*400.0+200.0
    y1 = -(z1.imag*400.0)+300.0
    x2 = z2.real*400.0+200.0
    y2 = -(z2.imag*400.0)+300.0
    canvas.create_line(x1, y1, x2, y2, fill=color)
    canvas.create_oval(x1-10, y1-10, x1+10, y1+10, fill=color)
    canvas.create_oval(x2-10, y2-10, x2+10, y2+10, fill=color)

def paint():
    canvas.delete(ALL)
    q1 = (p1[0] - 200.0)/400.0 - (p1[1] - 300.0)/400.0*(0+1j)
    q2 = (p2[0] - 200.0)/400.0 - (p2[1] - 300.0)/400.0*(0+1j)
    create_line(0,  q1, "red")
    create_line(q1, q2, "red")
    create_line(q2, 1,  "red")

def handler_button (event):
    global p1, p2
    if ((p1[0]-event.x)**2 + (p1[1]-event.y)**2 <
        (p2[0]-event.x)**2 + (p2[1]-event.y)**2):
        p1 = (event.x, event.y)
    else:
        p2 = (event.x, event.y)
    paint()

canvas.bind("<Button-1>",  handler_button)
canvas.bind("<B1-Motion>", handler_button)
canvas.focus_set()
canvas.pack(expand=YES, fill=BOTH)
paint()
canvas.mainloop()
```

To run the program, save the list into a file called, say, "sample.py", and double click the icon of the file (Windows) or type

```
% python sample.py
```

in the terminal (Mac OS X or Linux. Here, `%` is a prompt). Then a window appears (Figure 5 left). The user can move the middle vertices of degree two by mouse (Figure 5 right).

The overall structure of the program is the same as our final program *OptPy* in §5. It defines global variables, defines functions, calls several `canvas.`··· to set up the canvas, and stays in `canvas.mainloop()` forever. When the mouse is clicked or moved, `handler_button` is called. It

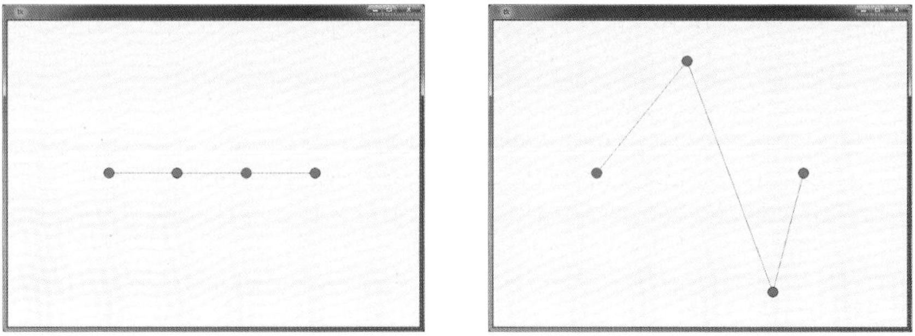

Fig. 5. Left: The initial window. Right: After some mouse movement.

change the value of the global variable `p1` or `p2` (location of the vertices) and calls `paint`. Using the value of the global variables, `paint` repaints the window.

3. Once Punctured Torus Groups

As mentioned in the introduction, we will make a very simple version of Wada's OPTi. In this section, we review T. Jørgensen's theory on once punctured torus groups. See [1] and [7] for detail.

3.1. *The deformation space of once punctured torus groups*

Let T be a once punctured torus. Its fundamental group $\pi_1(T)$ is the free group of rank two $F_2 = \langle X, Y \rangle$, generated by two elements X and Y corresponding to simple closed curves on T with geometric intersection number one. The $\mathrm{SL}(2,\mathbb{C})$ character variety $\mathcal{X} = \mathrm{Hom}(\pi_1(T), \mathrm{SL}(2,\mathbb{C}))// \mathrm{SL}(2,\mathbb{C})$ is identified with $\mathbb{C}^3$. The identification is given by

$$\iota : [\rho] \mapsto (x, y, z) = (\mathrm{tr}\,\rho(X), \mathrm{tr}\,\rho(Y), \mathrm{tr}\,\rho(XY)).$$

A $\mathrm{SL}(2,\mathbb{C})$ representation ρ of $\pi_1(T)$ is called *type preserving* if $\rho(XYX^{-1}Y^{-1}) = -2$. The set of characters which corresponds to type preserving representations is

$$\mathcal{X}_{-2} = \{(x, y, z) \in \mathcal{X} \mid x^2 + y^2 + z^2 = xyz\}.$$

$\mathcal{X}_{-2}$ is the *deformation space* of hyperbolic structures on once punctured torus.

We want to see the deformation by visualizing a fundamental domain of the corresponding representation. We refer to [7] for the pioneering work.

Later, it is studied by several people. See [1], [5] and [10] for example. As mentioned before, Wada made nice computer software named OPTi [13] which visualizes the fundamental domain. The user can deform the representation interactively by moving the mouse, and it shows the limit set, too. In the next section, we will create a very simplified version of Wada's OPTi by using Python introduced in the previous section.

3.2. *Ford domain*

We describe a type of fundamental domain, called Ford domain, which we will visualize.

For $T(z) = (az+b)/(cz+d)$ with $T(\infty) \neq \infty$ (that is $c \neq 0$), the *isometric circle* $Ic(T)$ of T is

$$Ic(T) := \{z \in \mathbb{C} \mid |T'(z)| = 1\} = \{z \in \mathbb{C} \mid |cz+d| = 1\}$$

and the *isometric hemisphere* $Ih(T)$ of T is the hyperplane of the upper half-space $\mathbb{H}^3$ bounded by $Ic(T)$. The center is $-d/c$ and the radius is $1/|c|$. $Eh(T) \subset \mathbb{H}^3$ denotes the closure of the exterior of the $Ih(T)$. Let Γ be a non-elementary Kleinian group such that the stabilizer Γ_∞ of ∞ consists of parabolic transformations. The *(extended) Ford domain* $F(\Gamma)$ of Γ is defined as follows:

$$F(\Gamma) = \bigcap \{Eh(A) \mid A \in \Gamma - \Gamma_\infty\}.$$

Then, $F(\Gamma)$ is a "fundamental polyhedron modulo Γ_∞".

3.3. *Jørgensen's normalization*

Let (x, y, z) be an element of $\mathcal{X}_{-2}$ with $y \neq 0$. Let ρ be the representation $\rho : \pi_1(T) \to \mathrm{SL}(2, \mathbb{C})$ defined by the following:

$$\rho(X) = \begin{pmatrix} x - z/y & x/y^2 \\ x & z/y \end{pmatrix}, \quad \rho(XY) = \begin{pmatrix} y & -1/y \\ y & 0 \end{pmatrix}, \tag{3.1}$$

$$\rho(Y) = \begin{pmatrix} z - x/y & -z/y^2 \\ -z & x/y \end{pmatrix}. \tag{3.2}$$

Then, ρ is type preserving and $\iota([\rho]) = (x, y, z)$. It is easy to see that $\rho(XYX^{-1}Y^{-1}) = \begin{pmatrix} -1 & -2 \\ -0 & -1 \end{pmatrix}$. Let $C(T)$ and $R(T)$ denote the center and the radius of $Ic(\rho(T))$ respectively. Then, by direct calculation, we have that

$$\begin{aligned} C(Y^{-1}) &= -y/xz - z/xy, & R(Y^{-1}) &= 1/|z| \\ C(X) &= -z/xy, & R(X) &= 1/|x| \\ C(XY) &= 0, & R(XY) &= 1/|y| \\ C(Y) &= x/yz, & R(Y) &= 1/|z| \\ C(X^{-1}) &= x/yz + y/zx, & R(X^{-1}) &= 1/|x| \\ C((XY)^{-1}) &= 1 & R((XY)^{-1}) &= 1/|y|\,. \end{aligned}$$

Define $a_1 = x/yz$, $a_2 = y/zx$ and $a_3 = z/xy$. Then we have $a_1+a_2+a_3 = 1$. Since $\rho(XYX^{-1}Y^{-1})(z) = z+2$, the above calculation shows that there are isometric hemispheres with (center, radius):

$$(n, 1/|y|),\ (n+a_1, 1/|z|),\ (n+a_1+a_2, 1/|x|) \quad (\text{for any } n \in \mathbb{Z}). \tag{3.3}$$

The circles in (3.4) are a part of the set of all isometric circles of $\Gamma - \Gamma_\infty$ that we (may) want to draw.

We will consider the "broken line" (or "polyline") given by connecting the consecutive centers of (3.4). That is, the union of line segments

$$\cup_{n\in\mathbb{Z}}\{L(n, n+a_1), L(n+a_1, n+a_1+a_2), L(n+a_1+a_2, n+1)\}, \tag{3.4}$$

where $L(z_1, z_2)$ is the line segment from z_1 to z_2 in $\mathbb{C}$. See "red broken line" in Figure 6.

Now, define "moves" from (x, y, z) as

$$(x_1, y_1, z_1) := (yz - x, z, y) \tag{3.5}$$
$$(x_2, y_2, z_2) := (z, zx - y, x) \tag{3.6}$$
$$(x_3, y_3, z_3) := (y, x, xy - z) \tag{3.7}$$

and

$$a_{i,1} := x_i/y_iz_i, \quad a_{i,2} := y_i/z_ix_i, \quad a_{i,3} := z_i/x_iy_i \quad (i = 1, 2, 3).$$

Then (x_i, y_i, z_i) is also in $\mathcal{X}_{-2}$ and we have $a_{i,1} + a_{i,2} + a_{i,3} = 1$. For $i = 1, 2, 3$, there are isometric hemispheres with (center, radius):

$$(s_i+n,\ 1/|y_i|),\ (s_i+n+a_{i,1},\ 1/|z_i|),\ (s_i+n+a_{i,1}+a_{i,2},\ 1/|x_i|), \tag{3.8}$$

for any $n \in \mathbb{Z}$, where $s_1 = a_1$, $s_2 = a_1 + a_{1,2}$, $s_3 = -a_3$. Also, we will consider the union of line segments connecting these centers:

$$\cup_{n\in\mathbb{Z}} \{L(s_i+n, s_i+n+a_{i,1}), L(s_i+n+a_{i,1}, s_i+n+a_{i,1}+a_{i,2}), \\ L(s_i+n+a_{i,1}+a_{i,2}, s_i+n+1)\}. \tag{3.9}$$

For (x_i, y_i, z_i), we can do the same "move" again, and this gives another infinite family of isometric hemispheres.

There is an action of the outer automorphism group $\mathrm{Out}(F_2)$ of F_2 on $\mathcal{X} \cong \mathbb{C}^3$ and $\mathcal{X}_{-2}$ is invariant under this action. $\mathrm{Out}(F_2)$ can be generated by three elements which induce polynomial automorphisms (3.5), (3.6), (3.7).

Let $\mathcal{F}$ be the Farey tessellation (triangulation) of $\mathbb{H}^2$. Each vertex corresponds to an isotopy class of simple nonperipheral curves on T and parameterized by $\widehat{\mathbb{Q}} := \mathbb{Q} \cup \{1/0\}$. The Markoff map $\phi_{(x,y,z)} : \widehat{\mathbb{Q}} \to \mathbb{C}$ associated with (x, y, z) is defined by the following conditions:

$$\phi_{(x,y,z)}(1/0) = x, \quad \phi_{(x,y,z)}(0/1) = y, \quad \phi_{(x,y,z)}(1/1) = z,$$

$$\phi_{(x,y,z)}(p) \times \phi_{(x,y,z)}(q) = \phi_{(x,y,z)}(r) + \phi_{(x,y,z)}(s), \quad \left(p, q, r, s \in \widehat{\mathbb{Q}}\right)$$

where p, q, r and p, q, s correspond to a pair of adjacent triangles in $\mathcal{F}$. If $p, q, r \in \widehat{\mathbb{Q}}$ correspond to a triangle in $\mathcal{F}$, then the triple of complex numbers $(\phi_{(x,y,z)}(p), \phi_{(x,y,z)}(q), \phi_{(x,y,z)}(r))$ is an orbit of (x, y, z) under the action of $\mathrm{Out}(F_2)$. For each triangle p, q, r in $\mathcal{F}$, an infinite family of isometric circles and an infinite broken line is defined using induction ((3.3),(3.4)(3.8),(3.9)). By abusing notation, we write $C(\phi_{(x,y,z)}(p), \phi_{(x,y,z)}(q), \phi_{(x,y,z)}(r))$ for the circles and $L(\phi_{(x,y,z)}(p), \phi_{(x,y,z)}(q), \phi_{(x,y,z)}(r))$ for the broken line.

3.4. *Jørgensen's method to construct the Ford domain*

The definition of the Ford domain says that we need to consider all the elements in $\Gamma - \Gamma_\infty$. But, most of the isometric hemispheres are "hidden" under other isometric hemispheres. Jørgensen [7] gave an algorithm which tells us how to find the "visible" ones, using the normalization in the previous subsection. The works in [1], [8], [13] are based on this theory. We outline this algorithm.

The input is (a_1, a_2, a_3). Then we calculate $x = 1/\sqrt{a_2 a_3}$, $y = 1/\sqrt{a_3 a_1}$, $z = 1/\sqrt{a_1 a_2}$.

Remark 3.1. We have to be careful which square root (for example $\pm\sqrt{a_1 a_2}$) we choose. This is done in our program in §5.

Step 1 (Go to "Sink"). We consider three cases.

- If $|yz - x| < |x|$, change (x, y, z) to $(yz - x, z, y)$.
- If $|zx - y| < |y|$, change (x, y, z) to $(z, zx - y, x)$.
- If $|xy - z| < |z|$, change (x, y, z) to $(y, x, xy - z)$.

Then, do this step 1 again until all three will fail.

Step 2 (Draw Isometric Circles). Draw isometric circles for $C(x, y, z)$, as in (3.4) or (3.9).

We say that x (resp. y, z) is *active* if $|x| > |y + zi|$ (resp. $|y| > |z + xi|$, $|z| > |x + yi|$).

(1) If x is active and y and z are not active,
change (x, y, z) to $(yz - x, z, y)$.
(2) If y is active and z and x are not active,
change (x, y, z) to $(z, zx - y, x)$.
(3) If z is active and x and y are not active,
change (x, y, z) to $(y, x, xy - z)$.
(4) If all x, y and z are not active, go to next step
(5) For the case where two of x, y, z is active and one is not active, there is an algorithm, but we omit here. (This is done in lines 68–81 in §5.)

Then, do this step 2 again until case (4) will happen.

Now, we set (x, y, z) as given at the end of step 1.

Step 3 (Draw Isometric Circles). Draw isometric circles for $C(x, y, z)$ as in (3.4) or (3.9). This step is almost the same as step 2 except one sign in the following definition of active. We say that x (resp. y, z) is *active* if $|x| > |y - zi|$ (resp. $|y| > |z - xi|$, $|z| > |x - yi|$).

(1) If x is active and y and z are not active,
change (x, y, z) to $(yz - x, z, y)$.
(2) If y is active and z and x are not active,
change (x, y, z) to $(z, zx - y, x)$.
(3) If z is active and x and y are not active,
change (x, y, z) to $(y, x, xy - z)$.
(4) If all x, y and z are not active, finish the process.
(5) For the case where two of x, y, z is active and one is not active, there is an algorithm, but we omit here. (This is done in lines 68–81 in §5.)

Then, do this step 3 again until case (4) will happen.

Remark 3.2. (1) By Jørgensen's inequality, if the absolute value of one of x, y, z becomes smaller than 1, we can stop our process and say that "indiscrete".

(2) Jørgensen's theory [7] tells us that if the group is quasi-fuchsian, then there exists a consecutive sequence of triangles $S = \{s_1, s_2, \ldots, s_m\}$ in the Farey tessellation $\mathcal{F}$ such that the corresponding family of isometric hemi-

spheres is equal to the boundary of the extended Ford domain $F(\rho(F_2))$ of the group $\rho(F_2)$ (Jørgensen's normalization), and the corresponding family of broken lines is the dual of the images of the boundaries (edges) of the faces of $F(\rho(F_2))$ projected to the ideal boundary $\mathbb{C}$. In **Step 1**, we try to find some triangle $t_i \in S$. In **Step 2** and **Step 3**, we have found triangles $\{t_{i-1}, \ldots, t_1\}$ and $\{t_{i+1}, \ldots, t_m\}$.

(3) If the infinite broken line $L(x, y, z)$ has a self intersection, than we stop our process. Typically, this happens when the group is indiscrete. But, this may happen even if the group is discrete, because the calculations in **Step 1** is not enough to find a triangle $t_i \in S$ mentioned above. The sink may not be unique in $\mathcal{F}$. Moreover, we don't know that S always contains a sink. In order to avoid this, we must change the step 1 using Bowditch's arguments. In [2], section 3, the notion $T(t)$ was introduced to show that Φ_Q (a subset of $\mathcal{X}$) is open. $T(t)$ is a connected subtree of the dual graph of $\mathcal{F}$. If $\rho(F_2)$ is quasi-fuchsian, then $[\rho]$ is in Φ_Q and, in this case, it was proved that $T(t)$ is finite and there is an algorithm which finds all the vertices in $T(t)$. We can find a triangle in S (starting point of **Step 2** and **Step 3**) from $T(t)$.

3.5. *Limit set*

The limit set Λ of a Kleinian group Γ is the set of all accumulation points of the orbits Γv for any $v \in \mathbb{H}^3$. Λ is a closed subset of the ideal boundary $\widehat{\mathbb{C}}$. If our group $\rho(F_2)$ is a quasi-fuchsian Kleinian group, Λ is a Jordan curve.

For the calculation, we will use matrices $\rho(X)$ and $\rho(Y)$ in (3.1, 3.2) and their inverses, Fix some point $v \in \mathbb{H}^3$. We will plot $(\rho(W))(v)$ for sufficiently long reduced words W over $\{X, X^{-1}, Y, Y^{-1}\}$.

4. OptPy

In this section, we will take a closer look at our example Python program. We named it "OptPy"— Once Punctured Torus Python. The program is listed in the next section. The reader can download the program from [14]. See Figure 6 for a screen shot. The circles are isometric circles and the blue points are the limit set.

At line 33, we define `canvas` with size 800×600 pixel. (See 2.6.) When started, we consider that this canvas corresponds to $[-0.5, 1.5] \times [-0.75, 0.75]$ in x-y coordinate. This is realized by the variables on lines 29–31 and two functions `create_line` (l. 87) and `create_circle` (l. 95). `create_line` takes two complex numbers `z1` and `z2` as inputs and draw

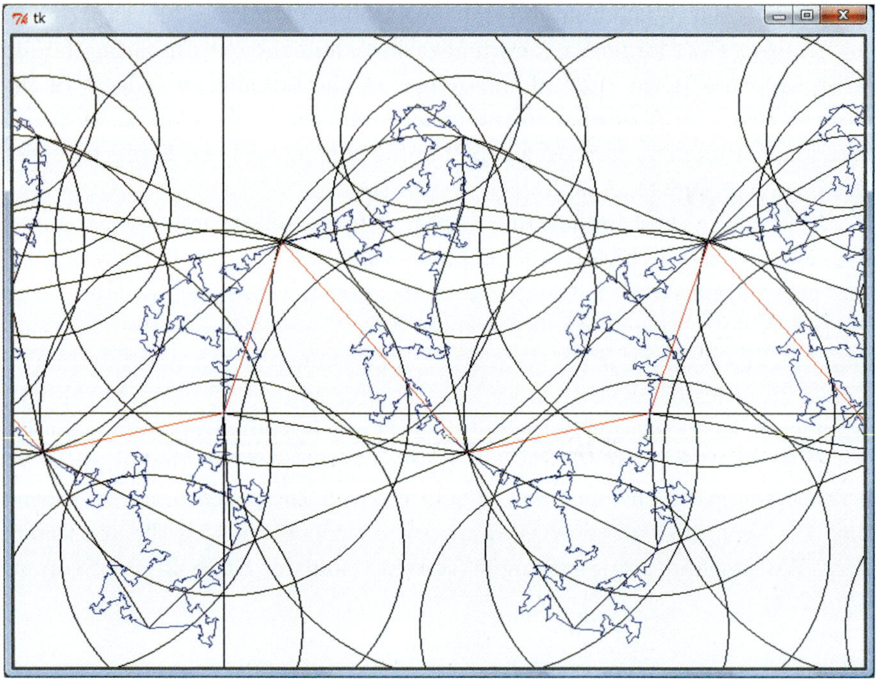

Fig. 6. OptPy.

the line segment connecting them. The arguments of `create_circle` are a complex number `center` and a real number `radius`.

By clicking or moving the mouse, the user can change $a_1 = x/yz$ and $a_1 + a_2 = x/yz + y/zx$ which results in the deformation of the group. The original position of these points are specified at line 13 and 14 in pixel coordinates. `p1` is the end point corresponding to a_1 and `p2` is the end point corresponding to $a_1 + a_2$. The comments on lines 16–27 give other (hopefully interesting) starting points.

`I` (l. 31) is used as $\sqrt{-1}$ (l. 32).

`oldpt` (l. 33) and `draw_limitset` (l.35) is used for drawing limit set.

The function `norm` (l. 37) returns $|\mathtt{x}|^2$ of complex number `x`.

The function `transition1` (l. 41) is used for step 1. The argument `x` is assumed to be a tuple of three elements (triple) which corresponds to (x, y, z). It returns

- -2 if the group is indiscrete
- 0, (resp. 1, 2) if x (resp. y, z) is active.

- -1 if there is no active, that is step 1 is finished.

`transition2` (l. 53)is used for step 2 and 3. The argument `x` is assumed to be a tuple of three elements (triple) which corresponds to (x, y, z). The next argument `direction` is set to be 1 when in step 2 and -1 when in step 3. It returns

- -2 if the group is indiscrete.
- -3 if there is a self intersection.
- 0, (resp. 1, 2) if x (resp. y, z) is active.
- -1 if there is no active, that is step 1 is finished.

`create_lines_and_circles` (l. 107) is used for step 2 and 3 to draw broken lines and isometric circles.

`next_stx` takes a triple `x` and returns the new triple if active = `ac`.

`limit set` is used to draw limit set. It is done by fairly simple recursive calls. See [9] for more detail.

The function `paint` (l. 165) is the most important. This function draws everything.

- clears the canvas (l. 168).
- draws axis (l. 170–171).
- calculate the triple `x`$(= (x, y, z))$ from `p1` and `p2` (l. 173–180).
- step 1 (l. 182–194).
- recalculate x and st (l. 196–204).
- step 2 (l. 209–216).
- step 3 (l. 218–228).
- draw limit set (l. 233–240).

After the line 243 are the handlers. See §2.6. Each handlers change some global variable and calls `paint()`.

5. Program List

```
1: #!/usr/bin/env python
2: # 31 Jan, 2011, Yasushi Yamashita
3:
4: from Tkinter import *
5: from cmath import *
6:
7: def matmult(a, b):
8:     return [ [ a[0][0]*b[0][0] + a[0][1]*b[1][0],
9:                a[0][0]*b[0][1] + a[0][1]*b[1][1] ],
```

```
             [ a[1][0]*b[0][0] + a[1][1]*b[1][0],
               a[1][0]*b[0][1] + a[1][1]*b[1][1] ] ]

# Fuchsian
p1 = (333, 300)
p2 = (466, 300)
# singly cusped
# p1 = (333.3333, 300)
# p2 = (466.6667, 150.9290)
# doubly cusped
# p1 = (400, 100)
# p2 = (500, 200)
# singly degenerate
# p1 = (279.3056, 149.2277)
# p2 = (490.6948,  92.1941)
# doubly degenerate
# p1 = (400, 184.5300)
# p2 = (400, 415.4700)

origin_x = 200.0
origin_y = 300.0
scale    = 400.0
I = 0+1j    # sqare root of -1
oldpt = 100 # used for drawing limit set
canvas = Canvas(width=800, height=600)
draw_limitset = True

def norm(x):
    return x.real*x.real + x.imag*x.imag

# Step 1 (Subsection 3.4)
def transition1(x):
    if norm(x[0]) < 1 or norm(x[1]) < 1 or norm(x[2]) < 1:
        return -2
    elif norm(x[0]) > norm(x[1]*x[2]-x[0]):
        return 0
    elif norm(x[1]) > norm(x[2]*x[0]-x[1]):
        return 1
    elif norm(x[2]) > norm(x[0]*x[1]-x[2]):
        return 2
    else:
        return -1

# Step 2(direction=1), 3(direction=-1) (Subsection 3.4)
def transition2(x, direction=1):
    # if not discrete
```

```
 if norm(x[0]) < 1 or norm(x[1]) < 1 or norm(x[2]) < 1:
   return -2
 # if not simple
 if (((x[0]*x[0])/(x[2]*x[2])).imag *
   ((x[1]*x[1])/(x[0]*x[0])).imag > 0 and
   ((x[1]*x[1])/(x[0]*x[0])).imag *
   ((x[2]*x[2])/(x[1]*x[1])).imag > 0):
   return -3
 active = [ norm(x[0]) > norm(x[1] + x[2]*I*direction),
            norm(x[1]) > norm(x[2] + x[0]*I*direction),
            norm(x[2]) > norm(x[0] + x[1]*I*direction) ]
 for i,j,k in [(0,1,2), (1,2,0), (2,0,1)]:
  if active[i] and not active[j] and not active[k]:
    return i
  if active[i] and active[j] and not active[k]:
    al = ( norm(x[k])-norm(x[i])+norm(x[j]) )/2
    am = ( norm(x[j])-norm(x[i])+norm(x[k]) )/2
    ar = ( norm(x[k])-norm(x[j])+norm(x[i]) )/2
    bl = (x[j]*x[j])/(x[k]*x[k])
    br = (x[i]*x[i])/(x[k]*x[k])
    cl = -(bl/abs(bl))*al
    cr = norm(x[k]) + (br/abs(br))*ar
    dl = ( bl.real*(cl.real-am) + bl.imag*cl.imag )/bl.imag
    dr = ( br.real*(cr.real-am) + br.imag*cr.imag )/br.imag
    if dl*direction < dr*direction:
      return i
    else:
      return j
 return -1

# draw line from z1 to z2 (z1, z2: complex number)
def create_line(z1, z2, color="black"):
  x1 = z1.real*scale+origin_x
  y1 = -(z1.imag*scale)+origin_y
  x2 = z2.real*scale+origin_x
  y2 = -(z2.imag*scale)+origin_y
  canvas.create_line(x1, y1, x2, y2, fill=color)

# draw circle (center: complex number, radius: real number)
def create_circle(center, radius):
  x1 = center.real - radius
  y1 = center.imag - radius
  x2 = center.real + radius
  y2 = center.imag + radius
  x1 = x1*scale+origin_x
  y1 = -(y1*scale)+origin_y
```

```
    x2 = x2*scale+origin_x
    y2 = -(y2*scale)+origin_y
    canvas.create_oval(x1, y1, x2, y2)

# for Step 2 and 3
def create_lines_and_circles(st, x, color="black"):
    a = (x[0]/(x[1]*x[2]), x[1]/(x[2]*x[0]), x[2]/(x[0]*x[1]))
    for i in range(-4, 4):
        create_line(st+i,             st+i+a[0], color)
        create_line(st+i+a[0],        st+i+a[0]+a[1], color)
        create_line(st+i+a[0]+a[1], st+i+a[0]+a[1]+a[2], color)
        create_circle(st+i, abs(1/x[1]))
        create_circle(st+i+a[0], abs(1/x[2]))
        create_circle(st+i+a[0]+a[1], abs(1/x[0]))

def nextstx(st, x, ac):
    if ac == 0:
        st = st + x[0]/(x[1]*x[2])
        x = (x[1]*x[2] - x[0], x[2], x[1])
    elif ac == 1:
        y = (x[2], x[2]*x[0] - x[1], x[0])
        st = st + x[0]/(x[1]*x[2]) + y[2]/(y[0]*y[1])
        x = y
    elif ac == 2:
        st = st - x[2]/(x[0]*x[1])
        x = (x[1], x[0], x[0]*x[1] - x[2])
    return st, x

# draws the limit set using depth first search
def limitset(T, lastlabel, level):
    global oldpt
    if lastlabel == 'E':
        limitset(matmult(T, ma), 'a', level+1)
        limitset(matmult(T, mB), 'B', level+1)
        limitset(matmult(T, mA), 'A', level+1)
        limitset(matmult(T, mb), 'b', level+1)
        return
    if norm(T[1][1]) > 0.00001:
        newpt = T[0][1]/T[1][1]
    else:
        newpt = 100j
    if (level > 4 and norm(newpt - oldpt) < 0.0001) or level > 9:
        create_line(oldpt, newpt, "blue")
        oldpt = newpt
        return
    if lastlabel == 'a':
```

```
            limitset(matmult(T, mb), 'b', level+1)
            limitset(matmult(T, ma), 'a', level+1)
            limitset(matmult(T, mB), 'B', level+1)
        if lastlabel == 'B':
            limitset(matmult(T, ma), 'a', level+1)
            limitset(matmult(T, mB), 'B', level+1)
            limitset(matmult(T, mA), 'A', level+1)
        if lastlabel == 'A':
            limitset(matmult(T, mB), 'B', level+1)
            limitset(matmult(T, mA), 'A', level+1)
            limitset(matmult(T, mb), 'b', level+1)
        if lastlabel == 'b':
            limitset(matmult(T, mA), 'A', level+1)
            limitset(matmult(T, mb), 'b', level+1)
            limitset(matmult(T, ma), 'a', level+1)

# This function draws everything
def paint():
    global ma, mA, mb, mB, oldpt
    # clears the canvas
    canvas.delete(ALL)
    # draws axis
    canvas.create_line(0, origin_y, 2000, origin_y)
    canvas.create_line(origin_x, 0, origin_x, 1000)
    # calculate the triple: x=(x,y,z)
    q1 = (p1[0] - origin_x)/scale - (p1[1] - origin_y)/scale*I
    q2 = (p2[0] - origin_x)/scale - (p2[1] - origin_y)/scale*I
    st = 0
    a = (q1, q2-q1, 1-q2)
    x = (1/sqrt(a[1]*a[2]), 1/sqrt(a[2]*a[0]), 1/sqrt(a[0]*a[1]))
    y = (x[0], x[1], -x[2])
    ax = (x[0]/(x[1]*x[2]), x[1]/(x[2]*x[0]), x[2]/(x[0]*x[1]))
    ay = (y[0]/(y[1]*y[2]), y[1]/(y[2]*y[0]), y[2]/(y[0]*y[1]))
    # Step 1
    if (norm(ax[0]-a[0]) + norm(ax[1]-a[1]) + norm(ax[2]-a[2]) >
        norm(ay[0]-a[0]) + norm(ay[1]-a[1]) + norm(ay[2]-a[2])):
        x = y
    discrete = True
    tmpx0 = x
    while True:
        ac = transition1(x)
        if ac < -1:
            discrete = False
        if ac < 0:
            break
        st, x = nextstx(st, x, ac)
```

```
        create_lines_and_circles(st, x, "green")
    # recalculate x and st
    a = (x[0]/(x[1]*x[2]), x[1]/(x[2]*x[0]), x[2]/(x[0]*x[1]))
    x = (x[0] if ((I/x[0])/a[1]).imag >= 0 else -x[0],
         x[1] if ((I/x[1])/a[2]).imag >= 0 else -x[1],
         x[2] if ((I/x[2])/a[0]).imag >= 0 else -x[2] )
    if (((x[0]*x[0])/(x[2]*x[2])).imag *
        ((x[1]*x[1])/(x[0]*x[0])).imag > 0 and
        ((x[1]*x[1])/(x[0]*x[0])).imag *
        ((x[2]*x[2])/(x[1]*x[1])).imag > 0):
        x = (x[0],x[1],-x[2])
    tmpx1 = x
    tmpst = st
    create_lines_and_circles(st, x, "black")
    # Step 2
    while True:
        ac = transition2(x, 1)
        if ac < -1:
            discrete = False
        if ac < 0:
            break
        st, x = nextstx(st, x, ac)
        create_lines_and_circles(st, x, "black")
    # Step 3
    x  = tmpx1
    st = tmpst
    while True:
        ac = transition2(x, -1)
        if ac < -1:
            discrete = False
        if ac < 0:
            break
        st, x = nextstx(st, x, ac)
        create_lines_and_circles(st, x, "black")
    create_lines_and_circles(0, tmpx0, "red")
    print q1
    print q2
    print "discrete" if discrete else "indiscrete"
    # draw limit set
    x = tmpx0
    oldpt = 100
    ma = [[x[0]-x[2]/x[1],  x[0]/(x[1]*x[1])], [ x[0], x[2]/x[1]]]
    mA = [[x[2]/x[1],-x[0]/(x[1]*x[1])], [-x[0], x[0]-x[2]/x[1] ]]
    mb = [[x[2]-x[0]/x[1], -x[2]/(x[1]*x[1])], [-x[2], x[0]/x[1]]]
    mB = [[x[0]/x[1], x[2]/(x[1]*x[1])], [ x[2], x[2]-x[0]/x[1] ]]
    if draw_limitset:
```

```
        limitset( [[1,0],[0,1]], 'E', 0 )

# handlers (See section 2.6)
def handler_button (event):
    global p1, p2
    if ((p1[0]-event.x)*(p1[0]-event.x) +
        (p1[1]-event.y)*(p1[1]-event.y) <
        (p2[0]-event.x)*(p2[0]-event.x) +
        (p2[1]-event.y)*(p2[1]-event.y)):
        p1 = (event.x, event.y)
    else:
        p2 = (event.x, event.y)
    paint()

def handler_left (event):
    global origin_x
    global p1
    global p2
    origin_x = origin_x + 20
    p1 = (p1[0]+20, p1[1])
    p2 = (p2[0]+20, p2[1])
    paint()

def handler_right (event):
    global origin_x
    global p1
    global p2
    origin_x = origin_x - 20
    p1 = (p1[0]-20, p1[1])
    p2 = (p2[0]-20, p2[1])
    paint()

def handler_up (event):
    global origin_y
    global p1
    global p2
    origin_y = origin_y + 20
    p1 = (p1[0], p1[1]+20)
    p2 = (p2[0], p2[1]+20)
    paint()

def handler_down (event):
    global origin_y
    global p1
    global p2
    origin_y = origin_y - 20
```

```
    p1 = (p1[0], p1[1]-20)
    p2 = (p2[0], p2[1]-20)
    paint()

def handler_l (event):
    global draw_limitset
    draw_limitset = not draw_limitset
    paint()

canvas.bind("<Button-1>",  handler_button)
canvas.bind("<B1-Motion>", handler_button)
canvas.bind("<Left>", handler_left)
canvas.bind("<Right>", handler_right)
canvas.bind("<Up>", handler_up)
canvas.bind("<Down>", handler_down)
canvas.bind("l", handler_l)

# This part is due to Ma Jia Jun.
# Type 's' to save the picture into a PostScript file.
import tkFileDialog
def saveps(event):
    filename = tkFileDialog.asksaveasfilename(
        defaultextension="ps",
        filetypes=[("PostScript","*.ps")],
        initialfile="OptPy.ps",
        title="Save to PostScript")
    canvas.postscript(
        {"file":filename
         })
canvas.bind("s",saveps)

canvas.focus_set()
canvas.pack(expand=YES, fill=BOTH)
paint()
canvas.mainloop()
```

6. Using OptPy

To use the program OptPy, create a file named `OptPy.py` with contents listed in the previous section. You can download this file from [14]. Double click this file, or type `python OptPy.py` in your terminal (Mac OS X or Linux). Then our program will start and a window appears. See Figure 6.

The initial broken line described at (3.4) is red. Other broken lines (3.9) calculated in steps 2 and 3 are black.

The user can change a_1 and a_2 by the mouse. Arrow keys can be used to translate the complex plane. The key '`l`' toggles whether draw the limit set or not. The key '`s`' saves the picture in PostScript format. The PostScript part is due to Ma Jia Jun (one of the audience). I would like to thank him for his contribution.

Let $(x, y, z) \in \mathcal{X}_2$ and $\phi_{(x,y,z)} : \widehat{\mathbb{Q}} \to \mathbb{C}$ be the associated Markoff map. (See subsection 3.3.) In [2] (Conjecture A), Bowditch conjectured that (x, y, z) corresponds to a quasifuchsian representation if and only if $\phi^{-1}_{(x,y,z)}([-2, 2]) = \emptyset$ and the number of elements $r \in \widehat{\mathbb{Q}}$ with $|\phi_{(x,y,z)}(r)| \leq 2$ is finite. This conjecture is still open. Quasifuchsian Kleinian groups might be better understood using this type of software.

Acknowledgements

I would like to express my thanks to Ser Peow Tan for organizing the program "Geometry, Topology and Dynamics of Character Varieties." I was supported as a visiting senior research fellow by the Department of Mathematics, National University of Singapore during the program. I am grateful to the IT staff of the Department of Mathematics for setting up software in the computer room needed for the lectures that I gave. I would also like to thank all the participants in the lectures. Finally, I wish to express my thanks to the referee for helpful comments.

References

1. H. Akiyoshi, M. Sakuma, M. Wada and Y. Yamashita, "Punctured torus groups and 2-bridge knot groups. I", *Lecture Notes in Mathematics*, **1909**, Springer, Berlin, 2007.
2. B. H. Bowditch, "Markoff triples and quasi-Fuchsian groups", *Proc. London Math. Soc.*, **77** (1998), 697–736.
3. B. Burton, "Regina", a normal surface theory calculator, `http://regina.sourceforge.net/`.
4. M. Culler, N. M. Dunfield and J. R. Weeks, "SnapPy", a computer program for studying the geometry and topology of 3-manifolds, `http://snappy.computop.org/`
5. F. Guéritaud, "Triangulated cores of punctured-torus groups", *J. Differential Geom.*, **81** (2009), 91–142.
6. T. Hamada and KNOPPIX/Math committers, "KNOPPIX/Math", a bootable live linux system with emphasis on mathematical software, `http://www.knoppix-math.org/`.
7. T. Jørgensen, "On pairs of once-punctured tori", *London Math. Soc. Lecture Note Ser.*, **299** (2003), 183–207.

8. Y. Komori, T. Sugawa, M. Wada and Y. Yamashita, "Drawing Bers embeddings of the Teichmüller space of once-punctured tori", *Experiment. Math.*, **15** (2006), 51–60.
9. D. Mumford, C. Series and D. Wright, "Indra's pearls. The vision of Felix Klein", Cambridge University Press, New York, 2002.
10. M. Lackenby, "The canonical decomposition of once-punctured torus bundles", *Comment. Math. Helv.*, **78** (2003), 363–384.
11. M. Lutz, "Learning Python", O'Reilly Media, Sebastopol CA, 2009.
12. K. Vlahos, "PyScripter", IDE for Python, `http://code.google.com/p/pyscripter/`
13. M. Wada, "OPTi's algorithm for discreteness determination", *Experiment. Math.*, **15** (2006), 61-66.
14. Y. Yamashita, "OptPy", Once punctured torus software using Python, `http://vivaldi.ics.nara-wu.ac.jp/~yamasita/OptPy/`.

TRACES IN COMPLEX HYPERBOLIC GEOMETRY

John R. Parker
Department of Mathematical Sciences
Durham University
Durham DH1 3LE, UK
j.r.parker@durham.ac.uk

In memory of Robert Miner

We discuss the relationship between the geometry of complex hyperbolic manifolds and orbifolds and the traces of elements of the corresponding subgroup of $\mathrm{SU}(2,1)$. We begin by showing how geometrical information about individual isometries is encoded by their trace. We then consider traces for groups Γ of isometries in two specific cases. First, we consider the case where Γ is a free group on two generators, which we view as the fundamental group of a three holed sphere. We indicate how to use this analysis to give complex hyperbolic Fenchel-Nielsen coordinates. Secondly, we consider the case where Γ is a triangle group generated by complex reflections in three complex lines. We keep in mind similar results from the more familiar setting of Fuchsian and Kleinian groups and we explain those examples from our point of view.

Keywords: Complex hyperbolic space, trace, invariants

Mathematics Subject Classification 2000: 51M10, 16R30, 22E40

1. Introduction

It is well known that any Riemann surface Σ of genus $g \geq 2$ may be written as $\Sigma = \mathbf{H}^2/\Gamma$ where Γ is a Fuchsian representation ρ of $\pi_1(\Sigma)$ to $\mathrm{PSL}(2,\mathbb{R})$, the orientation preserving isometries of the hyperbolic plane $\mathbf{H}^2$. One may lift ρ to a representation of $\pi_1(\Sigma)$ to $\mathrm{SL}(2,\mathbb{R})$. There is close relationship between the geometry of the surface and the representation $\Gamma = \rho\big(\pi_1(\Sigma)\big)$. For example, the lengths of closed geodesics on Σ may be written simply in terms of the traces of elements of Γ. Thus one may find useful geometric information on the possible hyperbolic metrics on the surface by studying either the representation variety or the character variety. This idea goes back to the work of Fricke in the nineteenth century.

Consider the *representation variety* $\mathrm{Hom}\big(\pi_1(\Sigma), \mathrm{SL}(2,\mathbb{R})\big)/\mathrm{SL}(2,\mathbb{R})$ of conjugacy classes of representations of $\pi_1(\Sigma)$ to $\mathrm{SL}(2,\mathbb{R})$. (One must take care with this quotient. The quotient by conjugation, that is by the group of inner automorphisms, is not Hausdorff and so in fact one takes the "maximal Hausdorff quotient space"; see the discussion on page 97 of Goldman [7].) One component of the representation variety is Teichmüller space $\mathcal{T}(\Sigma)$, which is the space of marked hyperbolic metrics on Σ (see page 98 of [7]). Fenchel-Nielsen coordinates [6] give a global set of parameters for $\mathcal{T}(\Sigma)$. These coordinates are the hyperbolic lengths of $3g-3$ simple closed geodesics on Σ and $3g-3$ twist parameters. The length parameters may be studied directly using traces but the twist parameters do not have an obvious direct interpretation in terms of the representation variety. A natural question is whether one can find $6g-6$ lengths (or traces) to give global parameters. A theorem proved by Okumura [18] and Schmutz [29] says that in fact one needs an extra length, or trace, to give global coordinates.

Similarly, one may consider the representation variety $\mathrm{Hom}\big(\pi_1(\Sigma), \mathrm{SL}(2,\mathbb{C})\big)/\mathrm{SL}(2,\mathbb{C})$ of conjugacy classes of representations of $\pi_1(\Sigma)$ to $\mathrm{SL}(2,\mathbb{C})$. Of particular interest are the *quasi-Fuchsian* representations. These are characterised by being discrete, faithful, type-preserving and geometrically finite. Complex Fenchel-Nielsen coordinates were defined by Tan [30] and Kourouniotis [12]. The main difference is that the lengths and twists are now complex numbers. For both the lengths and twists, the real part is just the ordinary hyperbolic length and the imaginary part measures how the normal plane is rotated as we move around the geodesic. This relationship is discussed in detail by Parker and Series [25]. A second key difference is that, while complex Fenchel Nielsen coordinates distinguish non-conjugate representations, it is in general not clear what subset of $\mathbb{C}^{6g-6}$ corresponds to quasi-Fuchsian space.

A triangle group is the group generated by reflections in the sides of a spherical, Euclidean or hyperbolic triangle. This triangle, and hence the group, is completely determined up to isometry (or similarity in the Euclidean case) by the internal angles of the triangle. These angles may be found using the trace of the product of reflections in the associated sides. So once again, traces lead to geometrical information and to a set of parameters for these groups. More specifically, one may write the Gram matrix G (or cosine matrix) of a triangle. This is a symmetric matrix whose diagonal entries are all 1 and the off diagonal entries are $-\cos(\theta_i)$ where θ_i for $i = 1, 2, 3$ are the internal angles; see Davis [3]. The Gram matrix G is positive definite, singular, of signature $(2,1)$ respectively if the sum of the internal angles is greater than, equal to or less than π respectively (see Theorem 6.8.12 of [3]). One may then write down a representation of the triangle group into $SO(G)$, the group of unimodular, orthogonal matrices preserving the bilinear form associated to G. When G has signature $(2, 1)$ then $\mathrm{SO}(G)$ is the isometry group of the hyperbolic plane $\mathbf{H}^2$. This representation of the triangle group is completely determined up to conjugacy by the three internal angles. In particular, one may write down the trace of any group element as an integer polynomial in the variables $2\cos(\theta_i)$.

The purpose of this article is to investigate how these ideas may be extended to complex hyperbolic space $\mathbf{H}^2_{\mathbb{C}}$, which is a natural generalisation of the hyperbolic plane to higher *complex* dimensions. As well as many similarities, there are many fascinating differences between the generalisation of the hyperbolic plane to higher real dimensions and to higher complex dimensions. Complex hyperbolic space has variable (quarter pinched) negative curvature, causing many things to become harder than in the constant curvature setting. However the complex structure simplifies many things. Particular instances of these similarities and differences will occur throughout these notes. We use results from real hyperbolic geometry to inspire us when formulating complex hyperbolic problems. Quite often this leads us to results in the complex world with a similar overall structure but whose details differ considerably from the analogous results from the real world.

The holomorphic isometry group of $\mathbf{H}^2_{\mathbb{C}}$ is the projective group $\mathrm{PU}(2, 1)$. It is more convenient to lift to a triple cover $\mathrm{SU}(2, 1)$. Hence we want to find out what types of geometrical information may be deduced from information about traces. In principle, this relationship is very similar to the connection between geometrical information about $\mathbf{H}^2$ or $\mathbf{H}^3$ and traces in $\mathrm{SL}(2, \mathbb{R})$ or $\mathrm{SL}(2, \mathbb{C})$. In practice this relationship is more subtle. Part of the subtlety arises from the fact that we are dealing with 3×3 matrices and part because $\mathbf{H}^2_{\mathbb{C}}$ has variable negative curvature.

The background material is quite standard. Readers may find more detail in the book of Goldman [8] and the forthcoming book of Parker [21]. The discussion of traces for two generator groups follows work of Lawton [13] and Will [33], [34]. The application of this work to Fenchel-Nielsen coordinates is related to work of Parker and Platis [23]. As such it is part of a wider area of complex hyperbolic quasi-Fuchsian groups. We will not discuss this topic in detail. Instead we refer readers to the survey article [24]. Finally, our treatment of traces for triangle groups follows Sandler [28] and Pratoussevitch [27]. It has applications for the construction of lattices, see Parker and Paupert [22].

2. Background

2.1. *Hermitian and unitary matrices*

We review well known facts about Hermitian linear algebra, and we show how to use this to construct complex hyperbolic space. The material in this section is completely standard; see Goldman [8], Chen and Greenberg [2] or Parker [21].

Let $A = (a_{ij})$ be a $k \times l$ complex matrix. The *Hermitian transpose* of A is the $l \times k$ complex matrix $A^* = (\overline{a}_{ji})$ formed by complex conjugating each entry of A and then taking the transpose. As with ordinary transpose, the Hermitian transpose of a product is the product of the Hermitian transposes in the reverse order. That is $(AB)^* = B^*A^*$. Clearly $(A^*)^* = A$ and, since $I^* = I$, if A is invertible we also have $(A^*)^{-1} = (A^{-1})^*$.

A $k \times k$ complex matrix H is said to be *Hermitian* if it equals its own Hermitian transpose $H = H^*$. Let H be a Hermitian matrix and μ an eigenvalue of H with eigenvector $\mathbf{z} \neq \mathbf{0}$. We claim that μ is real. In order to see this, observe that

$$\mu\mathbf{z}^*\mathbf{z} = \mathbf{z}^*(\mu\mathbf{z}) = \mathbf{z}^*H\mathbf{z} = \mathbf{z}^*H^*\mathbf{z} = (H\mathbf{z})^*\mathbf{z} = (\mu\mathbf{z})^*\mathbf{z} = \overline{\mu}\mathbf{z}^*\mathbf{z}.$$

Since $\mathbf{z}^*\mathbf{z}$ is real and non-zero, we see that μ is real. Suppose that H is a non-singular Hermitian matrix (that is, all its eigenvalues are non-zero) with p positive eigenvalues and q negative ones. Then we say that H has *signature* (p, q).

To each $k \times k$ Hermitian matrix H we can associate a *Hermitian form* $\langle \cdot, \cdot \rangle : \mathbb{C}^k \times \mathbb{C}^k \longrightarrow \mathbb{C}$ given by $\langle \mathbf{z}, \mathbf{w} \rangle = \mathbf{w}^*H\mathbf{z}$ (note that we change the order) where $\mathbf{w}$ and $\mathbf{z}$ are column vectors in $\mathbb{C}^k$. Hermitian forms are sesquilinear, that is they are linear in the first factor and conjugate linear in the second factor. In other words, for $\mathbf{z}$, $\mathbf{z}_1$, $\mathbf{z}_2$, $\mathbf{w}$ column vectors in $\mathbb{C}^k$ and λ a complex scalar, we have

$$\begin{aligned} \langle \mathbf{z}_1 + \mathbf{z}_2, \mathbf{w} \rangle &= \langle \mathbf{z}_1, \mathbf{w} \rangle + \langle \mathbf{z}_2, \mathbf{w} \rangle, \\ \langle \lambda\mathbf{z}, \mathbf{w} \rangle &= \lambda \langle \mathbf{z}, \mathbf{w} \rangle, \\ \langle \mathbf{w}, \mathbf{z} \rangle &= \overline{\langle \mathbf{z}, \mathbf{w} \rangle}. \end{aligned}$$

From these we see that

$$\begin{aligned} \langle \mathbf{z}, \mathbf{z} \rangle &\in \mathbb{R}, \\ \langle \mathbf{z}, \lambda\mathbf{w} \rangle &= \overline{\lambda} \langle \mathbf{z}, \mathbf{w} \rangle, \\ \langle \lambda\mathbf{z}, \lambda\mathbf{w} \rangle &= |\lambda|^2 \langle \mathbf{z}, \mathbf{w} \rangle. \end{aligned}$$

If H has signature (p, q) then we say that $\langle\cdot,\cdot\rangle$ also has signature (p, q). Let $\mathbb{C}^{p,q}$ be a complex vector space of (complex) dimension $p+q$ equipped with a non-degenerate Hermitian form $\langle\cdot,\cdot\rangle$ of signature (p, q). This means that $\langle\cdot,\cdot\rangle$ is given by a non-singular $(p+q) \times (p+q)$ Hermitian matrix H with p positive eigenvalues and q negative eigenvalues. Later we will see that choosing a different form of signature (p, q) leads to an isomorphic space. In what follows we will be interested in the case where $k = 3$, $p = 2$ and $q = 1$, but we will keep a running example of the case where $k = 2$ and $p = q = 1$. We will see that, depending on the choice of Hermitian form, this example yields the Poincaré disc or half plane model of $\mathbf{H}^2$. For a discussion of higher dimensional (real) hyperbolic space from the Hermitian point of view, see [19].

Example 2.1. Consider

$$H_0 = \begin{pmatrix} 1 & 0 \\ 0 & -1 \end{pmatrix}, \qquad H_0' = \begin{pmatrix} 0 & -i \\ i & 0 \end{pmatrix}. \tag{2.1}$$

It is clear that H_0 and H_0' are both Hermitian. Moreover, as H_0 is diagonal it is immediate that it has signature $(1, 1)$. It is not hard to check that H_0' also has signature $(1, 1)$.

Let H be a Hermitian form of signature (p, q) and let A be a matrix that preserves the corresponding Hermitian form. In other words, for all $\mathbf{v}$ and $\mathbf{w}$ in $\mathbb{C}^{p,q}$ we have

$$\mathbf{w}^* A^* H A \mathbf{v} = \langle A\mathbf{v},\, A\mathbf{w}\rangle = \langle \mathbf{v},\, \mathbf{w}\rangle = \mathbf{w}^* H \mathbf{v}. \tag{2.2}$$

We say that such a matrix A is *unitary* with respect to H. It is clear that the collection of all matrices that are unitary with respect to H form a group, the *unitary group of* H, denoted $\mathrm{U}(H)$. Sometimes we wish to consider unimodular unitary matrices and we denote the corresponding group by $\mathrm{SU}(H)$. By letting $\mathbf{v}$ and $\mathbf{w}$ run through a basis of of $\mathbb{C}^{p,q}$ we see that (2.2) implies $A^*HA = H$. In other words, $H^{-1}A^*HA = I$ and so $A^{-1} = H^{-1}A^*H$.

Example 2.2. Consider the Hermitian forms H_0 and H_0' in (2.1). Suppose that $A \in \mathrm{SU}(H_0)$. Then

$$\begin{pmatrix} d & -b \\ -c & a \end{pmatrix} = A^{-1} = H_0^{-1} A^* H_0 = \begin{pmatrix} \overline{a} & -\overline{c} \\ -\overline{b} & \overline{d} \end{pmatrix}.$$

Therefore $b = \overline{c}$ and $d = \overline{a}$. Hence $1 = ad - bc = |a|^2 - |c|^2$. Hence

$$\mathrm{SU}(H_0) = \left\{ \begin{pmatrix} a & \overline{c} \\ c & \overline{a} \end{pmatrix} \; : \; a,\, c \in \mathbb{C},\; |a|^2 - |c|^2 = 1 \right\}.$$

Similarly, suppose $A' \in \mathrm{SU}(H_0')$. Then

$$\begin{pmatrix} d & -b \\ -c & a \end{pmatrix} = A'^{-1} = H_0'^{-1} A'^* H_0' = \begin{pmatrix} \overline{d} & -\overline{b} \\ -\overline{c} & \overline{a} \end{pmatrix}.$$

Therefore a, b, c, d are all real. Hence

$$\mathrm{SU}(H_0') = \left\{ \begin{pmatrix} a & b \\ c & d \end{pmatrix} \; : \; a,\, b,\, c,\, d \in \mathbb{R},\; ad - bc = 1 \right\}.$$

That is, $\mathrm{SU}(H_0') = \mathrm{SL}(2, \mathbb{R})$.

Given two Hermitian forms H and H' of the same signature we can pass between them using a *Cayley transform* C. That is, we can write

$$H' = C^* H C.$$

The existence of the Cayley transform follows from Sylvester's law of inertia. The Cayley transform C is not unique for we may precompose and post-compose by any unitary matrix preserving the relevant Hermitian form. It is clear that if A is unitary with respect to H then $A' = C^{-1}AC$ is unitary with respect to H'. In order to see this, observe that, using $(C^{-1}AC)^* = C^* A^* C^{*-1}$, we have

$$A'^* H' A' = (C^{-1}AC)^* (C^* H C)(C^{-1}AC) = C^* A^* H A C = C^* H C = H'.$$

This means that one does not need to specify the form but only the signature. Hence we can talk about $\mathbb{C}^{p,q}$, which is the vector space $\mathbb{C}^{p+q}$ together with any Hermitian form of signature (p, q). The choice of a particular Hermitian form is equivalent to a choice of basis. Also, we write $\mathrm{U}(p, q)$ and $\mathrm{SU}(p, q)$ for the unitary and special unitary groups for a form of signature (p, q) and in doing so we do not need to specify a particular form.

Example 2.3. Consider H_0 and H_0' given by (2.1) and

$$C_0 = \frac{1}{\sqrt{2}} \begin{pmatrix} 1 & -i \\ -i & 1 \end{pmatrix}.$$

Then it is easy to check that $H_0' = C_0^* H_0 C_0$. Furthermore suppose

$$A = \begin{pmatrix} a & \overline{c} \\ c & \overline{a} \end{pmatrix} \in \mathrm{SU}(H_0).$$

Then

$$A' = C_0^{-1} A C_0 = \begin{pmatrix} \Re(a) - \Im(c) & \Im(a) + \Re(c) \\ -\Im(a) + \Re(c) & \Re(a) + \Im(c) \end{pmatrix} \in \mathrm{SU}(H_0') = \mathrm{SL}(2, \mathbb{R}).$$

2.2. *Complex hyperbolic space and its isometries*

We now define complex hyperbolic space. We first define the projective model and then go on to specialise to the unit ball model and the Siegel domain model. This is simultaneously a complex version of the projective and Klein-Beltrami models of ordinary (real) hyperbolic space and also a generalisation to higher complex dimensions of the Poincaré disc and half plane models of the hyperbolic plane. We restrict ourselves to the case of complex hyperbolic 2-space. These definitions can be generalised to higher dimensions in an obvious way.

Let H be a 3×3, non-singular Hermitian form of signature $(2, 1)$. If $\mathbf{z} \in \mathbb{C}^{2,1}$ then we know that $\langle \mathbf{z}, \mathbf{z} \rangle$ is real. Thus we may define subsets V_-, V_0 and V_+ of $\mathbb{C}^{2,1}$ by

$$V_- = \left\{ \mathbf{z} \in \mathbb{C}^{2,1} | \langle \mathbf{z}, \mathbf{z} \rangle < 0 \right\}, \tag{2.3}$$

$$V_0 = \left\{ \mathbf{z} \in \mathbb{C}^{2,1} - \{\mathbf{0}\} | \langle \mathbf{z}, \mathbf{z} \rangle = 0 \right\}, \tag{2.4}$$

$$V_+ = \left\{ \mathbf{z} \in \mathbb{C}^{2,1} | \langle \mathbf{z}, \mathbf{z} \rangle > 0 \right\}. \tag{2.5}$$

We say that $\mathbf{z} \in \mathbb{C}^{2,1}$ is *negative*, *null* or *positive* if $\mathbf{z}$ is in V_-, V_0 or V_+ respectively. Motivated by special relativity, these are sometimes called time-like, light-like and space-like.

Example 2.4. Consider $\mathbb{C}^{1,1}$ with the Hermitian form given by H_0 in (2.1). Then

$$V_- = \left\{ \begin{pmatrix} z_1 \\ z_2 \end{pmatrix} : |z_1|^2 < |z_2|^2 \right\}, \quad V_0 = \left\{ \begin{pmatrix} z_1 \\ z_2 \end{pmatrix} \neq \begin{pmatrix} 0 \\ 0 \end{pmatrix} : |z_1|^2 = |z_2|^2 \right\}.$$

Likewise for $\mathbb{C}^{1,1}$ with the Hermitian form given by H_0',

$$V_- = \left\{ \begin{pmatrix} z_1 \\ z_2 \end{pmatrix} : \Im(z_1 \overline{z}_2) > 0 \right\}, \quad V_0 = \left\{ \begin{pmatrix} z_1 \\ z_2 \end{pmatrix} \neq \begin{pmatrix} 0 \\ 0 \end{pmatrix} : \Im(z_1 \overline{z}_2) = 0 \right\}.$$

Define an equivalence relation on $\mathbb{C}^{2,1} - \{\mathbf{0}\}$ by $\mathbf{z} \sim \mathbf{w}$ if and only if there is a non-zero complex scalar λ so that $\mathbf{w} = \lambda \mathbf{z}$. Let $\mathbb{P} : \mathbb{C}^{2,1} - \{\mathbf{0}\} \longmapsto \mathbb{CP}^2$ denote the standard projection map defined by $\mathbb{P}(\mathbf{z}) = [\mathbf{z}]$ where $[\mathbf{z}]$ is the equivalence class of $\mathbf{z}$. On the chart of $\mathbb{C}^{2,1}$ with $z_3 \neq 0$ the projection map

$\mathbb{P}$ is given by

$$\mathbb{P} : \begin{bmatrix} z_1 \\ z_2 \\ z_3 \end{bmatrix} \longmapsto \begin{pmatrix} z_1/z_3 \\ z_2/z_3 \end{pmatrix} \in \mathbb{C}^2.$$

Because $\langle \lambda\mathbf{z}, \lambda\mathbf{z} \rangle = |\lambda|^2 \langle \mathbf{z}, \mathbf{z} \rangle$ we see that for any non-zero complex scalar λ the point $\lambda\mathbf{z}$ is negative, null or positive if and only if $\mathbf{z}$ is. The *projective model* of complex hyperbolic space is defined to be the collection of negative lines in $\mathbb{C}^{2,1}$ and its boundary is defined to be the collection of null lines. In other words $\mathbf{H}^2_{\mathbb{C}}$ is $\mathbb{P}V_-$ and $\partial\mathbf{H}^2_{\mathbb{C}}$ is $\mathbb{P}V_0$,

We define the other two standard models of complex hyperbolic space by considering two standard Hermitian forms on $\mathbb{C}^{2,1}$. We call these the first and second Hermitian forms. Let $\mathbf{z}$, $\mathbf{w}$ be the column vectors $(z_1, z_2, z_3)^t$ and $(w_1, w_2, w_3)^t$ respectively. The *first Hermitian form* is defined to be:

$$\langle \mathbf{z}, \mathbf{w} \rangle_1 = z_1\overline{w}_1 + z_2\overline{w}_2 - z_3\overline{w}_3. \tag{2.6}$$

It is given by the Hermitian matrix H_1:

$$H_1 = \begin{pmatrix} 1 & 0 & 0 \\ 0 & 1 & 0 \\ 0 & 0 & -1 \end{pmatrix}. \tag{2.7}$$

The *second Hermitian form* is defined to be:

$$\langle \mathbf{z}, \mathbf{w} \rangle_2 = z_1\overline{w}_3 + z_2\overline{w}_2 + z_3\overline{w}_1. \tag{2.8}$$

It is given by the Hermitian matrix H_2:

$$H_2 = \begin{pmatrix} 0 & 0 & 1 \\ 0 & 1 & 0 \\ 1 & 0 & 0 \end{pmatrix}. \tag{2.9}$$

Both of these forms have the property that each vector in V_- has non-zero third entry. Therefore, we can take the section defined by $z_3 = 1$. This gives a unique point on each complex line in V_-. In other words given $z = (z_1, z_2) \in \mathbb{C}^2$, we define its *standard lift* to $\mathbb{C}^{2,1}$ to be the column vector

$$\mathbf{z} = \begin{bmatrix} z_1 \\ z_2 \\ 1 \end{bmatrix}$$

in $\mathbb{C}^{2,1}$. Clearly $\mathbb{P}$ sends $\mathbf{z}$ back to z. We consider what it means for $\langle \mathbf{z}, \mathbf{z} \rangle$ to be negative for the first and second Hermitian forms, respectively.

For the first Hermitian form, we see that the standard lift of $z = (z_1, z_2) \in \mathbb{C}^2$ is in V_- if and only if $|z_1|^2 + |z_2|^2 < 1$. In other words, $\mathbb{P}V_- = \mathbf{H}^2_{\mathbb{C}}$ is the *unit ball* in $\mathbb{C}^2$. Likewise, the standard lift of z is in V_0 if and only if $|z_1|^2 + |z_2|^2 = 1$ and so $\mathbb{P}V_0 = \partial\mathbf{H}^2_{\mathbb{C}}$ is the unit 3-sphere S^3 in $\mathbb{C}^2$.

For the second Hermitian form, the standard lift of z is negative if and only if

$$z_1 + |z_2|^2 + \overline{z}_2 = 2\Re(z_1) + |z_2|^2 < 0.$$

Thus $\mathbb{P}V_-$ is a paraboloid in $\mathbb{C}^2$, called the *Siegel domain*. Similarly z in $\mathbb{P}V_0 \cap \mathbb{C}^2$ satisfies $2\Re(z_1) + |z_2|^2 = 0$. However, not all points in $\mathbb{P}(V_0)$ lie in $\mathbb{C}^2 \subset \mathbb{C}\mathbb{P}^2$. We have to add an extra point, denoted ∞, on the boundary of the Siegel domain. The standard lift of ∞ is

$$\begin{bmatrix} 1 \\ 0 \\ 0 \end{bmatrix}.$$

Of course ∞ is not the only point in $\mathbb{C}\mathbb{P}^2 - \mathbb{C}^2$, that is it is not the only point "at infinity". In this respect it is different from the point ∞ on the boundary of the upper half plane model of the hyperbolic plane, which is the only point of $\mathbb{C}\mathbb{P}^1$ that is not in $\mathbb{C}$.

Example 2.5. For $z \in \mathbb{C}$ the standard lift of z to $\mathbb{C}^{1,1}$ is

$$\mathbf{z} = \begin{bmatrix} z \\ 1 \end{bmatrix} \in \mathbb{C}^{1,1}.$$

If $\mathbb{C}^{1,1}$ has the Hermitian form given by H_0 from (2.1), then we see that $z \in \mathbb{P}V_-$ if and only if $|z| < 1$ and $z \in \mathbb{P}V_0$ if and only if $|z| = 1$. Thus $\mathbb{P}V_-$ is the unit disc and $\mathbb{P}V_0$ is the unit circle in $\mathbb{C}$.

Similarly for H_0', the point $z \in \mathbb{P}V_-$ if and only if $\Im(z) > 0$ and $z \in \mathbb{P}V_0 \cap \mathbb{C}$ if and only if z is real. We must add an extra point ∞ whose standard lift is

$$\begin{bmatrix} 1 \\ 0 \end{bmatrix}.$$

There are other Hermitian forms which are widely used in the literature. In particular, Chen and Greenberg (page 67 of [2]) give a close relative of the second Hermitian form, namely the one given by the matrix

$$\begin{pmatrix} 0 & -1 & 0 \\ -1 & 0 & 0 \\ 0 & 0 & 1 \end{pmatrix}.$$

When we consider complex hyperbolic triangle groups in Section 5 we will see that, following Mostow, it is most convenient to use the Gram matrix as the Hermitian form. This form will always be defined, but only for certain points in the parameter space will it have the correct signature.

So far we have only defined complex hyperbolic space as a set of points. In order to understand its geometry we must give it a metric. This metric is called the *Bergman metric*, given by the following element of arc length:

$$ds^2 = \frac{-4}{\langle \mathbf{z}, \mathbf{z} \rangle^2} \det \begin{pmatrix} \langle \mathbf{z}, \mathbf{z} \rangle & \langle d\mathbf{z}, \mathbf{z} \rangle \\ \langle \mathbf{z}, d\mathbf{z} \rangle & \langle d\mathbf{z}, d\mathbf{z} \rangle \end{pmatrix}. \tag{2.10}$$

The choice of the constant 4 in the above formula means that the holomorphic sectional curvature of $\mathbf{H}^2_{\mathbb{C}}$ is -1. Compare this with equation (3.3) on page 73 and Theorem 3.1.9 of Goldman [8]. The similar formula in equation (19.2) on page 135 of Mostow [16] has the constant 1 here and this leads to holomorphic sectional curvature -4.

Alternatively, the Bergman metric is given by the distance function $\rho(\cdot, \cdot)$ defined by the formula

$$\cosh^2\left(\frac{\rho(z, w)}{2}\right) = \frac{\langle \mathbf{z}, \mathbf{w} \rangle \langle \mathbf{w}, \mathbf{z} \rangle}{\langle \mathbf{z}, \mathbf{z} \rangle \langle \mathbf{w}, \mathbf{w} \rangle}.$$

See equation (3.4) on page 77 of Goldman [8]. The different normalisation given by Mostow means that $\rho/2$ is replaced with ρ in formula (19.4') on page 136 of [16].

For the ball model and Siegel domain model one can find the distance between points z and w by plugging their standard lifts $\mathbf{z}$ and $\mathbf{w}$ into the above formula. However, as may easily be seen, this formula is independent of which lifts $\mathbf{z}$ and $\mathbf{w}$ in $\mathbb{C}^{2,1}$ of z and w we choose. Thus the distance function is well defined in these two models.

Example 2.6. Consider $z \in \mathbb{C}$ with $|z| < 1$. We have seen that for the Hermitian form H_0 this point is in $z \in \mathbb{P}V_-$. Moreover, the standard lift of z and its derivative are

$$\mathbf{z} = \begin{bmatrix} z \\ 1 \end{bmatrix}, \quad d\mathbf{z} = \begin{bmatrix} dz \\ 0 \end{bmatrix}.$$

Plugging this vector and the Hermitian form given by H_0 into (2.10) gives

$$ds^2 = \frac{-4}{\left(|z|^2 - 1\right)^2} \det \begin{pmatrix} |z|^2 - 1 & \overline{z}\, dz \\ z\, d\overline{z} & dz\, d\overline{z} \end{pmatrix} = \frac{4|dz|^2}{\left(1 - |z|^2\right)^2}.$$

This is just the Poincaré metric on the unit disc.

Similarly consider $z \in \mathbb{C}$ with $\Im(z) > 0$. This is in $\mathbb{P}V_-$ for the Hermitian form H_0'. Plugging its standard lift into (2.10) gives

$$ds^2 = \frac{-4}{\left(-2\Im(z)\right)^2}\det\begin{pmatrix} -2\Im(z) & idz \\ -id\overline{z} & 0 \end{pmatrix} = \frac{|dz|^2}{\left(\Im(z)\right)^2}.$$

This is the Poincaré metric on the upper half plane.

Note that in both these examples we have the constant 4 and constant curvature -1.

Unitary matrices in $\mathrm{U}(2,1)$ act on $\mathbb{C}^{2,1}$ preserving V_+, V_0 and V_-. They also preserve the Bergman metric since it is given solely in terms of the Hermitian form. Therefore unitary matrices act as isometries on complex hyperbolic space. Let us see this action explicitly. Let $z = (z_1, z_2)$ be a point in $\mathbb{C}^2$ and let $\mathbf{z}$ be its standard lift to $\mathbb{C}^{2,1}$. Then $A \in \mathrm{U}(2,1)$ acts as follows:

$$A(z) = \mathbb{P}(A\mathbf{z}).$$

In other words, if

$$A = \begin{pmatrix} a & b & c \\ d & e & f \\ g & h & j \end{pmatrix}$$

then

$$A(z_1, z_2) = \left(\frac{az_1 + bz_2 + c}{gz_1 + hz_2 + j}, \frac{dz_1 + ez_2 + f}{gz_1 + hz_2 + j}\right).$$

This is just a linear fractional transformation in two variables.

Example 2.7. Consider $A \in SU(H_0)$:

$$A = \begin{pmatrix} a & \overline{c} \\ c & \overline{a} \end{pmatrix}.$$

Arguing as above, we see that A acts on the unit disc s the Möbius transformation in $\mathrm{PSU}(H_0)$

$$A(z) = \mathbb{P}\begin{pmatrix} a & \overline{c} \\ c & \overline{a} \end{pmatrix}\begin{pmatrix} z \\ 1 \end{pmatrix} = \frac{az + \overline{c}}{cz + \overline{a}}.$$

Similarly for H_0', the matrix $A \in \mathrm{SU}(H_0') = \mathrm{SL}(2, \mathbb{R})$ acts on the upper half plane as a Möbius transformation in $\mathrm{PSL}(2, \mathbb{R})$.

Any matrix in $\mathrm{U}(2,1)$ which is a (non-zero) complex scalar multiple of the identity maps each line through the origin in $\mathbb{C}^{2,1}$ to itself and so acts trivially on complex hyperbolic space. Since this matrix is unitary with respect to $\langle\cdot,\cdot\rangle$ then the scalar must have unit norm. Because of this, we define the projective unitary group $\mathrm{PU}(2,1) = \mathrm{U}(2,1)/\mathrm{U}(1)$ where $\mathrm{U}(1)$ is canonically identified with $\{e^{i\theta}I | 0 \leq \theta < 2\pi\}$, where I is the identity matrix in $\mathrm{U}(2,1)$. Sometimes it will be useful to consider $\mathrm{SU}(2,1)$, the group of matrices with determinant 1 which are unitary with respect to $\langle\cdot,\cdot\rangle$. The group $\mathrm{SU}(2,1)$ is a 3-fold covering of $\mathrm{PU}(2,1)$:

$$\mathrm{PU}(2,1) = \mathrm{SU}(2,1)/\{I,\, \omega I,\, \omega^2 I\}$$

where $\omega = (-1+i\sqrt{3})/2$ is a cube root of unity. This is completely analogous to the fact that $\mathrm{SL}(2,\mathbb{C})$ is a double cover of $\mathrm{PSL}(2,\mathbb{C}) = \mathrm{SL}(2,\mathbb{C})/\{I, -I\}$. Cube roots of unity are used because $\mathrm{SU}(2,1)$ consists of 3×3 matrices.

3. The Geometry of Isometries

3.1. *Introduction*

It is well known that the dynamical behaviour of hyperbolic isometries in $\mathrm{PSL}(2,\mathbb{C})$ may be classified as elliptic, parabolic or loxodromic (hyperbolic) and the trace of the corresponding matrix in $\mathrm{SL}(2,\mathbb{C})$ distinguishes between these classes. Moreover, for non-parabolic isometries, the geometry of the action in terms of rotation angle or hyperbolic translation length may be read off directly from the trace. Likewise, following Goldman [8], one may use the trace of an element of $\mathrm{SU}(2,1)$ to decide whether the corresponding complex hyperbolic isometry in $\mathrm{PU}(2,1)$ is elliptic, parabolic or loxodromic. Furthermore, one may deduce information about the geometry of the action of the isometry from this trace. For loxodromic isometries this is straightforward, but for elliptic and parabolic isometries there is some subtlety involved, which we will explain.

3.2. *Classification of elements of* $\mathbf{SU(2,1)}$ *by their trace*

In this section we will classify elements of $\mathrm{SU}(2,1)$ by their trace. Elements of $\mathrm{SU}(2,1)$ are holomorphic complex hyperbolic isometries and the familiar trichotomy from real hyperbolic geometry applies in this setting too; see Chen and Greenberg [2]. A holomorphic complex hyperbolic isometry A is said to be:

(i) *loxodromic* if it fixes exactly two points of $\partial\mathbf{H}^2_{\mathbb{C}}$;

(ii) *parabolic* if it fixes exactly one point of $\partial\mathbf{H}^2_{\mathbb{C}}$;
(iii) *elliptic* if it fixes at least one point of $\mathbf{H}^2_{\mathbb{C}}$.

Following Goldman, [8], we now show how $A \in \mathrm{SU}(2,1)$ may be classified as loxodromic, parabolic or elliptic using $\mathrm{tr}(A)$. First observe that, since $A^{-1} = H^{-1}A^*H$, we must have

$$\mathrm{tr}(A^{-1}) = \mathrm{tr}(A^*) = \overline{\mathrm{tr}(A)}.$$

Hence, if we denote the trace of A by $\mathrm{tr}(A) = \tau$, then $\mathrm{tr}(A^{-1}) = \overline{\tau}$. Putting this in the expression for the characteristic polynomial of $A \in \mathrm{SU}(2,1)$ from Lemma 4.5 below gives:

$$\mathrm{ch}_A(x) = x^3 - \mathrm{tr}(A)x^2 + \mathrm{tr}(A^{-1})x - 1 = x^3 - \tau x^2 + \overline{\tau}x - 1.$$

We want to find out when $A \in \mathrm{SU}(2,1)$ has repeated eigenvalues. In other words, we want to find conditions on τ for which $\mathrm{ch}_A(x) = 0$ has repeated solutions. This is true if and only if $\mathrm{ch}_A(x)$ and its derivative $\mathrm{ch}'_A(x)$ have a common root. Clearly

$$\mathrm{ch}'_A(x) = 3x^2 - 2\tau x + \overline{\tau}.$$

According to Lemma 3.3 on page 53 of Kirwan [10], two polynomials have a common root if and only if their resultant vanishes.

The *resultant* $R\big(p(x), q(x)\big)$ of two polynomials $p(x)$ and $q(x)$ of degree m and n, respectively, is the determinant of the $(m+n)\times(m+n)$ matrix defined as follows. Write the coefficients of $p(x)$ in the first row followed by $n-1$ zeros. In the next row the coefficients are displaced one place to the right, with one zero to the left and $n-2$ to the right. Continue in this fashion until the nth row is $n-1$ zeros followed by the coefficients of $p(x)$. For the last m rows we do the same thing with $p(x)$ and $q(x)$ interchanged. For details, see Definition 3.2 on page 52 of [10].

Since $\mathrm{ch}_A(x)$ and $\mathrm{ch}'_A(x)$ have degrees 3 and 2 respectively, the resultant is a 5×5 determinant. Applying the above recipe, we see that

$$R\big(\mathrm{ch}_A(x), \mathrm{ch}'_A(x)\big) = \det\begin{pmatrix} 1 & -\tau & \overline{\tau} & -1 & 0 \\ 0 & 1 & -\tau & \overline{\tau} & -1 \\ 3 & -2\tau & \overline{\tau} & 0 & 0 \\ 0 & 3 & -2\tau & \overline{\tau} & 0 \\ 0 & 0 & 3 & -2\tau & \overline{\tau} \end{pmatrix}.$$

Evaluating this determinant, we see that

$$R\big(\mathrm{ch}_A(x), \mathrm{ch}'_A(x)\big) = -|\tau|^4 + 8\Re(\tau^3) - 18|\tau|^2 + 27.$$

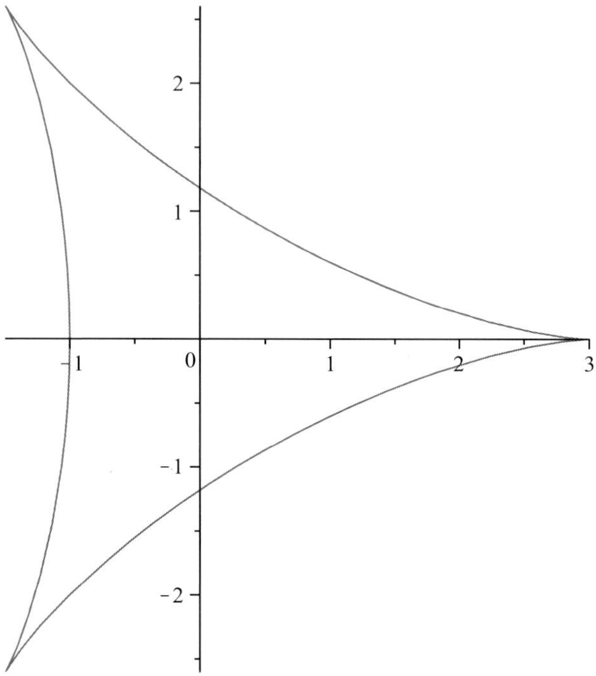

Fig. 1. *The deltoid given by $f(\tau) = 0$. The region with $f(\tau) < 0$ is inside and that with $f(\tau) > 0$ is outside.*

This leads to the following theorem of Goldman, Theorem 6.2.4 of [8]:

Theorem 3.1. *Let $f(\tau) = |\tau|^4 - 8\Re(\tau^3) + 18|\tau|^2 - 27$. Let $A \in \mathrm{SU}(2,1)$ then:*

(i) A has an eigenvalue λ with $|\lambda| \neq 1$ if and only if $f\big(\mathrm{tr}(A)\big) > 0$,
(ii) A has a repeated eigenvalue if and only if $f\big(\mathrm{tr}(A)\big) = 0$,
(iii) A has distinct eigenvalues of unit modulus if and only if $f\big(\mathrm{tr}(A)\big) < 0$.

The curve $f(\tau) = 0$ is a classical curve called a *deltoid*, see Chapter 8 of Lockwood [14] or page 26 of Kirwan [10] where it is written in terms of $x = \Re(\tau)$ and $y = \Im(\tau)$. The points outside correspond to case (i) in the theorem. This may be seen by considering A with eigenvalues r, r^{-1} and 1 where $r > 1$, which implies $\mathrm{tr}(A)$ lies in the interval $(3, \infty)$, and considering A with eigenvalues $e^{i\theta}$, $e^{-i\theta}$ and 1 whose trace lies in $(-1, 3)$. The rest follows by continuity.

It is easy to see that part (ii) of Theorem 3.1 follows from the construction of $f(\tau)$. We now look at the other cases separately.

Lemma 3.2. *Let $A \in \mathrm{SU}(2,1)$ and let λ be an eigenvalue of A. Then $\overline{\lambda}^{-1}$ is an eigenvalue of A.*

Proof. We know that A preserves the Hermitian form defined by H. Hence, $A^*HA = H$ and so $A = H^{-1}(A^*)^{-1}H$. Thus A has the same set of eigenvalues as $(A^*)^{-1}$ (they are conjugate). Since the characteristic polynomial of A^* is the complex conjugate of the characteristic polynomial of A, we see that if λ is an eigenvalue of A then $\overline{\lambda}$ is an eigenvalue of A^*. Therefore $\overline{\lambda}^{-1}$ is an eigenvalue of $(A^*)^{-1}$ and hence of A. □

Corollary 3.3. *Suppose $A \in \mathrm{SU}(2,1)$ has an eigenvalue λ with $|\lambda| \neq 1$. Then $\overline{\lambda}^{-1}$ is a distinct eigenvalue and the third eigenvalue is $\overline{\lambda}\lambda^{-1}$ of absolute value 1. Moreover, A is loxodromic.*

Proof. Using Lemma 3.2 we see $\overline{\lambda}^{-1}$ is an eigenvalue. Since $|\lambda| \neq 1$ it is not equal to λ. As the product of the eigenvalues is 1 we obtain the third eigenvalue.

If $\mathbf{v} \neq \mathbf{0}$ is an eigenvector corresponding to λ then

$$\langle \mathbf{v}, \mathbf{v} \rangle = \langle A\mathbf{v}, A\mathbf{v} \rangle = \langle \lambda\mathbf{v}, \lambda\mathbf{v} \rangle = |\lambda|^2 \langle \mathbf{v}, \mathbf{v} \rangle.$$

As $|\lambda| \neq 1$ we see that $\mathbf{v} \in V_0$ and $\mathbb{P}\mathbf{v}$ is a fixed point of A on $\partial\mathbf{H}^2_{\mathbb{C}}$. Similarly, a non-zero $\overline{\lambda}^{-1}$-eigenvector corresponds to a second fixed point on $\partial\mathbf{H}^2_{\mathbb{C}}$. Finally, a non-zero $\overline{\lambda}\lambda^{-1}$-eigenvector lies in V_+ and is a normal vector for the complex line through the two fixed points on $\partial\mathbf{H}^2_{\mathbb{C}}$. Hence A has precisely two fixed points on $\partial\mathbf{H}^2_{\mathbb{C}}$ and is loxodromic. □

Lemma 3.4. *Suppose that $A \in \mathrm{SU}(2,1)$ has an eigenvalue λ with $|\lambda| \neq 1$. Then $f\big(\mathrm{tr}(A)\big) > 0$.*

Proof. Suppose that $re^{i\theta}$ is an eigenvalue of A where r is positive and $r \neq 1$. Then by Corollary 3.3 we see that the other eigenvalues are $(\overline{re^{i\theta}})^{-1} = r^{-1}e^{i\theta}$ and $e^{-2i\theta}$. Therefore $\tau = re^{i\theta} + r^{-1}e^{i\theta} + e^{-2i\theta}$. Hence

$$\begin{aligned}|\tau|^2 &= (r + r^{-1})^2 + 2(r + r^{-1})\cos(3\theta) + 1,\\ \Re(\tau^3) &= (r + r^{-1})^3\cos(3\theta) + 3(r + r^{-1})^2 + 3(r + r^{-1})\cos(3\theta) + \cos(6\theta).\end{aligned}$$

From this it is easy to see that

$$f\big(re^{i\theta}+r^{-1}e^{i\theta}+e^{-2i\theta}\big)=(r-r^{-1})^2\big(r+r^{-1}-2\cos(3\theta)\big)^2>0. \qquad \square$$

Lemma 3.5. *Suppose that $A\in\mathrm{SU}(2,1)$ has three distinct eigenvalues, all of unit modulus. Then $f\big(\mathrm{tr}(A)\big)<0$.*

Proof. We write the eigenvalues as $e^{i\theta}$, $e^{i\phi}$, $e^{i\psi}$ where θ, ϕ and ψ are distinct and $e^{i\theta+i\phi+i\psi}=1$. Then $\tau=e^{i\theta}+e^{i\phi}+e^{i\psi}$ and

$$\begin{aligned}|\tau|^2&=3+2\cos(\theta-\phi)+2\cos(\phi-\psi)+2\cos(\psi-\theta),\\ \Re(\tau^3)&=\cos(3\theta)+\cos(3\phi)+\cos(3\psi)\\ &\qquad+6\cos(\theta-\phi)+6\cos(\phi-\psi)+6\cos(\psi-\theta)+6.\end{aligned}$$

But,

$$\begin{aligned}\cos(3\theta)&=\cos(2\theta-\phi-\psi)\\ &=\cos(\theta-\phi)\cos(\phi-\psi)+\sin(\theta-\phi)\sin(\phi-\psi).\end{aligned}$$

Hence

$$\begin{aligned}\Re(\tau^3)&=\cos(\theta-\phi)\cos(\phi-\psi)+\sin(\theta-\phi)\sin(\phi-\psi)\\ &\qquad+\cos(\phi-\psi)\cos(\psi-\theta)+\sin(\phi-\psi)\sin(\psi-\theta)\\ &\qquad+\cos(\psi-\theta)\cos(\theta-\phi)+\sin(\psi-\theta)\sin(\theta-\phi)\\ &\qquad+6\cos(\theta-\phi)+6\cos(\phi-\psi)+6\cos(\psi-\theta)+6.\end{aligned}$$

Using this we calculate

$$f\big(e^{i\theta}+e^{i\phi}+e^{i\psi}\big)=-4\Big(\sin(\theta-\phi)+\sin(\phi-\psi)+\sin(\phi-\psi)\Big)^2<0. \qquad \square$$

As these two lemmas exhaust all the possibilities when A has distinct eigenvalues, we have proved Theorem 3.1.

We will discuss loxodromic maps in the next two sections. We now briefly discuss elliptic maps. Suppose first that A has three distinct eigenvalues of unit modulus.

Proposition 3.6. *Suppose that $A\in\mathrm{SU}(2,1)$ has distinct eigenvalues $e^{i\theta}$, $e^{i\phi}$ and $e^{i\psi}$. Then A has a unique fixed point in $\mathbf{H}^2_{\mathbb{C}}$ corresponding to one of the eigenspaces. There are then three distinct conjugacy classes of elliptic maps with this trace.*

If the fixed point corresponds to the $e^{i\theta}$ eigenspace then A acts on the tangent space at this point by a unitary matrix with eigenvalues $e^{i\phi-i\theta}$ and $e^{i\psi-i\theta}$.

Proof. Since A is diagonalisable, we may consider its action on a basis of eigenvectors. The first part follows directly.

For the second part, we consider the action of $e^{-i\theta}A$ on $\mathbb{C}^{2,1}$ and restrict this to the tangent space. The result follows. $\square$

We now consider what happens if A has a repeated eigenvalue and so $\mathrm{tr}(A)$ lies on the deltoid. When all three eigenvalues are the same they must be a cube root of unity. These traces are the three vertices of the deltoid. Such maps are either parabolic or act as the identity on $\mathbf{H}^2_{\mathbb{C}}$. We will not discuss these maps here. We now consider the case where A has exactly two distinct eigenvalues.

Proposition 3.7. *Suppose that $A \in \mathrm{SU}(2,1)$ has two distinct eigenvalues, one of them repeated. Then the eigenvalues of A are $e^{i\psi}$, $e^{i\psi}$, $e^{-2i\psi}$ for some ψ with $3\psi \not\equiv 0$ mod 2π. Moreover, one of the following three possibilities arises:*

(i) A fixes a complex line L in $\mathbf{H}^2_{\mathbb{C}}$ and rotates a normal vector to L by -3ψ;
(ii) A fixes a point in $\mathbf{H}^2_{\mathbb{C}}$ and acts as $e^{3i\psi}I$ on the tangent space at this point;
(iii) A fixes a point on $\partial\mathbf{H}^2_{\mathbb{C}}$ and there is a complex line L with this point on its boundary so that A acts as a parabolic map on L and rotates a normal vector to L by -3ψ.

Proof. Suppose that A has a repeated eigenvalue λ. Since $\det(A) = 1$, it is clear that the third eigenvalue is λ^{-2}. Using Lemma 3.2 we see that $\{\lambda, \lambda, \lambda^{-2}\} = \{\overline{\lambda}^{-1}, \overline{\lambda}^{-1}, \overline{\lambda}^{2}\}$. This is a contradiction if $|\lambda| \neq 1$. Thus $|\lambda| = 1$ and the eigenvalues are $e^{i\psi}$, $e^{i\psi}$, $e^{-2i\psi}$ as claimed.

Now we discuss the possible conjugacy classes of A. Note that the $e^{-2i\psi}$-eigenspace of A is Hermitian orthogonal to the $e^{i\psi}$-eigenspace and so cannot be contained in V_0 (or else H would be singular). Suppose first that A is diagonalisable. First suppose that the $e^{-2i\psi}$-eigenspace is in V_+. Then the $e^{i\psi}$-eigenspace is indefinite, in that it contains vectors in V_-, V_0 and V_+. Its image under $\mathbb{P}$ is a complex line in $\mathbf{H}^2_{\mathbb{C}}$ fixed by A. Secondly, suppose that the $e^{-2i\psi}$-eigenspace is in V_- and so corresponds to an isolated fixed point of A in $\mathbf{H}^2_{\mathbb{C}}$. Then the $e^{i\psi}$-eigenspace is in V_+ and corresponds to the tangent space to $\mathbf{H}^2_{\mathbb{C}}$ at the fixed point. Thus A acts on this tangent space as $e^{3i\psi}I$.

Now suppose that A is not diagonalisable. Putting A into Jordan normal form, we see that there are non-zero vectors $\mathbf{v}$ and $\mathbf{u}$ so that $A\mathbf{v} = e^{i\psi}\mathbf{v}$ and $A\mathbf{u} = e^{i\psi}\mathbf{u} + \mathbf{v}$. Thus

$$\langle \mathbf{u}, \mathbf{v}\rangle = \langle A\mathbf{u}, A\mathbf{v}\rangle = \langle e^{i\psi}\mathbf{u} + \mathbf{v}, e^{i\psi}\mathbf{v}\rangle = \langle \mathbf{u}, \mathbf{v}\rangle + e^{-i\psi}\langle \mathbf{v}, \mathbf{v}\rangle.$$

Hence $\mathbf{v} \in V_0$ and corresponds to a fixed point of A on $\partial\mathbf{H}^2_{\mathbb{C}}$. The hyperplane spanned by $\mathbf{u}$ and $\mathbf{v}$ corresponds to a complex line L in $\mathbf{H}^2_{\mathbb{C}}$ and A acts on this line as a parabolic map with fixed point $\mathbb{P}\mathbf{v}$. The $e^{-2i\psi}$ eigenspace of A is spanned by a polar vector of A and so A acts on a normal vector to L as multiplication by $e^{-3i\psi}$. □

The maps described in Proposition 3.6 are called *regular elliptic*. Elliptic maps of the type given in Proposition 3.7 (i) are called *complex reflections in a line* and will be discussed in Section 5.2. Elliptic maps of the type given in Proposition 3.7 (ii) are called *complex reflections in a point*. The maps described in Proposition 3.7 (iii) are called *screw parabolic* or *ellipto-parabolic*.

3.3. *Traces and eigenvalues for loxodromic maps*

A loxodromic matrix A in $\mathrm{SL}(2,\mathbb{C})$ has eigenvalues λ and λ^{-1} where $|\lambda| > 1$. Furthermore writing $\mathrm{tr}(A) = \tau$ then $\tau = \lambda + \lambda^{-1}$. The map sending λ to τ is a conformal map from the exterior of the unit disc to the complex plane slit along the real axis from -2 to 2 (inclusive).

In this section we want to generalise this result to loxodromic maps in $\mathrm{SU}(2,1)$. The map from eigenvalues to traces is no longer holomorphic but, but we are able to show that it is a diffeomorphism from the exterior of the unit disc onto the set of points in $\mathbb{C}$ with $f(\tau) > 0$, that is the exterior of the deltoid, compare Theorem 3.1. If $\tau = \mathrm{tr}(A)$ then recall the discriminant function $f(\tau)$ of Theorem 3.1

$$f(\tau) = |\tau|^4 - 8\Re(\tau^3) + 18|\tau|^2 - 27. \tag{3.1}$$

Our main result is:

Proposition 3.8. *Let A be a loxodromic map in $\mathrm{SU}(2,1)$ with eigenvalue λ with $|\lambda| > 1$. Then the function Φ that gives the trace in terms of the eigenvalue*

$$\Phi : \{\lambda \in \mathbb{C} \ : \ |\lambda| > 1\} \longmapsto \{\tau \in \mathbb{C} \ : \ f(\tau) > 0\}$$

given by

$$\Phi(\lambda) = \tau = \lambda + \overline{\lambda}\lambda^{-1} + \overline{\lambda}^{-1}$$

is a diffeomorphism. Moreover, $\Phi(\omega\lambda) = \omega\Phi(\lambda)$, *where* ω *is a cube root of unity, and so this diffeomorphism is well defined for elements of* $\mathrm{PSU}(2,1)$.

Before we prove this result, we need to show that the map Φ is a surjection.

Lemma 3.9. *Suppose that* $\tau \in \mathbb{C}$ *satisfies* $f(\tau) > 0$ *then there exists* $\lambda \in \mathbb{C}$ *with* $|\lambda| > 1$ *so that* $\tau = \lambda + \overline{\lambda}\lambda^{-1} + \overline{\lambda}^{-1}$.

Proof. If we can find such a $\lambda = re^{i\theta}$ then, as in Lemma 3.4, we must find r and θ solving

$$|\tau|^2 = (r + r^{-1})^2 + 2(r + r^{-1})\cos(3\theta) + 1,$$
$$\Re(\tau^3) = (r + r^{-1})^3\cos(3\theta) + 3(r + r^{-1})^2 + 3(r + r^{-1})\cos(3\theta) + \cos(6\theta).$$

Eliminating $\cos(3\theta)$ from these equations, we must find $x = (r + r^{-1})^2 > 4$ solving $g(x) = 0$ where

$$g(x) = x^3 - \big(3 + |\tau|^2\big)x^2 + \big(3 + 2\Re(\tau^3) - |\tau|^2\big)x - \big(|\tau|^2 - 1\big)^2.$$

Moreover, since

$$\big((r + r^{-1}) - 1\big)^2 \le |\tau|^2 \le \big((r + r^{-1}) + 1\big)^2$$

such a solution must satisfy

$$\big(|\tau| - 1\big)^2 \le x = (r + r^{-1})^2 \le \big(|\tau| + 1\big)^2.$$

Note that since $f(\tau) > 0$ we must have $|\tau| > 1$ and so $\big(|\tau| + 1\big)^2 > 4$.

We now evaluate $g(x)$ at $x = 4$, $x = \big(|\tau| - 1\big)^2$ and $x = \big(|\tau| + 1\big)^2$:

$$g(4) = 27 - 18|\tau|^2 + 8\Re(\tau^3) - |\tau|^4 \;=\; -f(\tau) \;<\; 0,$$
$$g\big((|\tau| - 1)^2\big) = -2\big(|\tau| - 1\big)^2\big(|\tau^3| - \Re(\tau^3)\big) \;\le\; 0,$$
$$g\big((|\tau| + 1)^2\big) = 2\big(|\tau| + 1\big)^2\big(|\tau^3| + \Re(\tau^3)\big) \;\ge\; 0.$$

Therefore by the intermediate value theorem, there exists $x = x_0$ with $g(x_0) = 0$ so that

$$x_0 > 4, \quad x_0 \ge \big(|\tau| - 1\big)^2, \quad x_0 \le \big(|\tau| + 1\big)^2.$$

From this we can solve $(r + r^{-1})^2 = x_0$ to obtain

$$r = \frac{\sqrt{x_0} + \sqrt{x_0 - 4}}{2} > 1.$$

Substituting into the first of our equations, we obtain

$$\cos(3\theta) = \frac{|\tau|^2 - (r + r^{-1})^2 - 1}{2(r + r^{-1})} = \frac{|\tau|^2 - x_0 - 1}{2\sqrt{x_0}}.$$

The right hand side lies in $[-1, 1]$ by construction. So we can solve to find 3θ. Finally, by considering $\arg(\tau)$ we can solve for θ.

Writing $\lambda = re^{i\theta}$ gives the result. □

Proof of Proposition 3.8. Write

$$X = \{\lambda \in \mathbb{C} \ : \ |\lambda| > 1\}, \qquad Y = \{\tau \in \mathbb{C} \ : \ f(\tau) > 0\}.$$

From Lemma 3.4 we see that the image of X under Φ is contained in Y and by Lemma 3.9 we see that Φ maps X onto Y.

We calculate the Jacobian of $\tau(\lambda)$:

$$\begin{aligned} |J_\tau(\lambda)| &= \left|\frac{\partial \tau}{\partial \lambda}\right|^2 - \left|\frac{\partial \tau}{\partial \overline{\lambda}}\right|^2 \\ &= |1 - \overline{\lambda}\lambda^{-2}|^2 - |\lambda^{-1} - \overline{\lambda}^{-2}|^2 \\ &= \left(1 - |\lambda|^{-2}\right)\Big(1 - 2|\lambda|^{-1}\cos\big(3\arg(\lambda)\big) + |\lambda|^{-2}\Big). \end{aligned}$$

This is clearly different from 0 whenever $|\lambda| > 1$.

Therefore Φ is a local diffeomorphism from X onto Y. It is clear that, when $\lambda \in X$ then λ tends to infinity if and only if τ tends to infinity. Likewise, from the proof of Lemma 3.4, it is clear that Φ extends continuously to a map from the unit circle $\{\lambda \in \mathbb{C} \ : \ |\lambda| = 1\}$ to the set $\{\tau \in \mathbb{C} \ : \ f(\tau) = 0\}$. Hence Φ extends continuously to a map from $\overline{X}$ to $\overline{Y}$ and is therefore proper. Thus, by Ehresmann's fibration theorem we see that Φ is a locally trivial fibration (that is, when thought of as a map from an annulus to itself, it is a covering map). Because Φ is a bounded distance from the identity for large values of $|\lambda|$ we see that it has winding number 1 and so Φ is a global diffeomorphism. □

3.4. *Eigenvalues and complex displacement for loxodromic maps*

A loxodromic element A of $\mathrm{SL}(2, \mathbb{R})$ or $\mathrm{SU}(1, 1)$ with eigenvalues λ and λ^{-1} where $|\lambda| > 1$ corresponds to a hyperbolic isometry, which we also denote by A, in $\mathrm{PSL}(2,\mathbb{R})$ or $\mathrm{PU}(1,1)$, respectively. Since A is loxodromic, it has two fixed points on the boundary of the hyperbolic plane and these are the projections of the eigenspaces. The geodesic joining these two fixed points is called the *axis* of A, and is denoted $\tilde{\alpha}$. Then $\mathbf{H}^1_{\mathbb{C}}/\langle A\rangle$ is a hyperbolic cylinder (geometrically a catenoid) and $\alpha = \tilde{\alpha}/\langle A\rangle$ is the hyperbolic geodesic around its waist with hyperbolic length ℓ where

$$|\lambda| = e^{\ell/2}, \quad \big|\mathrm{tr}(A)\big| = 2\cosh\big(\ell/2\big).$$

In other words A translates along its axis by a hyperbolic translation length of ℓ. The ambiguity in the sign of $\mathrm{tr}(A)$ exactly corresponds to the choice of lift from $\mathrm{PSL}(2,\mathbb{R})$ to $\mathrm{SL}(2,\mathbb{R})$ or from $\mathrm{PSU}(1,1)$ to $\mathrm{SU}(1,1)$ respectively.

Similarly, when A is in $\mathrm{SL}(2,\mathbb{C})$ its trace corresponds to a complex length. More precisely, suppose $\mathrm{tr}(A) = \lambda + \lambda^{-1}$ where $|\lambda| > 1$. Then once again $|\lambda| = e^{\ell/2}$. To find the argument of λ, for any $z \in \tilde{\alpha}$, consider a tangent vector ξ in $T_z(\mathbf{H}^3)$ orthogonal to $\tilde{\alpha}$, the axis of A. Then A sends ξ in $T_z(\mathbf{H}^3)$ to a tangent vector $\xi e^{i\phi}$ in $T_{A(z)}(\mathbf{H}^3)$. In other words, A translates along $\tilde{\alpha}$ by a hyperbolic distance ℓ and rotates the tangent space by an angle ϕ. Then

$$\lambda = e^{\ell/2 + i\phi/2}, \quad \mathrm{tr}(A) = 2\cosh\big(\ell/2 + i\phi/2\big).$$

Since ϕ is defined mod 2π we see that the imaginary part of $\ell/2 + i\phi/2$ is defined mod π. This introduces an ambiguity of ± 1 in the trace and this corresponds exactly to the ambiguity introduced when lifting A from $\mathrm{PSL}(2,\mathbb{C})$ to $\mathrm{SL}(2,\mathbb{C})$; see Parker and Series [25].

In this section we illustrate how the geometric action of $A \in \mathrm{SU}(2,1)$ is recorded by $\mathrm{tr}(A)$. In principle the relationship is very similar to the case of $\mathrm{SL}(2,\mathbb{R})$ and $\mathrm{SL}(2,\mathbb{C})$ but the functions involved are more complicated. The main result of this section is:

Proposition 3.10. *Let $A \in \mathrm{SU}(2,1)$ be a loxodromic map with axis $\tilde{\alpha}$. Let $\lambda \in \mathbb{C}$ be the eigenvalue of A with $|\lambda| > 1$. Suppose that A has a Bergman translation length ℓ along $\tilde{\alpha}$ and rotates complex lines normal to $\tilde{\alpha}$ by an angle ϕ. Then*

$$\lambda = e^{\ell/2 - i\phi/3} \tag{3.2}$$

and

$$\mathrm{tr}(A) = 2\cosh\big(\ell/2\big)e^{-i\phi/3} + e^{2i\phi/3}. \tag{3.3}$$

Furthermore, since ϕ is defined mod 2π, the arguments of λ and τ are only given mod $2\pi/3$ and so these formulae are only well defined on $\mathrm{PU}(2,1)$.

Proof. It will be convenient to use the Hermitian form H_2 and to conjugate within $\mathrm{SU}(H_2)$ so that A is diagonal:

$$A = \begin{bmatrix} \lambda & 0 & 0 \\ 0 & \overline{\lambda}\lambda^{-1} & 0 \\ 0 & 0 & \overline{\lambda}^{-1} \end{bmatrix}.$$

The action of A on $\mathbf{H}^2_{\mathbb{C}}$ is given by

$$A(z_1, z_2) = \mathbb{P} \begin{bmatrix} \lambda & 0 & 0 \\ 0 & \overline{\lambda}\lambda^{-1} & 0 \\ 0 & 0 & \overline{\lambda}^{-1} \end{bmatrix} \begin{bmatrix} z_1 \\ z_2 \\ 1 \end{bmatrix} = \mathbb{P} \begin{bmatrix} \lambda z_1 \\ \overline{\lambda}\lambda^{-1} z_2 \\ \overline{\lambda}^{-1} \end{bmatrix} = \left(|\lambda|^2 z_1, \frac{\overline{\lambda}^3}{|\lambda|^2} z_2 \right).$$

The axis $\tilde{\alpha}$ of A is given by

$$\tilde{\alpha} = \big\{ (-x, 0) \in \mathbb{C}^2 \ : \ x > 0 \big\}.$$

Let $\mathbf{x}$ be the standard lift of $(-x, 0)$ in $\tilde{\alpha}$. Let ℓ be the Bergman translation length of A along its axis then

$$\begin{aligned} \cosh\big(\ell/2\big) &= \cosh\Big(\rho\big(A(-x,0), (-x,0)\big)/2\Big) \\ &= \left| \frac{\langle A\mathbf{x}, \mathbf{x} \rangle}{\langle \mathbf{x}, \mathbf{x} \rangle} \right| \\ &= \left| \frac{-\lambda x - \overline{\lambda}^{-1} x}{-2x} \right| \\ &= \big(|\lambda| + |\lambda|^{-1} \big)/2. \end{aligned}$$

Therefore, once again we have $|\lambda| = e^{\ell/2}$.

We now consider the argument of λ. The axis $\tilde{\alpha}$ is contained in a unique complex line, the *complex axis* $\tilde{\alpha}_{\mathbb{C}}$. With our normalisation,

$$\tilde{\alpha}_{\mathbb{C}} = \big\{ (z, 0) \in \mathbb{C}^2 \ : \ \Re(z) < 0 \big\}.$$

For any point $(-x, 0) \in \tilde{\alpha}$ let ξ be a tangent vector in $T_{(-x,0)}(\mathbf{H}^2_{\mathbb{C}})$ orthogonal to $\tilde{\alpha}_{\mathbb{C}}$. Since

$$A(z_1, z_2) = \big(|\lambda|^2 z_1, \overline{\lambda}^2 \lambda^{-1} z_2 \big),$$

we see that $\xi \in T_{(-x,0)}(\mathbf{H}^2_{\mathbb{C}})$ is sent to $\xi e^{i\phi}$ then

$$\phi = \arg\big(\overline{\lambda}^2 \lambda^{-1} \big) = -3 \arg(\lambda).$$

Hence $\arg(\lambda) = -\phi/3$. Thus we obtain (3.2).

Finally, since $\mathrm{tr}(A) = \lambda + \overline{\lambda}\lambda^{-1} + \overline{\lambda}^{-1}$, we obtain (3.3). □

Corollary 3.11. *Let A be as in Proposition 3.10. The function* (3.3) *relating ℓ and ϕ to* $\mathrm{tr}(A)$ *is a local diffeomorphism.*

Proof. Since it is clear that $\lambda = e^{\ell/2 - i\phi/3}$ is a local diffeomorphism the result follows by composing this map with the function relating λ and $\mathrm{tr}(A)$, and then using Proposition 3.8.

In fact it is jut as simple to calculate the Jacobian directly. Using (3.3), the real and imaginary parts of $\mathrm{tr}(A)$ are:

$$\begin{aligned}\mathrm{Re}\big(\mathrm{tr}(A)\big) &= 2\cosh(\ell/2)\cos(\phi/3) + \cos(2\phi/3),\\ \mathrm{Im}\big(\mathrm{tr}(A)\big) &= -2\cosh(\ell/2)\sin(\phi/3) + \sin(2\phi/3).\end{aligned}$$

Therefore:

$$\begin{aligned}\big|J_\tau(\ell,\phi)\big| &= \det\begin{pmatrix}\sinh(\frac{\ell}{2})\cos(\frac{\phi}{3}) & -\frac{2}{3}\cosh(\frac{\ell}{2})\sin(\frac{\phi}{3}) - \frac{2}{3}\sin(\frac{2\phi}{3})\\ -\sinh(\frac{\ell}{2})\sin(\frac{\phi}{3}) & -\frac{2}{3}\cosh(\frac{\ell}{2})\cos(\frac{\phi}{3}) + \frac{2}{3}\cos(\frac{2\phi}{3})\end{pmatrix}\\ &= -(2/3)\sinh(\ell/2)\big(\cosh(\ell/2) - \cos(\phi)\big).\end{aligned}$$

This is clearly non-zero when $\ell > 0$. □

4. Two Generator Groups and Fenchel-Nielsen Coordinates

4.1. *Introduction*

There is a long tradition of studying subgroups of $\mathrm{SL}(2,\mathbb{C})$ by relating the traces of group elements to their geometry. This goes back to work of Vogt and Fricke who showed that a non-elementary two generator subgroup of $\mathrm{SL}(2,\mathbb{C})$ is determined up to conjugation by the traces of the generators and their product. For a precise statement of this result see Theorem A of Goldman [9]. One goal of this section is to extend this result to two generator subgroups of $\mathrm{SU}(2,1)$ and our treatment follows work of Lawton [13] and Will [33], [34]. The method we use begins by discussing trace relations in $\mathrm{M}(3,\mathbb{C})$, then specialising to $\mathrm{SL}(3,\mathbb{C})$ before finally giving our results for $\mathrm{SU}(2,1)$.

We are also interested in the geometry of two generator subgroups of $\mathrm{SU}(2,1)$. In this section we concentrate on the case where the generators and their product are all loxodromic. The fundamental group of a three-holed sphere is a free group on two generators. The generators and their product correspond to the three boundary components. Since we require that these three elements are loxodromic, we can use the results of section 3.4 to give geometric information about the corresponding three-holed sphere. As an application, we discuss how to generalise Fenchel-Nielsen coordinates to complex hyperbolic representations of surface groups. This follows work of Parker and Platis [23].

4.2. *Trace identities in* $\mathbf{M(3,\mathbb{C})}$

In this section we derive some trace identities for 3×3 matrices. This follows Lawton [13] and Will [33], [34]. The first lemma follows by writing $\mathrm{tr}(A)$, $\mathrm{tr}(A^2)$ and $\mathrm{tr}(A^3)$ as homogeneous polynomials in the eigenvalues of A and then solving for the coefficients of the characteristic polynomial.

Lemma 4.1. *Let* $A \in \mathrm{M}(3,\mathbb{C})$. *Then the characteristic polynomial* $\mathrm{ch}_A(x)$ *of* A *is*

$$x^3 - \mathrm{tr}(A)x^2 + \frac{\mathrm{tr}(A)^2 - \mathrm{tr}(A^2)}{2}x - \frac{\mathrm{tr}(A)^3 - 3\mathrm{tr}(A)\mathrm{tr}(A^2) + 2\mathrm{tr}(A^3)}{6}.$$

For any $A \in \mathrm{M}(3,\mathbb{C})$ define $\mathrm{ch}(A)$ to be the following matrix (here I is the 3×3 identity matrix):

$$\begin{aligned}\mathrm{ch}(A) = A^3 - \mathrm{tr}(A)A^2 &+ \tfrac{1}{2}\big(\mathrm{tr}(A)^2 - \mathrm{tr}(A^2)\big)A \\ &-\tfrac{1}{6}\big(\mathrm{tr}(A)^3 - 3\mathrm{tr}(A)\mathrm{tr}(A^2) + 2\mathrm{tr}(A^3)\big)I.\end{aligned} \tag{4.1}$$

Then the Cayley-Hamilton theorem states that $\mathrm{ch}(A) = O$, the 3×3 zero matrix. We use a process known as trilinearisation on this identity to obtain the following:

Proposition 4.2. *Let* A, B, $C \in \mathrm{M}(3,\mathbb{C})$. *Then*

$$\begin{aligned} O = {} & ABC + ACB + BAC + BCA + CAB + CBA \\ & - \mathrm{tr}(A)\big(BC + CB\big) - \mathrm{tr}(B)\big(AC + CA\big) - \mathrm{tr}(C)\big(AB + BA\big) \\ & + \big(\mathrm{tr}(B)\mathrm{tr}(C) - \mathrm{tr}(BC)\big)A + \big(\mathrm{tr}(A)\mathrm{tr}(C) - \mathrm{tr}(AC)\big)B \\ & + \big(\mathrm{tr}(A)\mathrm{tr}(B) - \mathrm{tr}(AB)\big)C \\ & - \big(\mathrm{tr}(A)\mathrm{tr}(B)\mathrm{tr}(C) + \mathrm{tr}(ABC) + \mathrm{tr}(CBA)\big)I \\ & + \big(\mathrm{tr}(A)\mathrm{tr}(BC) + \mathrm{tr}(B)\mathrm{tr}(AC) + \mathrm{tr}(C)\mathrm{tr}(AB)\big)I. \end{aligned}$$

Proof. Using the Cayley-Hamilton theorem, as indicated above, for any A, B, $C \in \mathrm{M}(3,\mathbb{C})$ we have

$$\begin{aligned} O = {} & \mathrm{ch}(A+B+C) - \mathrm{ch}(A+B) - \mathrm{ch}(B+C) - \mathrm{ch}(A+C) \\ & + \mathrm{ch}(A) + \mathrm{ch}(B) + \mathrm{ch}(C). \end{aligned}$$

To obtain the result, we expand this expression and simplify, using $\mathrm{tr}(A+B) = \mathrm{tr}(A) + \mathrm{tr}(B)$ and $\mathrm{tr}(AB) = \mathrm{tr}(BA)$. □

Corollary 4.3. *For any A, $B \in \mathrm{M}(3,\mathbb{C})$ we have:*

$$\begin{aligned} O = {} & ABA^{-1} + B + A^{-1}BA \\ & - \mathrm{tr}(A)\big(BA^{-1} + A^{-1}B\big) - \mathrm{tr}(A^{-1})\big(AB + BA\big) + \mathrm{tr}(A)\mathrm{tr}(A^{-1})B \\ & + \big(\mathrm{tr}(B)\mathrm{tr}(A^{-1}) - \mathrm{tr}(BA^{-1})\big)A + \big(\mathrm{tr}(A)\mathrm{tr}(B) - \mathrm{tr}(AB)\big)A^{-1} \\ & - \big(\mathrm{tr}(A)\mathrm{tr}(B)\mathrm{tr}(A^{-1}) + \mathrm{tr}(B) - \mathrm{tr}(A)\mathrm{tr}(BA^{-1}) - \mathrm{tr}(A^{-1})\mathrm{tr}(AB)\big)I; \end{aligned}$$

$$\begin{aligned} O = {} & ABA + A^2B + BA^2 - \mathrm{tr}(A)\big(AB + BA\big) - \frac{1}{2}\mathrm{tr}(B)A^2 \\ & + \big(\mathrm{tr}(A)\mathrm{tr}(B) - \mathrm{tr}(AB)\big)A + \frac{1}{2}\big(\mathrm{tr}(A)^2 - \mathrm{tr}(A^2)\big)B \\ & - \frac{1}{2}\big(\mathrm{tr}(A)^2 - \mathrm{tr}(A^2)\big)\mathrm{tr}(B)I + \big(\mathrm{tr}(A)\mathrm{tr}(AB) - \mathrm{tr}(A^2B)\big)I. \end{aligned}$$

Proof. For the first identity put $C = A^{-1}$ into the expression from Proposition 4.2 and use $\mathrm{tr}(I) = 3$. For the second put $C = A$ into Proposition 4.2 and divide by 2. □

Corollary 4.4. *For any A, $B \in \mathrm{M}(3,\mathbb{C})$ we have*

$$\begin{aligned} & \mathrm{tr}[A,B] + \mathrm{tr}[A^{-1},B] \\ & = \mathrm{tr}(A)\mathrm{tr}(A^{-1}) + \mathrm{tr}(B)\mathrm{tr}(B^{-1}) + \mathrm{tr}(A)\mathrm{tr}(A^{-1})\mathrm{tr}(B)\mathrm{tr}(B^{-1}) - 3 \\ & \quad + \mathrm{tr}(AB)\mathrm{tr}(A^{-1}B^{-1}) - \mathrm{tr}(A)\mathrm{tr}(B)\mathrm{tr}(A^{-1}B^{-1}) \\ & \quad - \mathrm{tr}(A^{-1})\mathrm{tr}(B^{-1})\mathrm{tr}(AB) + \mathrm{tr}(A^{-1}B)\mathrm{tr}(AB^{-1}) \\ & \quad - \mathrm{tr}(A^{-1})\mathrm{tr}(B)\mathrm{tr}(AB^{-1}) - \mathrm{tr}(A)\mathrm{tr}(B^{-1})\mathrm{tr}(A^{-1}B). \end{aligned}$$

Proof. Multiplying the first expression from Corollary 4.3 on the right by B^{-1} gives

$$\begin{aligned} O = {} & ABA^{-1}B^{-1} + I + A^{-1}BAB^{-1} \\ & - \mathrm{tr}(A)\big(BA^{-1}B^{-1} + A^{-1}\big) - \mathrm{tr}(A^{-1})\big(A + BAB^{-1}\big) \\ & + \mathrm{tr}(A)\mathrm{tr}(A^{-1})I + \big(\mathrm{tr}(B)\mathrm{tr}(A^{-1}) - \mathrm{tr}(BA^{-1})\big)AB^{-1} \\ & + \big(\mathrm{tr}(A)\mathrm{tr}(B) - \mathrm{tr}(AB)\big)A^{-1}B^{-1} \\ & - \big(\mathrm{tr}(A)\mathrm{tr}(B)\mathrm{tr}(A^{-1}) + \mathrm{tr}(B)\big)B^{-1} \\ & + \big((\mathrm{tr}(A)\mathrm{tr}(BA^{-1})\mathrm{tr}(A^{-1})\mathrm{tr}(AB)\big)B^{-1}. \end{aligned}$$

Taking traces gives the result. □

4.3. *Trace identities in* $\mathbf{SL(3,\mathbb{C})}$

When A is in $\mathrm{SL}(3,\mathbb{C})$ its characteristic polynomial may be written in a form that is somewhat simpler the form given in Lemma 4.1.

Lemma 4.5. *Let* $A \in \mathrm{SL}(3,\mathbb{C})$. *The characteristic polynomial of* A *is*

$$\mathrm{ch}_A(x) = x^3 - \mathrm{tr}(A)x^2 + \mathrm{tr}(A^{-1})x - 1.$$

Proof. Let λ_1, λ_2, λ_3 be the eigenvalues of A. Then $\lambda_1\lambda_2\lambda_3 = \det(A) = 1$. This gives the constant term in $\mathrm{ch}_A(x)$. Observe that λ_1^{-1}, λ_2^{-1}, λ_3^{-1} are the eigenvalues of A^{-1}. Thus, using both of these facts, we see that the linear term in $\mathrm{ch}_A(x)$ is

$$\lambda_2\lambda_3 + \lambda_1\lambda_3 + \lambda_1\lambda_2 = \lambda_1^{-1} + \lambda_2^{-1} + \lambda_3^{-1} = \mathrm{tr}(A^{-1}).$$ □

Again, using the Cayley-Hamilton theorem, we see that for $A \in \mathrm{SL}(3,\mathbb{R})$ we have

$$O = A^3 - \mathrm{tr}(A)A^2 + \mathrm{tr}(A^{-1})A - I. \tag{4.2}$$

Lemma 4.6. *Let* $A \in \mathrm{SU}(2,1)$. *Then*

(i) $\mathrm{tr}(A^2) = \big(\mathrm{tr}(A)\big)^2 - 2\mathrm{tr}(A^{-1})$;
(ii) $\mathrm{tr}(A^3) = \big(\mathrm{tr}(A)\big)^3 - 3\mathrm{tr}(A)\mathrm{tr}(A^{-1}) + 3$.

Proof. Multiplying equation (4.2) by A^{-1} gives:

$$A^2 = \mathrm{tr}(A)A - \mathrm{tr}(A^{-1})I + A^{-1}.$$

Taking traces we see that

$$\mathrm{tr}(A^2) = \mathrm{tr}(A)\mathrm{tr}(A) - 3\mathrm{tr}(A^{-1}) + \mathrm{tr}(A^{-1}) = \big(\mathrm{tr}(A)\big)^2 - 2\mathrm{tr}(A^{-1}),$$

which gives part (i).

Taking traces in equation (4.2) and then substituting for $\mathrm{tr}(A^2)$ gives

$$\begin{aligned}\mathrm{tr}(A^3) &= \mathrm{tr}(A)\mathrm{tr}(A^2) - \mathrm{tr}(A^{-1})\mathrm{tr}(A) + 3\\ &= \mathrm{tr}(A)\Big(\big(\mathrm{tr}(A)\big)^2 - 2\mathrm{tr}(A^{-1})\Big) - \mathrm{tr}(A^{-1})\mathrm{tr}(A) + 3\\ &= \big(\mathrm{tr}(A)\big)^3 - 3\mathrm{tr}(A)\mathrm{tr}(A^{-1}) + 3,\end{aligned}$$

which gives part (ii). □

Proposition 4.7. *Let* A, $B \in \mathrm{SL}(3,\mathbb{C})$. *Then* $\mathrm{tr}[A,B]\mathrm{tr}[B,A]$ *may be expressed as a polynomial function of the traces of* A, B, AB, $A^{-1}B$ *and their inverses.*

The precise polynomial is given by Lawton in equation (18) of [13] or by Will on page 58 of [33]. In the case of $A \in \mathrm{SU}(2,1)$ we give a version of this polynomial in Proposition 4.10 below.

Sketch Proof. This follows Will [33]; Lawton [13] has a different proof.

Write $A = MN$ and $B = NM$ in the expression in Corollary 4.4. This gives

$$\begin{aligned}
&\mathrm{tr}[MN, NM] + \mathrm{tr}[NM, MN] \\
&= 2\mathrm{tr}(MN)\mathrm{tr}(M^{-1}N^{-1}) + \mathrm{tr}(MN)^2\mathrm{tr}(M^{-1}N^{-1})^2 - 3 \\
&\quad + \mathrm{tr}(M^2N^2)\mathrm{tr}(M^{-2}N^{-2}) - \mathrm{tr}(MN)^2\mathrm{tr}(M^{-2}N^{-2}) \\
&\quad - \mathrm{tr}(N^{-1}M^{-1})^2\mathrm{tr}(M^2N^2) + \mathrm{tr}[M,N]\mathrm{tr}[N,M] \\
&\quad - \mathrm{tr}(MN)\mathrm{tr}(M^{-1}N^{-1})\big(\mathrm{tr}[M,N] + \mathrm{tr}[M^{-1},N]\big).
\end{aligned}$$

Using Corollary 4.4 we can express $\mathrm{tr}[M,N] + \mathrm{tr}[M^{-1},N]$ in terms of the traces of M, N, MN, $M^{-1}N$ and their inverses.

If M and N are in $\mathrm{SL}(3,\mathbb{C})$ we can use their characteristic polynomials to write

$$M^2 = \mathrm{tr}(M)M - \mathrm{tr}(M^{-1})I + M^{-1}, \quad N^2 = \mathrm{tr}(N)N - \mathrm{tr}(N^{-1})I + N^{-1}.$$

Hence

$$M^2N^2 = \big(\mathrm{tr}(M)M - \mathrm{tr}(M^{-1})I + M^{-1}\big)\big(\mathrm{tr}(N)N - \mathrm{tr}(N^{-1})I + N^{-1}\big).$$

From this we can express the trace of M^2N^2 in terms of the traces of M, N, MN, MN^{-1} and their inverses. Likewise we can express the traces of M^2N^{-2}, $M^{-2}N^2$ and $M^{-2}N^{-2}$ in terms of these traces.

Thus it suffices to express the trace of $[MN, NM]$ and $[NM, MN]$ in terms of these other traces. To do so, first write

$$[MN, NM] = MN^2MN^{-1}M^{-2}N^{-1}$$

and substitute for N^2 and M^{-2} as above. Then use Corollary 4.3 to substitute for words such as MNM, MNM^{-1}. At each stage, we write the words in terms of shorter words in the group. Eventually we obtain the desired expression. □

4.4. *Trace parameters for two generator groups of* SU(2, 1)

Let Y be a three holed sphere (also known as a "pair of pants"). Let $\pi_1 = \pi_1(Y)$ be the fundamental group of Y. If $[\alpha]$, $[\beta]$ and $[\gamma]$ are the homotopy classes in π_1 representing the boundary curves then $\alpha\beta\gamma = Id$. In fact, π_1 is a free group generated by any two of $[\alpha]$, $[\beta]$, $[\gamma]$.

It is well known that for $\mathrm{SL}(2,\mathbb{R})$ or $\mathrm{SU}(1,1)$ (the holomorphic, hyperbolic isometry groups of the upper plane and Poincaré disc respectively) then the group generated by A and B is completely determined up to conjugacy by $\mathrm{tr}(A)$, $\mathrm{tr}(B)$ and $\mathrm{tr}(AB)$. Geometrically, under mild hypotheses, $\langle A, B\rangle$ corresponds to a representation ρ_0 of π_1 which gives Y a hyperbolic metric. The mild hypotheses are that $\langle A, B\rangle$ should be discrete, faithful (so free), totally loxodromic and that the axes of A, B and AB should bound a common region in the hyperbolic plane. We may choose geodesic representatives for the three boundary homotopy classes, and we denote these geodesics by α, β and γ respectively. Then by construction we have $\rho(\alpha) = A$, $\rho(\beta) = B$ and $\rho(\gamma) = C = B^{-1}A^{-1}$ with lengths given by

$$\begin{aligned} \big|\mathrm{tr}(A)\big| &= 2\cosh\big(\ell(\alpha)/2\big), \\ \big|\mathrm{tr}(B)\big| &= 2\cosh\big(\ell(\beta)/2\big), \\ \big|\mathrm{tr}(C)\big| &= 2\cosh\big(\ell(\gamma)/2\big). \end{aligned}$$

In fact, our mild hypotheses about the axes of A, B and C imply that

$$\mathrm{tr}(A)\mathrm{tr}(B)\mathrm{tr}(C) < 0$$

and so we may choose a lift from $\mathrm{PSL}(2,\mathbb{R})$ to $\mathrm{SL}(2,\mathbb{R})$ where all three traces are negative. Conversely, given $\ell(\alpha)$, $\ell(\beta)$, $\ell(\gamma)$ in $\mathbb{R}_+$ we can construct a hyperbolic metric on Y whose boundary geodesics have these lengths. This in turn gives rise to a group $\langle A, B\rangle$ satisfying $\big|\mathrm{tr}(A)\big| = 2\cosh\big(\ell(\alpha)/2\big)$ et cetera.

Similarly, if $\langle A, B\rangle$ is a discrete, free, geometrically finite and totally loxodromic subgroup of $\mathrm{SL}(2,\mathbb{C})$ then we have a similar picture, but the lengths of the boundary curves are now complex, as discussed in the introduction to Section 3.4. The main difference here is that, not all triples of complex lengths give rise to a discrete, free, totally loxodromic, geometrically finite group.

We now want to play a similar game using complex hyperbolic representations of $\pi_1(Y)$. Again the representations we will be interested in will be discrete, free, totally loxodromic and geometrically finite; for a discussion of the latter see Bowditch [1]. We will also add the hypothesis that $\langle A, B\rangle$ is Zariski dense. A subgroup of $\mathrm{PSU}(2,1)$ is Zariski dense if and only if its action on $\mathbb{CP}^2$ does not have a global fixed point (see Remark 10 of Will [34]). Equivalently, it does not fix a point on $\overline{\mathbf{H}}^2_{\mathbb{C}}$ or preserve a complex line in $\mathbf{H}^2_{\mathbb{C}}$. Consider $\rho : \pi_1(Y) \longrightarrow \mathrm{SU}(2,1)$. Then ρ is irreducible if and only if its image is Zariski dense.

The main question is what are the data we need to completely determine $\langle A, B\rangle$ up to conjugation. Our first observation is that $\mathrm{SU}(2,1)$ has complex dimension four and so we do not expect to be able to determine $\langle A, B\rangle$ using only three complex numbers. The following theorem is a special case of a result of Wen [32], see also Lemma 5 of Lawton [13] or Corollary 3.10 of Will [33], Proposition 9 of [34]:

Theorem 4.8. *Suppose that* A, $B \in \mathrm{SU}(2,1)$ *and that* $\langle A, B\rangle$ *is Zariski dense. Then* $\langle A, B\rangle$ *is determined up to conjugation within* $\mathrm{SU}(2,1)$ *by*

$$\mathrm{tr}(A), \quad \mathrm{tr}(B), \quad \mathrm{tr}(AB), \quad \mathrm{tr}(A^{-1}B), \quad \mathrm{tr}[A,B].$$

We remark that Wen's theorem refers to A and B in $\mathrm{SL}(3,\mathbb{C})$ and also requires the traces of A^{-1}, B^{-1}, $A^{-1}B^{-1}$ and AB^{-1}. Also, this theorem has a similar flavour to the the theorem of Okumura [18] and Schmutz [29]. Namely, one would expect to only need to use four traces to describe $\langle A, B\rangle$. In fact ones needs an extra one, $\mathrm{tr}[A,B]$, and this satisfies relations with the other traces.

In what follows we want to consider A, B, $C \in \mathrm{SU}(2,1)$ with $ABC = I$. It is clear that $\mathrm{tr}(AB) = \mathrm{tr}(C^{-1}) = \overline{\mathrm{tr}(C)}$. We want to express the other parameters in a way that is symmetrical with respect to cyclic permutations of A, B and C. First we consider the trace of $A^{-1}B$.

Lemma 4.9. *Let* A, B, C *be elements of* $\mathrm{SU}(2,1)$ *so that* $ABC = I$. *Then*

$$\begin{aligned}\mathrm{tr}(A^{-1}B) - \mathrm{tr}(A^{-1})\mathrm{tr}(B) &= \mathrm{tr}(B^{-1}C) - \mathrm{tr}(B^{-1})\mathrm{tr}(C)\\ &= \mathrm{tr}(C^{-1}A) - \mathrm{tr}(C^{-1})\mathrm{tr}(A).\end{aligned}$$

Proof. We have already seen that

$$A^3 - \mathrm{tr}(A)A^2 + \mathrm{tr}(A^{-1})A - I = O.$$

Multiplying on the right by $A^{-1}B$ gives

$$A^2B - \mathrm{tr}(A)AB = A^{-1}B - \mathrm{tr}(A^{-1})B.$$

Taking traces and using $AB = C^{-1}$ gives

$$\mathrm{tr}(C^{-1}A) - \mathrm{tr}(C^{-1})\mathrm{tr}(A) = \mathrm{tr}(A^{-1}B) - \mathrm{tr}(A^{-1})\mathrm{tr}(B).$$

This shows equality between the first and third expressions. Cyclically permuting A, B and C gives the second as well. □

Therefore by using $\mathrm{tr}(A^{-1}B) - \mathrm{tr}(A^{-1})\mathrm{tr}(B)$ instead of $\mathrm{tr}(A^{-1}B)$ we can give symmetric parameters. Furthermore, trivially we have

$$\mathrm{tr}[A,B] = \mathrm{tr}[B,C] = \mathrm{tr}[C,A] = \overline{\mathrm{tr}[B,A]} = \overline{\mathrm{tr}[C,B]} = \overline{\mathrm{tr}[A,C]}.$$

As we have seen above in Corollary 4.4 and Proposition 4.7, the real part and absolute value of $\mathrm{tr}[A,B]$ are determined by the other parameters. We now make this explicit.

Proposition 4.10. *Let A, B, $C \in \mathrm{SU}(2,1)$ with $ABC = I$. Let*

$$a = \mathrm{tr}(A), \quad b = \mathrm{tr}(B), \quad c = \mathrm{tr}(C), \quad d = \mathrm{tr}(A^{-1}B) - \mathrm{tr}(A^{-1})\mathrm{tr}(B).$$

Then

$$2\Re\big(\mathrm{tr}[A,B]\big) = |a|^2 + |b|^2 + |c|^2 + |d|^2 - abc - \overline{a}\overline{b}\overline{c} - 3$$

and

$$\begin{aligned}
&\big|\mathrm{tr}[A,B]\big|^2 \\
&= |a|^2|b|^2|c|^2 + a^2b^2\overline{c} + \overline{a}^2\overline{b}^2c + a^2\overline{b}c^2 + \overline{a}^2b\overline{c}^2 + \overline{a}b^2c^2 + a\overline{b}^2\overline{c}^2 \\
&\quad +|a|^2|b|^2 + |b|^2|c|^2 + |a|^2|c|^2 \\
&\quad -2ab\overline{c}^2 - 2\overline{a}\overline{b}c^2 - 2a\overline{b}^2c - 2\overline{a}b^2\overline{c} - 2\overline{a}^2bc - 2a^2\overline{b}\overline{c} \\
&\quad +a^3 + \overline{a}^3 + b^3 + \overline{b}^3 + c^3 + \overline{c}^3 + 3abc + 3\overline{a}\overline{b}\overline{c} - 6|a|^2 - 6|b|^2 - 6|c|^2 \\
&\quad +d\big(|a|^2b\overline{c} + a\overline{b}|c|^2 + \overline{a}|b|^2c + ab^2 + \overline{a}^2\overline{b} + a^2c + \overline{a}\,\overline{c}^2 + bc^2 + \overline{b}^2\overline{c}\big) \\
&\quad +\overline{d}\big(|a|^2\overline{b}c + \overline{a}b|c|^2 + a|b|^2\overline{c} + \overline{a}\overline{b}^2 + a^2b + \overline{a}^2\overline{c} + ac^2 + \overline{b}\overline{c}^2 + b^2c\big) \\
&\quad +(d^2 - 3\overline{d})\big(\overline{a}b + \overline{b}c + a\overline{c}\big) + (\overline{d}^2 - 3d)\big(a\overline{b} + b\overline{c} + \overline{a}c\big) \\
&\quad +|d|^2\big(|a|^2 + |b|^2 + |c|^2 - 6\big) + d^3 + \overline{d}^3 + 9.
\end{aligned}$$

Proof. Using $\mathrm{tr}(A^{-1}) = \overline{\mathrm{tr}(A)} = \overline{a}$ et cetera and also $\mathrm{tr}(A^{-1}B) = d + \overline{a}b$ in the expression of Corollary 4.4 gives:

$$\begin{aligned}
2\Re\big(\mathrm{tr}[A,B]\big) &= |a|^2|b|^2 + |a|^2 + |b|^2 + |c|^2 + |d + \overline{a}b|^2 \\
&\qquad -\overline{a}b(\overline{d} + a\overline{b}) - a\overline{b}(d + \overline{a}b) - abc - \overline{a}\overline{b}\overline{c} - 3 \\
&= |a|^2 + |b|^2 + |c|^2 + |d|^2 - abc - \overline{a}\overline{b}\overline{c} - 3.
\end{aligned}$$

Similarly, using the expression for $\big|\mathrm{tr}[A,B]\big|^2$ given by Lawton in equation (18) of [13] or by Will on page 58 of [33] (see Proposition 4.7) gives the second expression. □

Putting this together gives:

Proposition 4.11. *Let A, B, C be elements of* $\mathrm{SU}(2,1)$ *with* $ABC = I$. *Then, if* $\langle A, B, C\rangle$ *is Zariski dense, it is determined up to conjugacy by*

$$\mathrm{tr}(A), \quad \mathrm{tr}(B), \quad \mathrm{tr}(C), \quad \mathrm{tr}(A^{-1}B) - \mathrm{tr}(A^{-1})\mathrm{tr}(B), \quad \mathrm{tr}[A, B].$$

Also, the last two of these expressions remain unchanged under cyclic permutations of A, B and C.

Moreover, the group is determined by $\mathrm{tr}(A)$, $\mathrm{tr}(B)$, $\mathrm{tr}(C)$ *and* $\mathrm{tr}(A^{-1}B) - \mathrm{tr}(A^{-1})\mathrm{tr}(B)$ *together with the sign of the imaginary part of* $\mathrm{tr}[A, B]$.

Example 4.12. We now give an example that shows the traces of A, B, AB and $A^{-1}B$ do not determine the imaginary part of $\mathrm{tr}[A, B]$.

For $\theta \in (-\pi/2, \pi/2)$ let $Q(\theta) \in \mathrm{SU}(2,1)$ be the matrix

$$Q(\theta) = \frac{-e^{-i\theta/6}}{2\cos(\theta/2)} \begin{bmatrix} 1 & \sqrt{2\cos(\theta)} & -e^{i\theta} \\ \sqrt{2\cos(\theta)} & e^{-i\theta} - 1 & \sqrt{2\cos(\theta)} \\ -e^{i\theta} & \sqrt{2\cos(\theta)} & 1 \end{bmatrix}.$$

Note that $Q(\theta)^{-1} = Q(-\theta)$. For $r > 1$ and $s > 1$, define A, $B_\theta \in \mathrm{SU}(2,1)$ by

$$A = \begin{bmatrix} re^{i\phi} & 0 & 0 \\ 0 & e^{-2i\phi} & 0 \\ 0 & 0 & r^{-1}e^{i\phi} \end{bmatrix}, \quad B_\theta = Q(\theta) \begin{bmatrix} se^{i\psi} & 0 & 0 \\ 0 & e^{-2i\psi} & 0 \\ 0 & 0 & s^{-1}e^{i\psi} \end{bmatrix} Q(-\theta).$$

Then we have

$$\begin{aligned}
\mathrm{tr}(A) &= (r + r^{-1})e^{i\phi} + e^{-2i\phi}, \\
\mathrm{tr}(B_\theta) &= (s + s^{-1})e^{i\psi} + e^{-2i\psi}, \\
\mathrm{tr}(AB_\theta) &= \frac{1}{2 + 2\cos(\theta)} \Big((r + r^{-1})e^{i\phi} + 2\cos(\theta)e^{-2i\phi} \Big) \\
&\quad \cdot \Big((s + s^{-1})e^{i\psi} + 2\cos(\theta)e^{-2i\psi} \Big) \\
&\quad - \frac{1}{2 + 2\cos(\theta)} e^{-2i\phi - 2i\psi} \Big(2\cos(\theta) + 2\cos(2\theta) \Big), \\
\mathrm{tr}(A^{-1}B_\theta) &= \frac{1}{2 + 2\cos(\theta)} \Big((r + r^{-1})e^{-i\phi} + 2\cos(\theta)e^{2i\phi} \Big) \\
&\quad \cdot \Big((s + s^{-1})e^{i\psi} + 2\cos(\theta)e^{-2i\psi} \Big) \\
&\quad - \frac{1}{2 + 2\cos(\theta)} e^{2i\phi - 2i\psi} \Big(2\cos(\theta) + 2\cos(2\theta) \Big),
\end{aligned}$$

$$\begin{aligned}
&\mathrm{tr}[A, B_\theta] \\
&= 3 - \frac{1}{(2+2\cos(\theta))^2}(r-r^{-1})^2(s-s^{-1})^2 \\
&\quad - \frac{8\cos(\theta)}{(2+2\cos(\theta))^2}\Big(2-(r+r^{-1})\cos(3\phi)\Big)\Big(2-(s+s^{-1})\cos(3\psi)\Big) \\
&\quad + \frac{2\cos(\theta)}{(2+2\cos(\theta))^2}\Big(e^{i\theta}(rs^{-1}+r^{-1}s)+e^{-i\theta}(rs+r^{-1}s^{-1})\Big) \\
&\qquad \cdot\Big(r+r^{-1}-2\cos(3\phi)\Big)\Big(s+s^{-1}-2\cos(3\psi)\Big).
\end{aligned}$$

Then it is easy to see that

$$\mathrm{tr}(B_\theta) = \mathrm{tr}(B_{-\theta}), \quad \mathrm{tr}(AB_\theta) = \mathrm{tr}(AB_{-\theta}), \quad \mathrm{tr}(A^{-1}B_\theta) = \mathrm{tr}(A^{-1}B_{-\theta})$$

but $\mathrm{tr}[A, B_\theta] \neq \mathrm{tr}[A, B_{-\theta}]$.

4.5. *Cross-ratios*

In [23] Parker and Platis used cross-ratios to describe the complex hyperbolic three holed spheres. In this section we will briefly outline their construction and relate them to the trace coordinates of the previous section.

Choose a signature-$(2,1)$ Hermitian form on $\mathbb{C}^{2,1}$. Let V_- and V_0 be the associated subspaces given by (2.3) and (2.4). Then $\mathbb{P}V_- = \mathbf{H}^2_{\mathbb{C}}$ and $\mathbb{P}V_0 = \partial\mathbf{H}^2_{\mathbb{C}}$. Let z_1, z_2, z_3, z_4 be four distinct points on $\partial\mathbf{H}^2_{\mathbb{C}}$ and choose lifts of them $\mathbf{z}_1$, $\mathbf{z}_2$, $\mathbf{z}_3$, $\mathbf{z}_4$ in V_0. Following Korányi and Reimann [11], we define the *complex cross-ratio* of these four points to be

$$\mathbb{X} = [z_1, z_2, z_3, z_4] = \frac{\langle \mathbf{z}_3, \mathbf{z}_1\rangle\langle \mathbf{z}_4, \mathbf{z}_2\rangle}{\langle \mathbf{z}_4, \mathbf{z}_1\rangle\langle \mathbf{z}_3, \mathbf{z}_2\rangle}.$$

Since the z_i are distinct we see that $\mathbb{X}$ is finite and non-zero. We note that $\mathbb{X}$ is independent of the choice of the lifts $\mathbf{z}_j$ and is invariant under $\mathrm{SU}(2,1)$.

By choosing different orderings of our four points we may define other cross-ratios. There are some symmetries associated to certain permutations, see Property 5 on page 225 of [8]. After taking these into account, there are only three cross-ratios that remain. Given distinct points $z_1, \ldots, z_4 \in \partial\mathbf{H}^2_{\mathbb{C}}$, we define

$$\mathbb{X}_1 = [z_1, z_2, z_3, z_4], \quad \mathbb{X}_2 = [z_1, z_3, z_2, z_4], \quad \mathbb{X}_3 = [z_2, z_3, z_1, z_4]. \tag{4.3}$$

Then the three complex numbers $\mathbb{X}_1$, $\mathbb{X}_2$ and $\mathbb{X}_3$ satisfy the following identities (see Proposition 5.2 of [23])

$$|\mathbb{X}_2| = |\mathbb{X}_1|\,|\mathbb{X}_3|, \tag{4.4}$$

$$2|\mathbb{X}_1|^2\Re(\mathbb{X}_3) = |\mathbb{X}_1|^2 + |\mathbb{X}_2|^2 + 1 - 2\Re(\mathbb{X}_1 + \mathbb{X}_2). \tag{4.5}$$

Note that the norm and real part of $\mathbb{X}_3$ are determined by $\mathbb{X}_1$ and $\mathbb{X}_2$, but that the sign of $\Im(\mathbb{X}_3)$ is not determined. Falbel has given a different parametrisation of the space of cross-ratios Proposition 2.3 of [5].

Suppose we are given loxodromic maps A and B in $\mathrm{SU}(2,1)$ with distinct fixed points. Let the attracting fixed points of A and B be a_A, a_B and the repelling fixed points r_A, r_B respectively. Suppose that these fixed points correspond to attractive eigenvectors $\mathbf{a}_A$, $\mathbf{a}_B$ and repulsive eigenvectors $\mathbf{r}_A$, $\mathbf{r}_B$ respectively. Following (4.3), we define the first, second and third cross-ratios of the loxodromic maps A and B to be

$$\mathbb{X}_1(A,B) = [a_B, a_A, r_A, r_B] = \frac{\langle \mathbf{r}_A, \mathbf{a}_B\rangle\langle \mathbf{r}_B, \mathbf{a}_A\rangle}{\langle \mathbf{r}_B, \mathbf{a}_B\rangle\langle \mathbf{r}_A, \mathbf{a}_A\rangle}, \tag{4.6}$$

$$\mathbb{X}_2(A,B) = [a_B, r_A, a_A, r_B] = \frac{\langle \mathbf{a}_A, \mathbf{a}_B\rangle\langle \mathbf{r}_B, \mathbf{r}_A\rangle}{\langle \mathbf{r}_B, \mathbf{a}_B\rangle\langle \mathbf{a}_A, \mathbf{r}_A\rangle}, \tag{4.7}$$

$$\mathbb{X}_3(A,B) = [a_A, r_A, a_B, r_B] = \frac{\langle \mathbf{a}_B, \mathbf{a}_A\rangle\langle \mathbf{r}_B, \mathbf{r}_A\rangle}{\langle \mathbf{r}_B, \mathbf{a}_A\rangle\langle \mathbf{a}_B, \mathbf{r}_A\rangle}. \tag{4.8}$$

We can parametrise Zariski dense subgroups of $\mathrm{SU}(2,1)$ generated by a pair of loxodromic maps by their traces and the cross-ratios of their fixed points. This is Theorem 7.1 of [23]:

Theorem 4.13. *Suppose that A and B are loxodromic elements of* $\mathrm{SU}(2,1)$ *with distinct fixed points. Suppose also that $\langle A, B\rangle$ does not preserve a complex line. Then the group $\langle A, B\rangle$ is determined up to conjugation in* $\mathrm{SU}(2,1)$ *by:* $\mathrm{tr}(A)$, $\mathrm{tr}(B)$, $\mathbb{X}_1(A,B)$, $\mathbb{X}_2(A,B)$ *and* $\mathbb{X}_3(A,B)$.

Obviously, this result is asymmetrical in that it depends on a choice of two of the boundary curves. In order to get around this difficulty, Parker and Platis, Theorem 7.2 of [23], show that choosing a different pair of boundary coordinates amounts to a real change of coordinates.

Proposition 4.14. *Let A, B and C be loxodromic elements of* $\mathrm{SU}(2,1)$ *with $ABC = I$. Then* $\mathrm{tr}(C)$, $\mathbb{X}_1(A,C)$, $\mathbb{X}_2(A,C)$ *and* $\mathbb{X}_3(A,C)$ *may be expressed as real analytic functions of* $\mathrm{tr}(A)$, $\mathrm{tr}(B)$, $\mathbb{X}_1(A,B)$, $\mathbb{X}_2(A,B)$ *and* $\mathbb{X}_3(A,B)$.

We conclude this section by showing how these mixed trace and cross-ratio coordinates are related to the trace coordinates we found in the previous section; see Propositions 6.4 and 7.6 of [23].

Proposition 4.15. *Let A and B be loxodromic maps in* $\mathrm{SU}(2,1)$ *with* $\mathrm{tr}(A) = \lambda + \overline{\lambda}\lambda^{-1} + \overline{\lambda}^{-1}$ *and* $\mathrm{tr}(B) = \mu + \overline{\mu}\mu^{-1} + \overline{\mu}^{-1}$. *Let* $\mathbb{X}_1$, $\mathbb{X}_2$, $\mathbb{X}_3$

be the cross-ratios of their fixed points given by (4.6), (4.7) *and* (4.8). *Then the traces of* AB, $A^{-1}B$ *and* $[A,B]$ *are given by*

$$\begin{aligned}
&\mathrm{tr}(AB)\\
&= (\lambda+\overline{\lambda}^{-1})\overline{\mu}\mu^{-1} + \overline{\lambda}\lambda^{-1}(\mu+\overline{\mu}^{-1}) - \overline{\lambda}\lambda^{-1}\overline{\mu}\mu^{-1}\\
&\qquad + \mathbb{X}_1\,(\overline{\lambda}^{-1} - \overline{\lambda}\lambda^{-1})(\overline{\mu}^{-1} - \overline{\mu}\mu^{-1}) + \overline{\mathbb{X}}_1\,(\lambda - \overline{\lambda}\lambda^{-1})(\mu - \overline{\mu}\mu^{-1})\\
&\qquad + \mathbb{X}_2\,(\lambda - \overline{\lambda}\lambda^{-1})(\overline{\mu}^{-1} - \overline{\mu}\mu^{-1}) + \overline{\mathbb{X}}_2\,(\overline{\lambda}^{-1} - \overline{\lambda}\lambda^{-1})(\mu - \overline{\mu}\mu^{-1}),\\
&\mathrm{tr}(A^{-1}B)\\
&= (\lambda^{-1}+\overline{\lambda})\overline{\mu}\mu^{-1} + \lambda\overline{\lambda}^{-1}(\mu+\overline{\mu}^{-1}) - \lambda\overline{\lambda}^{-1}\overline{\mu}\mu^{-1}\\
&\qquad + \mathbb{X}_1\,(\overline{\lambda} - \lambda\overline{\lambda}^{-1})(\overline{\mu}^{-1} - \overline{\mu}\mu^{-1}) + \overline{\mathbb{X}}_1\,(\lambda^{-1} - \lambda\overline{\lambda}^{-1})(\mu - \overline{\mu}\mu^{-1})\\
&\qquad + \mathbb{X}_2\,(\lambda^{-1} - \lambda\overline{\lambda}^{-1})(\overline{\mu}^{-1} - \overline{\mu}\mu^{-1}) + \overline{\mathbb{X}}_2\,(\overline{\lambda} - \lambda\overline{\lambda}^{-1})(\mu - \overline{\mu}\mu^{-1})
\end{aligned}$$

and

$$\begin{aligned}
&\mathrm{tr}[A,B]\\
&= 3 - 2\Re\Big(\mathbb{X}_1\,(\overline{\lambda} - \lambda\overline{\lambda}^{-1})(\overline{\lambda}^{-1} - \overline{\lambda}\lambda^{-1})(\overline{\mu} - \mu\overline{\mu}^{-1})(\overline{\mu}^{-1} - \overline{\mu}\mu^{-1})\Big)\\
&\qquad - 2\Re\Big(\mathbb{X}_2\,(\lambda - \overline{\lambda}\lambda^{-1})(\lambda^{-1} - \lambda\overline{\lambda}^{-1})(\overline{\mu} - \mu\overline{\mu}^{-1})(\overline{\mu}^{-1} - \overline{\mu}\mu^{-1})\Big)\\
&\qquad + \Big(1 - 2\Re(\mathbb{X}_1 + \mathbb{X}_2)\Big)\\
&\qquad\quad \cdot\Big(\big|(\lambda - \overline{\lambda}\lambda^{-1})(\mu - \overline{\mu}\mu^{-1})\big|^2 + \big|(\lambda^{-1} - \lambda\overline{\lambda}^{-1})(\mu^{-1} - \mu\overline{\mu}^{-1})\big|^2\Big)\\
&\qquad + \left|\begin{matrix} \mathbb{X}_1\,(\overline{\lambda} - \lambda\overline{\lambda}^{-1})(\overline{\mu} - \mu\overline{\mu}^{-1}) + \overline{\mathbb{X}}_1\,(\lambda^{-1} - \lambda\overline{\lambda}^{-1})(\mu^{-1} - \mu\overline{\mu}^{-1}) \\ + \mathbb{X}_2\,(\lambda^{-1} - \lambda\overline{\lambda}^{-1})(\overline{\mu} - \mu\overline{\mu}^{-1}) + \overline{\mathbb{X}}_2\,(\overline{\lambda} - \lambda\overline{\lambda}^{-1})(\mu^{-1} - \mu\overline{\mu}^{-1})\end{matrix}\right|^2\\
&\qquad + \Big(|\mathbb{X}_2|^2 - |\mathbb{X}_1|^2\mathbb{X}_3\Big)\\
&\qquad\quad \cdot\Big(\big|\lambda - \overline{\lambda}\lambda^{-1}\big|^2 - \big|\lambda^{-1} - \lambda\overline{\lambda}^{-1}\big|^2\Big)\Big(\big|\mu - \overline{\mu}\mu^{-1}\big|^2 - \big|\mu^{-1} - \mu\overline{\mu}^{-1}\big|^2\Big).
\end{aligned}$$

We can use the expressions from Proposition 4.15 to express $\mathbb{X}_1$ and $\mathbb{X}_2$ in terms of λ, μ, $\mathrm{tr}(AB)$ and $\mathrm{tr}(A^{-1}B)$. This means that these coordinates are "equivalent" to the trace coordinates we found earlier. Furthermore, the real part and modulus of $\mathrm{tr}[A,B]$ may be expressed in terms of λ, μ, $\mathbb{X}_1$ and $\mathbb{X}_2$ and the ambiguity in the sign of $\Im\big(\mathrm{tr}[A,B]\big)$ is exactly the same as the ambiguity in the sign of $\Im(\mathbb{X}_3)$. In Example 4.12 we have $\mathbb{X}_1 = \mathbb{X}_2 = -1/(2-2\cos(\theta))$ and $\mathbb{X}_3 = e^{-2i\theta}$.

4.6. *Twist-bend parameters*

Let Σ be a surface of genus $g \geq 2$. We may decompose Σ into three holed spheres and use this decomposition to define complex hyperbolic Fenchel-Nielsen coordinates. The trace coordinates from Proposition 4.11 give the Fenchel-Nielsen complex lengths via Proposition 3.10. In this section we give a brief sketch of how to define Fenchel-Nielsen twist-bends. Details may be found in Parker and Platis [23].

For hyperbolic surfaces, a Fenchel-Nielsen *twist* about a simple, closed, oriented curve α involves cutting the surface along α and then re-attaching so that points on one side are moved a hyperbolic distance k relative to the other side. It is often useful to think about doing this with the lift $\tilde{\alpha}$ of α to the hyperbolic plane. If $\tilde{\alpha}$ is the geodesic in the upper halfplane with endpoints 0 and ∞, then this process involves applying the dilation $K : z \longmapsto e^k z$ to the part of the hyperbolic plane on one side of $\tilde{\alpha}$, say the part with $\mathrm{Re}(z) > 0$. We allow k to be negative and this corresponds to moving in the opposite direction relative to the orientation of α. If we twist by a hyperbolic distance k equal to the length of α then the Fenchel-Nielsen twist is the same as a Dehn twist.

We can generalise the definition of Fenchel-Nielsen twist to twist-bends in hyperbolic 3-space $\mathbf{H}^2_{\mathbb{R}}$, following Tan [30] and Kourouniotis [12]. The easiest way to describe this is to suppose that the universal cover of the surface is a hyperbolic plane inside $\mathbf{H}^3_{\mathbb{R}}$ and that $\tilde{\alpha}$ is again the geodesic with endpoints 0 and ∞. As well as applying a Fenchel-Nielsen twist, we may also rotate through an angle θ in the plane normal to $\tilde{\alpha}$. This is called a *bend* and corresponds to applying the rotation $K : z \longmapsto e^{i\theta} z$. Doing a twist through distance k and a bend through angle θ gives a *twist-bend* with parameter $k + i\theta$. It corresponds to applying the loxodromic map $K : z \longmapsto e^{k+i\theta} z$. Doing this process to a hyperbolic surface results in a pleated surface where the bending locus is the geodesic α. The relationship between traces and bending for such surfaces is explored by Parker and Series in [25].

This description can be extended to the case of complex hyperbolic representations of surface groups even though there is no longer a hyperbolic surface. However, the local picture is the same. Namely, a twist will be a hyperbolic translation of distance k along a geodesic α (or $\tilde{\alpha}$) and a bend through angle θ in the plane normal to the complex line containing $\tilde{\alpha}$.

We can describe twist-bends on the level of the fundamental group. This works for all of the cases described above, but we only give details in the case of SU(2, 1). There are two cases, (a) when the parts of the surface on either side of α are in different three holed spheres and (b) when they are in the same three holed sphere. In terms group theory, (a) corresponds to a free product with amalgamation and (b) to an HNN extension. The definition of twist bends in each case are similar but not quite the same.

Let Y be a three holed sphere with oriented boundary geodesics α, β, γ and let $\rho_0 : \pi_1 \longmapsto \mathrm{SU}(2,1)$ be a corresponding representation with $\rho_0([\alpha]) = A$, $\rho_0([\beta]) = B$ and $\rho_0([\gamma]) = C$ where $ABC = I$. Suppose that Y' is another such surface. We have α', β', γ', ρ_0', A', B', C' all as above. We must decide when we can attach Y and Y' along α and α'. We can do so when α' has the same length as α but the opposite orientation. Hence A' must be conjugate to A^{-1}.

There are two cases to consider. First we must investigate what happens when we attach two distinct three holed spheres along a common boundary; see Section 8.2 of [23]. For our initial configuration, we suppose $A' = A^{-1}$. Attaching Y and Y' is then the same as taking the free product of $\langle A, B\rangle$ and $\langle A', B'\rangle$ with amalgamation along the common subgroup $\langle A\rangle = \langle A'\rangle$. The resulting group is then

$$\langle A, B\rangle *_{\langle A\rangle} \langle A', B'\rangle = \langle A, B, A', B' \,|A' = A^{-1}\rangle = \langle A, B, B'\rangle.$$

In this case a twist-bend consists of fixing the surface corresponding to $\langle A, B\rangle$ and moving the surface corresponding to $\langle A', B'\rangle$ by a hyperbolic translation along the axis of A (the twist) and a rotation around the complex axis of A (the bend). In other words, we take a map K that commutes with $A' = A^{-1}$ and we conjugate $\langle A', B'\rangle$ by K. Thus the new group is

$$\begin{aligned}\langle A, B\rangle *_{\langle A\rangle} K\langle A', B'\rangle K^{-1} &= \langle A, B\rangle *_{\langle A\rangle} \langle KA'K^{-1}, KB'K^{-1}\rangle \\ &= \langle A, B, KA'K^{-1}, KB'K^{-1} \,|KA'K^{-1} = A^{-1}\rangle \\ &= \langle A, B, KB'K^{-1}\rangle.\end{aligned}$$

Note that if we swap the roles of Y and Y' then the same process yields a twist-bend associated to the matrix K^{-1}. That is, the new group is $\langle A', B', K^{-1}BK\rangle$ which is conjugate, via K^{-1}, to $\langle A, B, KB'K^{-1}\rangle$.

Secondly, we must consider the case where we close a handle; see Section 8.3 of [23]. In this case we consider Y and we want to glue two of its boundary components. Suppose one of them is represented by A then the other must be conjugate to A^{-1}, say it is $BA^{-1}B^{-1}$. Note that if A and $BA^{-1}B^{-1}$ correspond to boundary components of the same three holed sphere, this means the third boundary component is $C = (BA^{-1}B^{-1})^{-1}A^{-1} = [B, A]$. Then in order to close the handle we take HNN extension associated to the isomorphism $\phi : \langle A \rangle \longrightarrow \langle BAB^{-1} \rangle$ given by $\phi(A) = BA^{-1}B^{-1}$. It is clear that we may do this by adjoining the stable letter B to obtain

$$\langle A, BA^{-1}B^{-1} \rangle *_\phi = \langle A, BA^{-1}B^{-1}, B \mid BA^{-1}B^{-1} = \phi(A) \rangle = \langle A, B \rangle.$$

Because we want to keep track of the stable letter, we will write this extension as

$$\langle A, BA^{-1}B^{-1} \rangle *_\phi (B).$$

If K is a map that commutes with A then we also have $\phi(A) = (BK)A^{-1}(BK)^{-1}$. Therefore we can take an isomorphic HNN extension by adding the stable letter BK instead of B:

$$\langle A, (BK)A^{-1}(BK)^{-1} \rangle *_\phi (BK) = \langle A, BK \rangle.$$

Thus, in the case of closing a handle performing a complex twist associated to a map K that commutes with A involves changing the stable letter of the HNN extension from B to BK.

In both cases, the geometry of the complex twist is recorded by $\mathrm{tr}(K)$ in exactly the same way that $\mathrm{tr}(A)$ is related to $\ell(\alpha) + i\phi(\alpha)$ as described in Proposition 3.10. Therefore if K corresponds to a twist through distance $k \in \mathbb{R}$ and bend through angle $\theta \in (-\pi, \pi]$ then we have

$$\mathrm{tr}(K) = 2\cosh(k/2)e^{-i\theta/3} + e^{2i\theta/3}.$$

We note that there are subtleties about the direction of twist and the sign of $\Re(\kappa)$; see [23].

We need to find a conjugation invariant way of measuring the twist-bend parameter κ from data associated to the representation of the fundamental group π_1. In [23] this was done using cross-ratios. We will not discuss this here.

5. Traces for Triangle Groups

5.1. *Introduction*

In this section we consider groups generated by three complex reflections. Such groups are important because of their connection with Deligne-Mostow groups. In [17] Mostow showed that certain groups generated by three complex reflections are non-arithmetic lattices. Later Livné, Mostow, Deligne, Thurston and Deraux constructed more examples of lattices that are generated by three complex reflections. See [20] for a survey of this area, including a discussion of arithmeticity. Recently, Deraux, Parker and Paupert [4] and Thompson [31] have found new candidates for non-arithmetic lattices from triangle groups.

Suppose that Δ is a group generated by three complex reflections in $\mathrm{SU}(2,1)$ all with the same angle. The goal of this section is to give combinatorial formulae for the traces of elements of Δ. These formulae are due to Pratoussevitch [27] generalising earlier work of Sandler [28]. These formulae are useful when deciding arithmeticity and commensurability in the case where Δ is a lattice; see Paupert [26] for more details. In particular, one may use this technique for the groups constructed by Mostow in [17].

5.2. *Reflections*

Consider $\mathbb{R}^{n+1}$ with the standard inner product. Let Π be a hyperplane through the origin and let $\mathbf{n}$ be a normal vector to Π. Thus the orthogonal complement $\Pi^\perp$ of Π is spanned by $\mathbf{n}$. We may decompose a vector $\mathbf{x} \in \mathbb{R}^{n+1}$ into a component in Π and a component in $\Pi^\perp$. Reflection R in Π is obtained by multiplying the component in $\Pi^\perp$ by -1. Specifically,

$$\mathbf{x} = \left(\mathbf{x} - \frac{\mathbf{x}\cdot\mathbf{n}}{\mathbf{n}\cdot\mathbf{n}}\mathbf{n}\right) + \frac{\mathbf{x}\cdot\mathbf{n}}{\mathbf{n}\cdot\mathbf{n}}\mathbf{n},$$

where

$$\left(\mathbf{x} - \frac{\mathbf{x}\cdot\mathbf{n}}{\mathbf{n}\cdot\mathbf{n}}\mathbf{n}\right) \in \Pi, \qquad \frac{\mathbf{x}\cdot\mathbf{n}}{\mathbf{n}\cdot\mathbf{n}}\mathbf{n} \in \Pi^\perp.$$

Then

$$R(\mathbf{x}) = \left(\mathbf{x} - \frac{\mathbf{x}\cdot\mathbf{n}}{\mathbf{n}\cdot\mathbf{n}}\mathbf{n}\right) + (-1)\frac{\mathbf{x}\cdot\mathbf{n}}{\mathbf{n}\cdot\mathbf{n}}\mathbf{n} = \mathbf{x} - 2\frac{\mathbf{x}\cdot\mathbf{n}}{\mathbf{n}\cdot\mathbf{n}}\mathbf{n}.$$

Then R is given by a matrix in $\mathrm{O}(n+1)$ with determinant -1. The unit sphere S^n in $\mathbb{R}^{n+1}$ is given by all vectors with norm 1. Any reflection in great hypersphere on S^n may be obtained by restricting the above formula to S^n.

We may repeat this construction for Minkowski space $\mathbb{R}^{n,1}$ and (real) hyperbolic n-space. Specifically, consider a non-degenerate bilinear form of signature $(n,1)$ given by the symmetric matrix B:

$$(\mathbf{x},\mathbf{y}) = \mathbf{y}^t B\mathbf{x}.$$

Suppose that Π is a real hyperplane through the origin with space-like normal vector $\mathbf{n}$. That is, for given $\mathbf{n} \in \mathbb{R}^{n,1}$ with $(\mathbf{n},\mathbf{n}) > 0$

$$\Pi = \left\{\mathbf{x} \in \mathbb{R}^{n,1} \ : \ (\mathbf{x},\mathbf{n}) = 0\right\}.$$

Also $\Pi^{\perp}$ is the line spanned by $\mathbf{n}$. We may generalise the formula for reflection in Π given above. That is we can write any $\mathbf{x} \in \mathbb{R}^{n,1}$ as

$$\mathbf{x} = \left(\mathbf{x} - \frac{(\mathbf{x},\mathbf{n})}{(\mathbf{n},\mathbf{n})}\mathbf{n}\right) + \frac{(\mathbf{x},\mathbf{n})}{(\mathbf{n},\mathbf{n})}\mathbf{n}.$$

Then

$$R(\mathbf{x}) = \left(\mathbf{x} - \frac{(\mathbf{x},\mathbf{n})}{(\mathbf{n},\mathbf{n})}\mathbf{n}\right) + (-1)\frac{(\mathbf{x},\mathbf{n})}{(\mathbf{n},\mathbf{n})}\mathbf{n} = \mathbf{x} - 2\frac{(\mathbf{x},\mathbf{n})}{(\mathbf{n},\mathbf{n})}\mathbf{n}.$$

This is represented by a matrix in $\mathrm{O}(n,1)$ with determinant -1. The hyperboloid model of (real) hyperbolic n-space is given by

$$\left\{\mathbf{x} \in \mathbb{R}^{n,1} \ : \ (\mathbf{x},\mathbf{x}) = -1, \ x_{n+1} > 0\right\}.$$

Then R maps this hyperboloid to itself. Alternatively, we may want to obtain the Klein-Beltrami model of $\mathbf{H}^n_{\mathbb{R}}$ by taking a projective action on the points with $x_{n+1} = 1$.

We want to generalise this whole setup from the real world to the complex world. To do so, consider $\mathbb{C}^{n+1}$ with a Hermitian form H, which we assume to be non-degenerate but, at this stage, without restrictions on its signature. We use angle brackets to denote the Hermitian form associated to H as usual: $\mathbf{w}^* H\mathbf{z} = \langle \mathbf{z},\mathbf{w}\rangle$ for $\mathbf{z}$ and $\mathbf{w}$ in $\mathbb{C}^{n+1}$. Since we are interested in complex hyperbolic space $\mathbf{H}^n_{\mathbb{C}}$ we mainly think of the case where the form H has signature $(n,1)$ but this is not necessary for the definition of complex reflections.

Suppose that Π is a complex hyperplane in $\mathbb{C}^{n+1}$. That is $\Pi^{\perp}$ is spanned by $\mathbf{n} \in \mathbb{C}^{n+1}$ and so we have $\Pi = \left\{\mathbf{z} \in \mathbb{C}^{n+1} \ : \ \langle \mathbf{z},\mathbf{n}\rangle = 0\right\}$. Any $\mathbf{z} \in \mathbb{C}^{n+1}$ may then be decomposed into components in Π and $\Pi^{\perp}$:

$$\mathbf{z} = \left(\mathbf{z} - \frac{\langle \mathbf{z},\mathbf{n}\rangle}{\langle \mathbf{n},\mathbf{n}\rangle}\mathbf{n}\right) + \frac{\langle \mathbf{z},\mathbf{n}\rangle}{\langle \mathbf{n},\mathbf{n}\rangle}\mathbf{n},$$

where

$$\left(\mathbf{z} - \frac{\langle \mathbf{z},\mathbf{n}\rangle}{\langle \mathbf{n},\mathbf{n}\rangle}\mathbf{n}\right) \in \Pi, \quad \frac{\langle \mathbf{z},\mathbf{n}\rangle}{\langle \mathbf{n}\cdot\mathbf{n}\rangle}\mathbf{n} \in \Pi^{\perp}.$$

At this point we come to most crucial difference between real and complex reflections. Once again a complex reflection will preserve the decomposition of $\mathbf{z}$ into components in Π and $\Pi^{\perp}$, will pointwise fix the component in Π and will preserve the norm of $\mathbf{z}$. However, since $\Pi^{\perp}$ is a complex line we have greater freedom than we did before: we may multiply the component in

$\Pi^\perp$ by any complex number with modulus 1. Hence we define *the reflection in* Π *with angle* ψ to be the map $R(\mathbf{z})$ given by

$$R(\mathbf{z}) = \left(\mathbf{z} - \frac{\langle \mathbf{z}, \mathbf{n}\rangle}{\langle \mathbf{n}, \mathbf{n}\rangle}\mathbf{n}\right) + e^{i\psi}\frac{\langle \mathbf{z}, \mathbf{n}\rangle}{\langle \mathbf{n}, \mathbf{n}\rangle}\mathbf{n} = \mathbf{z} + (e^{i\psi} - 1)\frac{\langle \mathbf{z}, \mathbf{n}\rangle}{\langle \mathbf{n}, \mathbf{n}\rangle}\mathbf{n}. \qquad (5.1)$$

The map R is given by a matrix in $U(H)$ with n eigenvalues $+1$ and one eigenvalue $e^{i\psi}$. Hence its determinant is $e^{i\psi}$. In order to obtain a map in $\mathrm{SU}(H)$ we must multiply this matrix by $e^{-i\psi/(n+1)}$.

In what follows, we will mainly be interested in the case where $n = 2$, H has signature $(2,1)$ and $\mathbf{n} \in V_+$. This means that, in terms of its action of $\mathbf{H}^2_{\mathbb{C}}$, the reflection R fixes a complex line $L = \mathbb{P}\Pi \cap \mathbf{H}^2_{\mathbb{C}}$. However, it will be useful to consider the space of groups generated by three complex reflections (all with the same angle) for a Hermitian form H and then consider the subspace where H has the correct signature.

We now give two examples in signature $(1,1)$ that show that in this case complex reflections are just rotations, that is elliptic matrices.

Example 5.1.

(i) Consider $\mathbb{C}^{1,1}$ where the Hermitian form is given by H_0, as given in (2.1). Let Π be the complex line in $\mathbb{C}^{1,1}$ with polar vector $\mathbf{n}$ where

$$\mathbf{n} = \begin{pmatrix} 1 \\ 0 \end{pmatrix}.$$

Then using (5.1) the reflection in Π with angle ψ is given by

$$\begin{aligned} R(z) &= \begin{pmatrix} z_1 \\ z_2 \end{pmatrix} + (e^{i\psi} - 1)\frac{z_1}{1}\begin{pmatrix} 1 \\ 0 \end{pmatrix} \\ &= \begin{pmatrix} e^{i\psi} & 0 \\ 0 & 1 \end{pmatrix}\begin{pmatrix} z_1 \\ z_2 \end{pmatrix}. \end{aligned}$$

The matrix in the last line is in $U(1,1)$. In order to obtain a matrix in $\mathrm{SU}(1,1)$ we must multiply by $e^{-i\psi/2}$ to get

$$R = \begin{pmatrix} e^{i\psi/2} & 0 \\ 0 & e^{-i\psi/2} \end{pmatrix}.$$

(ii) Consider $\mathbb{C}^{1,1}$ where the Hermitian form is given by H_0'. Let Π be the complex line in $\mathbb{C}^{1,1}$ with polar vector $\mathbf{n}$ where

$$\mathbf{n} = \begin{pmatrix} -i \\ 1 \end{pmatrix}.$$

Then $\langle \mathbf{n}, \mathbf{n} \rangle = 2$. Then, arguing as before, the reflection in Π with angle ψ is

$$\begin{aligned} R(z) &= \begin{pmatrix} z_1 \\ z_2 \end{pmatrix} + (e^{i\psi} - 1)\frac{iz_1 + z_2}{2}\begin{pmatrix} -i \\ 1 \end{pmatrix} \\ &= \frac{1}{2}\begin{pmatrix} (e^{i\psi}+1)z_1 - i(e^{i\psi}-1)z_2 \\ i(e^{i\psi}-1)z_1 + (e^{i\psi}+1)z_2 \end{pmatrix} \\ &= \begin{pmatrix} e^{i\psi/2}\cos(\psi/2) & e^{i\psi/2}\sin(\psi) \\ -e^{i\psi/2}\sin(\psi/2) & e^{i\psi/2}\cos(\psi/2) \end{pmatrix}\begin{pmatrix} z_1 \\ z_2 \end{pmatrix}. \end{aligned}$$

In order to obtain a matrix with determinant 1 we must multiply by $e^{-i\psi/2}$ to get

$$R = \begin{pmatrix} \cos(\psi/2) & \sin(\psi/2) \\ -\sin(\psi/2) & \cos(\psi/2) \end{pmatrix} \in \mathrm{SL}(2, \mathbb{R}).$$

5.3. *Complex reflections in* SU(2, 1)

We now give more details for the case we are most interested in, namely where $n = 2$ and H has signature $(2, 1)$. Let Π be a complex hyperplane in $\mathbb{C}^{2,1}$ with normal vector $\mathbf{n} \in \mathbb{C}^{2,1}$ with $\langle \mathbf{n}, \mathbf{n} \rangle > 0$. The complex reflection with angle ψ fixing Π is given by (5.1). In order that R is represented by a matrix in $\mathrm{SU}(2, 1)$, we multiply this formula by $e^{-i\psi/3}$. This gives

$$R(\mathbf{z}) = e^{-i\psi/3}\left(\mathbf{z} + (e^{i\psi}-1)\frac{\langle \mathbf{z}, \mathbf{n} \rangle}{\langle \mathbf{n}, \mathbf{n} \rangle}\mathbf{n}\right) = e^{-i\psi/3}\mathbf{z} + (e^{2i\psi/3} - e^{-\psi/3})\frac{\langle \mathbf{z}, \mathbf{n} \rangle}{\langle \mathbf{n}, \mathbf{n} \rangle}\mathbf{n}. \tag{5.2}$$

If R is as above, for any $A \in \mathrm{SU}(2, 1)$ we now relate $\mathrm{tr}(RA)$ and $\mathrm{tr}(A)$.

Lemma 5.2. *Let R be complex reflection in the hyperplane orthogonal to* $\mathbf{n}$ *with angle ψ given by (5.2). Let A be any element of* $\mathrm{SU}(2, 1)$. *Then*

$$\mathrm{tr}(RA) = e^{-i\psi/3}\mathrm{tr}(A) + (e^{2i\psi/3} - e^{-i\psi/3})\frac{\langle A\mathbf{n}, \mathbf{n} \rangle}{\langle \mathbf{n}, \mathbf{n} \rangle}.$$

Proof. We have

$$\begin{aligned} RA(\mathbf{z}) &= e^{-i\psi/3}A\mathbf{z} + \frac{(e^{2i\psi/3} - e^{-\psi/3})}{\langle \mathbf{n}, \mathbf{n} \rangle}\mathbf{n}\langle A\mathbf{z}, \mathbf{n} \rangle \\ &= e^{-i\psi/3}A\mathbf{z} + \frac{(e^{2i\psi/3} - e^{-\psi/3})}{\langle \mathbf{n}, \mathbf{n} \rangle}\mathbf{n}\mathbf{n}^* HA\mathbf{z}. \end{aligned}$$

Therefore, the matrix of RA is

$$e^{-i\psi/3}A + \frac{(e^{2i\psi/3} - e^{-\psi/3})}{\langle \mathbf{n}, \mathbf{n} \rangle}\mathbf{n}\mathbf{n}^* HA.$$

Now if a matrix can be written in the form $\mathbf{u}\mathbf{v}^*$ for column vectors $\mathbf{u}$ and $\mathbf{v}$, then its trace is just $\mathbf{v}^*\mathbf{u}$. Thus

$$\mathrm{tr}(\mathbf{n}\mathbf{n}^* HA) = \mathrm{tr}\big(\mathbf{n}(A^*H\mathbf{n})^*\big) = (A^*H\mathbf{n})^*\mathbf{n} = \mathbf{n}^*HA\mathbf{n} = \langle A(\mathbf{n}), \mathbf{n}\rangle.$$

Hence

$$\begin{aligned}\mathrm{tr}(RA) &= e^{-i\psi/3}\mathrm{tr}(A) + \frac{(e^{2i\psi/3} - e^{-\psi/3})}{\langle \mathbf{n}, \mathbf{n}\rangle}\mathrm{tr}(\mathbf{n}\mathbf{n}^* H)\\ &= e^{-i\psi/3}\mathrm{tr}(A) + (e^{2i\psi/3} - e^{-\psi/3})\frac{\langle A\mathbf{n}, \mathbf{n}\rangle}{\langle \mathbf{n}, \mathbf{n}\rangle}.\end{aligned}$$

□

Setting A to be the identity matrix, we see the fact (which we already knew from our consideration of eigenvalues, Proposition 3.7) that $\mathrm{tr}(R) = 3e^{-i\psi/3} + (e^{2i\psi/3} - e^{-i\psi/3}) = e^{2i\psi/3} + 2e^{-i\psi/3}$.

5.4. *Equilateral triangle groups*

Suppose that we are given three complex lines L_1, L_2 and L_3 in $\mathbf{H}^2_{\mathbb{C}}$. These correspond to hyperplanes Π_1, Π_2 and Π_3 in $\mathbb{C}^{2,1}$ with normal vectors $\mathbf{n}_1$, $\mathbf{n}_2$ and $\mathbf{n}_3$ with $\langle \mathbf{n}_j, \mathbf{n}_j\rangle > 0$. For $j = 1, 2, 3$, consider complex reflections R_j with angle ψ about complex lines with polar vectors $\mathbf{n}_j$. Using (5.2) that is

$$R_j(\mathbf{z}) = e^{-i\psi/3}\mathbf{z} + (e^{2i\psi/3} - e^{-\psi/3})\frac{\langle \mathbf{z}, \mathbf{n}_j\rangle}{\langle \mathbf{n}_j, \mathbf{n}_j\rangle}\mathbf{n}_j. \qquad (5.3)$$

Note that this formula is preserved if $\mathbf{n}_j$ is sent to $\lambda\mathbf{n}_j$ for any $\lambda \in \mathbb{C} - \{0\}$.

Suppose first that the three complex lines L_1, L_2, L_3 form an equilateral triangle. That is, there is a J map of order 3 cyclically permuting them. In other words $J \in \mathrm{SU}(2,1)$ satisfies $\Pi_2 = J\Pi_1$, $\Pi_3 = J\Pi_2 = J^{-1}\Pi_1$ and $\mathbf{n}_2 = J\mathbf{n}_1$, $\mathbf{n}_3 = J\mathbf{n}_2 = J^{-1}\mathbf{n}_1$. Thus

$$\langle \mathbf{n}_1, \mathbf{n}_1\rangle = \langle \mathbf{n}_2, \mathbf{n}_2\rangle = \langle \mathbf{n}_3, \mathbf{n}_3\rangle, \quad \langle \mathbf{n}_2, \mathbf{n}_1\rangle = \langle \mathbf{n}_3, \mathbf{n}_2\rangle = \langle \mathbf{n}_1, \mathbf{n}_3\rangle.$$

Note that if ω is a cube root of unity, all these formulae remain valid if, for $j = 1, 2, 3$, we send $\mathbf{n}_j$ to $\omega^j\mathbf{n}_j$.

The map J will have eigenvalues 1, ω and $\overline{\omega}$ and so $\mathrm{tr}(J) = 0$. Using this fact, the following result is an easy corollary of Lemma 5.2.

Lemma 5.3. *Let R be a complex reflection with angle ψ fixing a complex line L with polar vector $\mathbf{n}$. Let $J \in \mathrm{SU}(2,1)$ be a regular elliptic map of order 3. Then*

$$\mathrm{tr}(RJ) = (e^{2i\psi/3} - e^{-i\psi/3})\frac{\langle J\mathbf{n}, \mathbf{n}\rangle}{\langle \mathbf{n}, \mathbf{n}\rangle}.$$

Motivated by Lemma 5.3, we define the variable τ to be this trace (where indices are taken cyclically):

$$\tau = \mathrm{tr}(R_j J) = (e^{2i\psi/3} - e^{-i\psi/3})\frac{\langle J\mathbf{n}_j, \mathbf{n}_j\rangle}{\langle \mathbf{n}_j, \mathbf{n}_j\rangle} = (e^{2i\psi/3} - e^{-i\psi/3})\frac{\langle \mathbf{n}_{j+1}, \mathbf{n}_j\rangle}{\langle \mathbf{n}_j, \mathbf{n}_j\rangle}. \tag{5.4}$$

Sending $\mathbf{n}_j$ to $\omega^j \mathbf{n}_j$ means that τ is multiplied by ω. Therefore given R_1, R_2 and R_3 the map τ is only defined up to multiplication by a cube root of unity.

Furthermore, following Section 3.3.2 of Goldman [8], if L_j and L_{j+1} meet with angle θ (by symmetry this is the same for all three pairs of lines) then

$$\cos(\theta) = \frac{\left|\langle \mathbf{n}_{j+1}, \mathbf{n}_j\rangle\right|}{\langle \mathbf{n}_j, \mathbf{n}_j\rangle} = \frac{|\tau|}{2\sin(\psi/2)}. \tag{5.5}$$

This shows that once again the traces lead to geometrical information about the group.

All of this has been defined without reference to any particular Hermitian form. Following Mostow [17], we choose $\mathbf{n}_1$, $\mathbf{n}_2$ and $\mathbf{n}_3$ to be the standard basis vectors of $\mathbb{C}^{2,1}$; see Parker and Paupert [22]. Thus

$$\mathbf{n}_1 = \begin{bmatrix} 1 \\ 0 \\ 0 \end{bmatrix}, \qquad \mathbf{n}_2 = \begin{bmatrix} 0 \\ 1 \\ 0 \end{bmatrix}, \qquad \mathbf{n}_3 = \begin{bmatrix} 0 \\ 0 \\ 1 \end{bmatrix}. \tag{5.6}$$

An immediate consequence is that the permuting matrix is

$$J = \begin{bmatrix} 0 & 0 & 1 \\ 1 & 0 & 0 \\ 0 & 1 & 0 \end{bmatrix}. \tag{5.7}$$

Sending $\mathbf{n}_j$ to $\omega^j \mathbf{n}_j$ means that J is multiplied by ω. Therefore given complex reflections R_1, R_2, R_3 the symmetry map J is only defined up to multiplication by a cube root of unity.

The matrix defining the Hermitian form is then (a multiple of) the Gram matrix. This makes sense only when these three vectors are linearly independent or equivalently the group generated by R_1, R_2 and R_3 does not preserve any lower dimensional complex subspace, that is it is Zariski dense (see Remark 10 of Will [34]). Moreover, (5.6) determines the Hermitian form H up to a real multiple. In order to avoid denominators, we choose

$$\langle \mathbf{n}_j, \mathbf{n}_j\rangle = |e^{2i\psi/3} - e^{-i\psi/3}|^2 = 2 - e^{i\psi} - e^{-i\psi}.$$

This means that

$$\langle \mathbf{n}_{j+1}, \mathbf{n}_j \rangle = \frac{\langle \mathbf{n}_j, \mathbf{n}_j \rangle}{e^{2i\psi/3} - e^{-i\psi/3}}\,\tau = (e^{-2i\psi/3} - e^{i\psi/3})\tau,$$

and so

$$H = \begin{bmatrix} 2 - e^{i\psi} - e^{-i\psi} & (e^{-2i\psi/3} - e^{i\psi/3})\tau & (e^{2i\psi/3} - e^{-i\psi/3})\overline{\tau} \\ (e^{2i\psi/3} - e^{-i\psi/3})\overline{\tau} & 2 - e^{i\psi} - e^{-i\psi} & (e^{-2i\psi/3} - e^{i\psi/3})\tau \\ (e^{-2i\psi/3} - e^{i\psi/3})\tau & (e^{2i\psi/3} - e^{-i\psi/3})\overline{\tau} & 2 - e^{i\psi} - e^{-i\psi} \end{bmatrix}. \quad (5.8)$$

As we indicated in Section 5.2 the construction of R_1, R_2 and R_2 in terms of $\mathbf{n}_1$, $\mathbf{n}_2$ and $\mathbf{n}_3$ works whatever the signature of H. It is only when H has signature $(2,1)$ that these reflections act on complex hyperbolic space as complex reflections in complex lines L_1, L_2 and L_3. We now discuss the geometry when H has other signatures.

Since the trace of H is positive, we see that it must have at least one positive eigenvalue. If H has signature $(3,0)$ then our triangle lies on $\mathbb{CP}^2$ and $\langle R_1, R_2, R_3 \rangle$ is a subgroup of SU(3). If H has signature $(2,1)$, which is the case we are interested in, then $\langle R_1, R_2, R_3 \rangle$ is generated by reflections in complex lines in complex hyperbolic space. If H has signature $(1,2)$ then $\langle R_1, R_2, R_3 \rangle$ is generated by reflections in points in complex hyperbolic space, generalising Theorem 6.8.12 of Davis [3]. We now give a criterion for determining when H has signature $(2,1)$.

Lemma 5.4. *The signature of the matrix H given by (5.8) is $(2,1)$ if and only if*

$$\begin{aligned} 0 < 3(2 - e^{i\psi} - e^{-i\psi})|\tau|^2 - (1 - e^{-i\psi})\tau^3 - (1 - e^{i\psi})\overline{\tau}^3 \\ - (2 - e^{i\psi} - e^{-i\psi})^2. \end{aligned}$$

Proof. We must find when H has two eigenvalues that are positive and one that is negative. Since the sum of the eigenvalues of H is

$$\mathrm{tr}(H) = 3(2 - e^{i\psi} - e^{-i\psi}) > 0,$$

it is easy to see that all three eigenvalues cannot be negative simultaneously. This means we only need to check when H has negative determinant. Hence

$$\begin{aligned} 0 &> \det(H) \\ &= (2 - e^{i\psi} - e^{-i\psi})^3 + (e^{-2i\psi/3} - e^{i\psi/3})^3\tau^3 + (e^{2i\psi/3} - e^{-i\psi/3})^3\overline{\tau}^3 \\ &\quad - 3(2 - e^{i\psi} - e^{-i\psi})(e^{-2i\psi/3} - e^{i\psi/3})(e^{2i\psi/3} - e^{-i\psi/3})^3|\tau|^2 \\ &= (2 - e^{i\psi} - e^{-i\psi})^3 \\ &\quad + (2 - e^{i\psi} - e^{-i\psi})(1 - e^{-i\psi})\tau^3 + (2 - e^{i\psi} - e^{-i\psi})(1 - e^{i\psi})\overline{\tau}^3 \\ &\quad - 3(2 - e^{i\psi} - e^{-i\psi})^2|\tau|^2. \end{aligned}$$

The result follows since $2 - e^{i\psi} - e^{-i\psi} > 0$ □

Corollary 5.5. *Suppose that the matrix H given by (5.8) has signature $(2,1)$. For $j = 1, 2, 3$ let $\mathbf{n}_j$ be given by (5.6) and let L_j be the complex line with polar vector $\mathbf{n}_j$. If L_j and L_{j+1} intersect in $\mathbf{H}^2_{\mathbb{C}}$ then they meet at an angle of less than $\pi/3$.*

Proof. Since H has signature $(2,1)$ Lemma 5.4 implies

$$\begin{aligned} 0 &< 3(2 - e^{i\psi} - e^{-i\psi})|\tau|^2 - (1 - e^{-i\psi})\tau^3 - (1 - e^{i\psi})\overline{\tau}^3 \\ &\quad - (2 - e^{i\psi} - e^{-i\psi})^2 \\ &\le 4\sin(\psi/2)|\tau|^3 + 12\sin^2(\psi/2)|\tau|^2 - 16\sin^4(\psi/2) \\ &= 4\sin(\psi/2)\big(|\tau| - \sin(\psi/2)\big)\big(|\tau| + 2\sin(\psi/2)\big)^2. \end{aligned}$$

This implies that $|\tau| > \sin(\psi/2)$. Note that the converse of this inequality is not necessarily true, since in the second line we used $2\mathrm{Re}\big(-(1 - e^{-i\psi})\tau^3\big) \le 2|1 - e^{-i\psi}|\,|\tau^3| = 4\sin(\psi/2)|\tau|^3$.

If L_j and L_{j+1} intersect in $\mathbf{H}^2_{\mathbb{C}}$ then, from (5.5), the angle θ between L_j and L_{j+1} is given by:

$$\cos(\theta) = \frac{\big|\langle \mathbf{n}_{j+1}, \mathbf{n}_j \rangle\big|}{\langle \mathbf{n}_j, \mathbf{n}_j \rangle} = \frac{|\tau|}{2\sin(\psi/2)} > \frac{1}{2}.$$

Therefore $\theta < \pi/3$ as claimed. □

Using H and the formula (5.3) we have

$$R_1 = \begin{bmatrix} e^{2i\psi/3} & \tau & -e^{i\psi/3}\overline{\tau} \\ 0 & e^{-i\psi/3} & 0 \\ 0 & 0 & e^{-i\psi/3} \end{bmatrix}, \tag{5.9}$$

$$R_2 = JR_1J^{-1} = \begin{bmatrix} e^{-i\psi/3} & 0 & 0 \\ -e^{i\psi/3}\overline{\tau} & e^{2i\psi/3} & \tau \\ 0 & 0 & e^{-i\psi/3} \end{bmatrix}, \tag{5.10}$$

$$R_3 = J^{-1}R_1J = \begin{bmatrix} e^{-i\psi/3} & 0 & 0 \\ 0 & e^{-i\psi/3} & 0 \\ \tau & -e^{i\psi/3}\overline{\tau} & e^{2i\psi/3} \end{bmatrix}. \tag{5.11}$$

The parameter τ completely determines the group $\langle R_1, R_2, R_3 \rangle$, or equivalently the normal subgroup $\langle R_1, J \rangle$, up to conjugation. Therefore, in principle, the trace of any element of $\langle R_1, R_2, R_3 \rangle$ may be given as a function of τ. Pratoussevitch [27] has given some beautiful combinatorial formulae for these traces. This generalises earlier work of Sandler [28] for the case where $\psi/3 = \pi$ and so R_j has order 2. With a little more work, Pratoussevitch gives similar formulae for the non-equilateral case. We derive these formulae in the Section 5.6.

5.5. *General triangle groups*

In this section we consider three complex lines in general position and the group generated by complex reflections of angle ψ in their sides. Let L_1, L_2 and L_3 be three complex lines in $\mathbf{H}^2_{\mathbb{C}}$ with normal vectors $\mathbf{n}_1$, $\mathbf{n}_2$ and $\mathbf{n}_3$. Suppose that $\langle \mathbf{n}_1, \mathbf{n}_1 \rangle = \langle \mathbf{n}_2, \mathbf{n}_2 \rangle = \langle \mathbf{n}_3, \mathbf{n}_3 \rangle > 0$. Define

$$\rho = (e^{2i\psi/3} - e^{-i\psi/3}) \frac{\langle \mathbf{n}_2, \mathbf{n}_1 \rangle}{\langle \mathbf{n}_1, \mathbf{n}_1 \rangle},$$
$$\sigma = (e^{2i\psi/3} - e^{-i\psi/3}) \frac{\langle \mathbf{n}_3, \mathbf{n}_2 \rangle}{\langle \mathbf{n}_2, \mathbf{n}_2 \rangle},$$
$$\tau = (e^{2i\psi/3} - e^{-i\psi/3}) \frac{\langle \mathbf{n}_1, \mathbf{n}_3 \rangle}{\langle \mathbf{n}_3, \mathbf{n}_3 \rangle}.$$

These formulae generalise (5.4) but now, since we no longer have the symmetry J, they are not the trace of any group elements. Using Proposition 5.10 below, we will be able to relate them to other traces.

As before, using Section 3.3.2 of Goldman [8], if L_j and L_k meet with angle θ_{jk} then

$$\cos(\theta_{12}) = \frac{|\rho|}{2\sin(\psi/2)}, \quad \cos(\theta_{23}) = \frac{|\sigma|}{2\sin(\psi/2)}, \quad \cos(\theta_{31}) = \frac{|\tau|}{2\sin(\psi/2)}. \tag{5.12}$$

We can use ρ, σ, τ to define a Hermitian form. Once again we normalise so that $\langle \mathbf{n}_j, \mathbf{n}_j \rangle = 2 - e^{i\psi} - e^{-i\psi} > 0$. Then

$$H = \begin{bmatrix} 2 - e^{i\psi} - e^{-i\psi} & (e^{-2i\psi/3} - e^{i\psi/3})\rho & (e^{2i\psi/3} - e^{-i\psi/3})\overline{\tau} \\ (e^{2i\psi/3} - e^{-i\psi/3})\overline{\rho} & 2 - e^{i\psi} - e^{-i\psi} & (e^{-2i\psi/3} - e^{i\psi/3})\sigma \\ (e^{-2i\psi/3} - e^{i\psi/3})\tau & (e^{2i\psi/3} - e^{-i\psi/3})\overline{\sigma} & 2 - e^{i\psi} - e^{-i\psi} \end{bmatrix}. \tag{5.13}$$

We require that H should have signature $(2, 1)$. Since its trace is positive, the same argument we used before shows that this is equivalent to $\det(H) < 0$. Doing so and then arguing in a similar fashion to the proof of Lemma 5.4, we find:

Lemma 5.6. *The matrix H given by* (5.13) *has signature* $(2, 1)$ *if and only* *if:*

$$0 < (2 - e^{i\psi} - e^{-i\psi})\big(|\rho|^2 + |\sigma|^2 + |\tau|^2\big) \\ - (1 - e^{-i\psi})\rho\sigma\tau - (1 - e^{i\psi})\overline{\rho}\,\overline{\sigma}\,\overline{\tau} - (2 - e^{i\psi} - e^{-i\psi})^2.$$

This criterion is equivalent to the one given by Pratoussevitch in Proposition 1 of [27]. A simple geometric consequence of this inequality, generalising Corollary 5.5, is:

Corollary 5.7. *The angles* θ_{jk} *from* (5.12) *satisfy* $\theta_{12} + \theta_{23} + \theta_{31} < \pi$.

Proof. Using the inequality from Lemma 5.6 we find:

$$\begin{aligned} 0 &< (2 - e^{i\psi} - e^{-i\psi})\big(|\rho|^2 + |\sigma|^2 + |\tau|^2\big) \\ &\quad - (1 - e^{-i\psi})\rho\sigma\tau - (1 - e^{i\psi})\overline{\rho}\,\overline{\sigma}\,\overline{\tau} - (2 - e^{i\psi} - e^{-i\psi})^2 \\ &\le 4\sin^2(\psi/2)\big(|\rho|^2 + |\sigma|^2 + |\tau|^2\big) \\ &\quad + 4\sin(\psi/2)|\rho||\sigma||\tau| - 16\sin^4(\psi/2) \\ &= 16\sin^4(\psi/2)\big(\cos^2(\theta_{12}) + \cos^2(\theta_{23}) + \cos^2(\theta_{31})\big) \\ &\quad + 32\sin^4(\psi/2)\cos(\theta_{12})\cos(\theta_{23})\cos(\theta_{31}) - 16\sin^4(\psi/2) \\ &= 16\sin^4(\psi/2)\big(\cos(\theta_{12})\cos(\theta_{23}) + \cos(\theta_{31})\big)^2 \\ &\quad - 16\sin^4(\psi/2)\sin^2(\theta_{12})\sin^2(\theta_{23}) \\ &= 16\sin^4(\psi/2)\big(\cos(\theta_{12} - \theta_{23}) + \cos(\theta_{31})\big)\big(\cos(\theta_{12} + \theta_{23}) + \cos(\theta_{31})\big). \end{aligned}$$

Since $\theta_{jk} \in (0, \pi/2)$ we see that $\cos(\theta_{12} - \theta_{23})$ and $\cos(\theta_{31})$ are both positive. Thus we must have

$$\cos(\theta_{31}) > -\cos(\theta_{12} + \theta_{23}) = \cos(\pi - \theta_{12} - \theta_{23}).$$

Hence $\theta_{31} < \pi - \theta_{12} - \theta_{23}$ as required. □

Matrices for the reflections R_1, R_2, R_3 can the be obtained by using H in the formula (5.3):

$$R_1 = \begin{bmatrix} e^{2i\psi/3} & \rho & -e^{i\psi/3}\overline{\tau} \\ 0 & e^{-i\psi/3} & 0 \\ 0 & 0 & e^{-i\psi/3} \end{bmatrix}, \tag{5.14}$$

$$R_2 = \begin{bmatrix} e^{-i\psi/3} & 0 & 0 \\ -e^{i\psi/3}\overline{\rho} & e^{2i\psi/3} & \sigma \\ 0 & 0 & e^{-i\psi/3} \end{bmatrix}, \tag{5.15}$$

$$R_3 = \begin{bmatrix} e^{-i\psi/3} & 0 & 0 \\ 0 & e^{-i\psi/3} & 0 \\ \tau & -e^{i\psi/3}\overline{\sigma} & e^{2i\psi/3} \end{bmatrix}. \tag{5.16}$$

5.6. *Traces in general triangle groups*

Let R_1, R_2 and R_3 be as in (5.14), (5.15) and (5.16). We will be interested in finding a formula for the trace of each element of $\Delta = \langle R_1, R_2, R_3\rangle$, written as a word in R_1, R_2, R_3 and their inverses. Since cyclic permutation does not affect the trace it will be easier for us to consider cyclic words. Consider an element $R_{a_1}^{\epsilon_1}\dots R_{a_r}^{\epsilon_r}$ of $\langle R_1, R_2, R_3\rangle$ where $a_j \in \{1,2,3\}$ and $\epsilon_j \in \{1,-1\}$. For ease of notation, we make the canonical identification between this word and the sequences $a = (a_1 \dots a_r)$ and $\epsilon = (\epsilon_1 \dots \epsilon_r)$. We shall regard these indices as being defined cyclically, that is $a_{r+1} = a_1$ and $\epsilon_{r+1} = \epsilon_1$.

We need to introduce some notation. For the sequence $a = (a_1 \dots a_r)$ as above and for $j = 1, 2, 3$ taken cyclically (so when $j = 3$ we have $j+1 = 1$) we define

$$z_j(a) = \#\Big\{k \in \{1, \dots, r\} \ : \ a_{k+1} = a_k = j\Big\}, \tag{5.17}$$

$$p_j(a) = \#\Big\{k \in \{1, \dots, r\} \ : \ a_{k+1} = j+1,\ a_k = j\Big\}, \tag{5.18}$$

$$n_j(a) = \#\Big\{k \in \{1, \dots, r\} \ : \ a_{k+1} = j,\ a_k = j+1\Big\}. \tag{5.19}$$

It is easy to see that

$$\#\Big\{k \in \{1, \dots, r\} \ : \ a_k = j\Big\} = z_j(a) + p_j(a) + n_{j-1}(a),$$
$$\#\Big\{k \in \{1, \dots, r\} \ : \ a_{k+1} = j\Big\} = z_j(a) + p_{j-1}(a) + n_j(a).$$

By relabelling the sequence a, it is clear that these numbers must be the same. That is $z_j(a) + p_j(a) + n_{j-1}(a) = z_j(a) + p_{j-1}(a) + n_j(a)$. Therefore we have

$$p_1(a) - n_1(a) = p_2(a) - n_2(a) = p_3(a) - n_3(a).$$

Following Sandler [28], see also Pratoussevitch [27], we define the *winding number* $w(a)$ of the sequence $a = (a_1 \dots a_r)$ to be

$$w(a) = p_j(a) - n_j(a). \tag{5.20}$$

Similarly, for $\epsilon = (\epsilon_1 \dots \epsilon_r)$ define

$$m_+(\epsilon) = \#\Big\{k \in \{1, \dots, r\} \ : \ \epsilon_k = 1\Big\}, \tag{5.21}$$

$$m_-(\epsilon) = \#\Big\{k \in \{1, \dots, r\} \ : \ \epsilon_k = -1\Big\}. \tag{5.22}$$

We now give the main result for computing traces which is due to Pratoussevitch, see Theorems 4 and 10 of [27].

Proposition 5.8. *Let $a = (a_1 \dots a_r)$ be a cyclic word with $a_k \in \{1, 2, 3\}$. Let $\epsilon = (\epsilon_1 \dots \epsilon_r)$ with $\epsilon_k \in \{1, -1\}$. Let $E = \sum_{j=1}^{n} \epsilon_j$. Then*

$$\begin{aligned}
&\mathrm{tr}(R_{a_1}^{\epsilon_1} \cdots R_{a_r}^{\epsilon_r}) \\
&= (e^{i\psi})^{-E/3}\left(3 + \sum_{S} \frac{(e^{i\psi}-1)^z(-e^{i\psi})^n(e^{i\psi})^w}{(-e^{i\psi})^{m_-}} \rho^{p_1}\overline{\rho}^{n_1}\sigma^{p_2}\overline{\sigma}^{n_2}\tau^{p_3}\overline{\tau}^{n_3}\right)
\end{aligned}$$

where the sum is taken over all non-empty subsets $S = \{k_1, \dots, k_m\}$ of the set $\{1, \dots, r\}$. Such a subset determines a subset $a_S = (a_{k_1} \dots a_{k_m})$ of a and $\epsilon_S = (\epsilon_{k_1} \dots \epsilon_{k_m})$ of ϵ. The numbers p_j, n_j, $w = p_j - n_j$, $z = z_1 + z_2 + z_3$, $n = n_1 + n_2 + n_3$ are determined from a_S by (5.17), (5.18), (5.19) *and* (5.20). *Finally, m_- is determined from ϵ_S by* (5.22).

Proof. Let $S = \{k_1, \dots, k_m\}$ be a non-empty subset of $\{1, \dots, r\}$ and denote the corresponding subsets of a and ϵ by $a_S = (a_{k_1} \dots a_{k_m})$ and $\epsilon_S = (\epsilon_{k_1} \dots \epsilon_{k_m})$. Write $a_{k_l} = b_l$ for $l = 1, \dots, m$.

Using the expression for R_j given in equation (5.3), we have

$$\begin{aligned}
e^{i\psi/3}R_j\mathbf{z} &= \mathbf{z} + (e^{i\psi}-1)\frac{\langle \mathbf{z}, \mathbf{n}_j\rangle}{\langle \mathbf{n}_j, \mathbf{n}_j\rangle}\mathbf{n}_j = \left(I + (e^{i\psi}-1)\frac{\mathbf{n}_j\mathbf{n}_j^* H}{\langle \mathbf{n}_j, \mathbf{n}_j\rangle}\right)\mathbf{z}, \\
e^{-i\psi/3}R_j^{-1}\mathbf{z} &= \mathbf{z} + (e^{-i\psi}-1)\frac{\langle \mathbf{z}, \mathbf{n}_j\rangle}{\langle \mathbf{n}_j, \mathbf{n}_j\rangle}\mathbf{n}_j = \left(I + (e^{-i\psi}-1)\frac{\mathbf{n}_j\mathbf{n}_j^* H}{\langle \mathbf{n}_j, \mathbf{n}_j\rangle}\right)\mathbf{z}.
\end{aligned}$$

Therefore

$$\begin{aligned}
&(e^{i\psi/3})^{\epsilon_1}R_{a_1}^{\epsilon_1} \cdots (e^{i\psi/3})^{\epsilon_r}R_{a_r}^{\epsilon_r} \\
&= \left(I + (e^{\epsilon_1 i\psi}-1)\frac{\mathbf{n}_{a_1}\mathbf{n}_{a_1}^* H}{\langle \mathbf{n}_{a_1}, \mathbf{n}_{a_1}\rangle}\right) \cdots \left(I + (e^{\epsilon_r i\psi}-1)\frac{\mathbf{n}_{a_r}\mathbf{n}_{a_r}^* H}{\langle \mathbf{n}_{a_r}, \mathbf{n}_{a_r}\rangle}\right) \\
&= I + \sum_{S\neq\emptyset}(e^{i\psi}-1)^{m_+}(e^{-i\psi}-1)^{m_-}\frac{\mathbf{n}_{b_1}\mathbf{n}_{b_1}^* H\mathbf{n}_{b_2}\cdots\mathbf{n}_{b_{m-1}}^* H\mathbf{n}_{b_m}\mathbf{n}_{b_m}^* H}{\langle \mathbf{n}_{b_1}, \mathbf{n}_{b_1}\rangle\cdots\langle \mathbf{n}_{b_m}, \mathbf{n}_{b_m}\rangle} \\
&= I + \sum_{S\neq\emptyset}\frac{(e^{i\psi}-1)^{|S|}}{(-e^{i\psi})^{m_-}}\frac{\mathbf{n}_{b_1}\langle \mathbf{n}_{b_2}, \mathbf{n}_{b_1}\rangle\cdots\langle \mathbf{n}_{b_m}, \mathbf{n}_{b_{m-1}}\rangle\mathbf{n}_{b_m}^* H}{\langle \mathbf{n}_{b_1}, \mathbf{n}_{b_1}\rangle\cdots\langle \mathbf{n}_{b_m}, \mathbf{n}_{b_m}\rangle}.
\end{aligned}$$

We can gather together the powers of $e^{i\psi}$ on the left hand side to obtain $(e^{i\psi/3})^E = (e^{i\psi})^{E/3}$. Arguing as in the proof of Lemma 5.2, we have

$\mathrm{tr}\left(\mathbf{n}_{b_1}\mathbf{n}_{b_m}^* H\right) = \langle \mathbf{n}_{b_1}, \mathbf{n}_{b_m}\rangle$. Hence

$$\begin{aligned}
&(e^{i\psi})^{E/3}\mathrm{tr}(R_{a_1}^{\epsilon_1}\cdots R_{a_r}^{\epsilon_r})\\
&= 3 + \sum_{S\neq\emptyset} \frac{(e^{i\psi}-1)^{|S|}}{(-e^{i\psi})^{m_-}} \frac{\langle \mathbf{n}_{b_2}, \mathbf{n}_{b_1}\rangle}{\langle \mathbf{n}_{b_1}, \mathbf{n}_{b_1}\rangle} \cdots \frac{\langle \mathbf{n}_{b_m}, \mathbf{n}_{b_{m-1}}\rangle}{\langle \mathbf{n}_{b_{m-1}}, \mathbf{n}_{b_{m-1}}\rangle} \cdot \frac{\langle \mathbf{n}_{b_1}, \mathbf{n}_{b_m}\rangle}{\langle \mathbf{n}_{b_m}, \mathbf{n}_{b_m}\rangle}\\
&= 3 + \sum_{S\neq\emptyset} (-e^{-i\psi})^{m_-} \frac{(e^{i\psi}-1)\langle \mathbf{n}_{b_2}, \mathbf{n}_{b_1}\rangle}{\langle \mathbf{n}_{b_1}, \mathbf{n}_{b_1}\rangle} \cdots \frac{(e^{i\psi}-1)\langle \mathbf{n}_{b_1}, \mathbf{n}_{b_m}\rangle}{\langle \mathbf{n}_{b_m}, \mathbf{n}_{b_m}\rangle}.
\end{aligned}$$

From the definitions of ρ, σ and τ we have

$$\frac{(e^{i\psi}-1)\langle \mathbf{n}_{b_{k+1}}, \mathbf{n}_{b_k}\rangle}{\langle \mathbf{n}_{b_k}, \mathbf{n}_{b_k}\rangle} = \begin{cases}
e^{i\psi}-1 & \text{if } b_{k+1} = b_k;\\
e^{i\psi/3}\rho & \text{if } b_{k+1} = 2,\ b_k = 1;\\
(-e^{i\psi})e^{-i\psi/3}\overline{\rho} & \text{if } b_{k+1} = 1,\ b_k = 2;\\
e^{i\psi/3}\sigma & \text{if } b_{k+1} = 3,\ b_k = 2;\\
(-e^{i\psi})e^{-i\psi/3}\overline{\sigma} & \text{if } b_{k+1} = 2,\ b_k = 3;\\
e^{i\psi/3}\tau & \text{if } b_{k+1} = 1,\ b_k = 3;\\
(-e^{i\psi})e^{-i\psi/3}\overline{\tau} & \text{if } b_{k+1} = 3,\ b_k = 1.
\end{cases}$$

Thus for each sum $S \neq \emptyset$ we have:

$$\begin{aligned}
&(-e^{-i\psi})^{m_-} \frac{(e^{i\psi}-1)\langle \mathbf{n}_{b_2}, \mathbf{n}_{b_1}\rangle}{\langle \mathbf{n}_{b_1}, \mathbf{n}_{b_1}\rangle} \cdots \frac{(e^{i\psi}-1)\langle \mathbf{n}_{b_1}, \mathbf{n}_{b_m}\rangle}{\langle \mathbf{n}_{b_m}, \mathbf{n}_{b_m}\rangle}\\
&= (-e^{-i\psi})^{m_-}\left(e^{i\psi}-1\right)^{z_1}\left(e^{i\psi/3}\rho\right)^{p_1}\left((-e^{i\psi})e^{-i\psi/3}\overline{\rho}\right)^{n_1}\\
&\qquad\cdot\left(e^{i\psi}-1\right)^{z_2}\left(e^{i\psi/3}\sigma\right)^{p_2}\left((-e^{i\psi})e^{-i\psi/3}\overline{\sigma}\right)^{n_2}\\
&\qquad\cdot\left(e^{i\psi}-1\right)^{z_3}\left(e^{i\psi/3}\tau\right)^{p_3}\left((-e^{i\psi})e^{-i\psi/3}\overline{\tau}\right)^{n_3}\\
&= \left(-e^{-i\psi}\right)^{m_-}\left(e^{i\psi}-1\right)^{z_1+z_2+z_3}\left(-e^{i\psi}\right)^{n_1+n_2+n_3}\\
&\qquad\cdot\left(e^{i\psi/3}\right)^{p_1-n_1}\rho^{p_1}\overline{\rho}^{n_1}\left(e^{i\psi/3}\right)^{p_2-n_2}\sigma^{p_2}\overline{\sigma}^{n_2}\left(e^{i\psi/3}\right)^{p_3-n_3}\tau^{p_3}\overline{\tau}^{n_3}\\
&= \left(-e^{-i\psi}\right)^{m_-}\left(e^{i\psi}-1\right)^{z}\left(-e^{i\psi}\right)^{n}\left(e^{i\psi}\right)^{w}\rho^{p_1}\overline{\rho}^{n_1}\sigma^{p_2}\overline{\sigma}^{n_2}\tau^{p_3}\overline{\tau}^{n_3}
\end{aligned}$$

where, in the last line, we have used $w = p_1 - n_1 = p_2 - n_2 = p_3 - n_2$, $z = z_1 + z_2 + z_3$ and $n = n_1 + n_2 + n_3$. □

Note $|S| = m = m_+ + m_- = z + p + n$ where $z = z_1 + z_2 + z_3$, $p = p_1 + p_2 + p_3$ and $n = n_1 + n_2 + n_3$. This means that if we consider

$$(R_{a_1}^{\epsilon_1}\cdots R_{a_r}^{\epsilon_r})^{-1} = R_{a_r}^{-\epsilon_r}\cdots R_{a_1}^{-\epsilon_1}$$

then we must send E to $-E$ and swap m_+ and m_-; p_j and n_j. Using the formula of Proposition 5.8 we can deduce

$$\mathrm{tr}\big((R_{a_1}^{\epsilon_1}\cdots R_{a_r}^{\epsilon_r})^{-1}\big) = \overline{\mathrm{tr}(R_{a_1}^{\epsilon_1}\cdots R_{a_r}^{\epsilon_r})}.$$

This is a good check that the formula might be correct.

An immediately consequence of Proposition 5.8 is the following result, which enables us find the trace field of a triangle group; see Pratoussevitch [27] Theorem 9 for the case where the generators are involutions.

Corollary 5.9. *The trace of any element of Δ may be written as a power of $e^{i\psi/3}$ times a polynomial in $|\rho|^2$, $|\sigma|^2$, $|\tau|^2$, $\rho\sigma\tau$ and $\overline{\rho}\,\overline{\sigma}\,\overline{\tau}$ with coefficients in $\mathbb{Z}[e^{i\psi}, e^{-i\psi}]$. In particular, when ψ is a rational multiple of π then the coefficients may be written in $\mathbb{Z}[e^{i\psi}]$.*

Proof. We examine the term coming from $S \neq \emptyset$ as in the proof of Proposition 5.8. First we have $p_j - n_j = w$ and so when $w \geq 0$ we have $p_j \geq n_j$. Thus writing $p_j = w + n_j$ we have

$$\begin{aligned}
\rho^{p_1}\overline{\rho}^{n_1} &= \rho^{w+n_1}\overline{\rho}^{n_1} &&= \rho^w\big(|\rho|^2\big)^{n_1(s)},\\
\sigma^{p_2}\overline{\sigma}^{n_2} &= \sigma^{w+n_2}\overline{\sigma}^{n_2} &&= \sigma^w\big(|\sigma|^2\big)^{n_2(s)},\\
\tau^{p_3}\overline{\tau}^{n_3} &= \tau^{w+n_3}\overline{\tau}^{n_3} &&= \tau^w\big(|\tau|^2\big)^{n_3(s)}
\end{aligned}$$

and so

$$\rho^{p_1}\overline{\rho}^{n_1}\sigma^{p_2}\overline{\sigma}^{n_2}\tau^{p_3}\overline{\tau}^{n_3} = \big(|\rho|^2\big)^{n_1}\big(|\sigma|^2\big)^{n_2}\big(|\tau|^2\big)^{n_3}(\rho\sigma\tau)^{|w|}.$$

Likewise, when $w \leq 0$, writing $n_j = p_j - w_j$ we have

$$\rho^{p_1}\overline{\rho}^{n_1}\sigma^{p_2}\overline{\sigma}^{n_2}\tau^{p_3}\overline{\tau}^{n_3} = \big(|\rho|^2\big)^{p_1}\big(|\sigma|^2\big)^{p_2}\big(|\tau|^2\big)^{p_3}(\overline{\rho}\,\overline{\sigma}\,\overline{\tau})^{|w|}.$$

In each case this is a monomial in $|\rho|^2$, $|\sigma|^2$, $|\tau|^2$, $\rho\sigma\tau$ and $\overline{\rho}\,\overline{\sigma}\,\overline{\tau}$. □

To make this construction explicit, we now go through it with some particularly important traces; see also Section 8 of Pratoussevitch [27].

Proposition 5.10. *Let R_1, R_2 and R_3 be as above. Then for any distinct j, k, $l \in \{1, 2, 3\}$ we have*

$$\begin{aligned}
\mathrm{tr}(R_1R_2) &= e^{i\psi/3}\big(2 - |\rho|^2\big) + e^{-2i\psi/3},\\
\mathrm{tr}(R_1R_2^{-1}) &= 1 + 2\cos(\psi) + |\rho|^2,\\
\mathrm{tr}(R_1R_2R_3) &= 3 - |\rho|^2 - |\sigma|^2 - |\tau|^2 + \rho\sigma\tau,\\
\mathrm{tr}(R_3R_2R_1) &= 3 - |\rho|^2 - |\sigma|^2 - |\tau|^2 - e^{i\psi}\overline{\rho}\,\overline{\sigma}\,\overline{\tau},\\
\mathrm{tr}(R_1R_2R_3R_2^{-1}) &= e^{i\psi/3}\big(2 - |\rho\sigma - \overline{\tau}|^2\big) + e^{-2i\psi/3}.
\end{aligned}$$

Proof. First consider R_1R_2. We now enumerate all non-empty subsets, their index and winding number, and the contribution they make to the trace. For R_1R_2 the terms are given in the following table:

a_S	ϵ_S	m_-	z	p_1	n_1	p_2	n_2	p_3	n_3	w	term
$\{1\}$	$\{+\}$	0	1	0	0	0	0	0	0	0	$e^{i\psi}-1$
$\{2\}$	$\{+\}$	0	1	0	0	0	0	0	0	0	$e^{i\psi}-1$
$\{1,2\}$	$\{+,+\}$	0	0	1	1	0	0	0	0	0	$-e^{i\psi}\|\rho\|^2$

From this we see that

$$\mathrm{tr}(R_1R_2)=e^{-2i\psi/3}\big(3+e^{i\psi}-1+e^{i\psi}-1-e^{i\psi}|\rho|^2\big)=e^{i\psi/3}\big(2-|\rho|^2\big)+e^{-2i\psi/3}.$$

For $R_1R_2^{-1}$ this table becomes:

a_S	ϵ_S	m_-	z	p_1	n_1	p_2	n_2	p_3	n_3	w	term
$\{1\}$	$\{+\}$	0	1	0	0	0	0	0	0	0	$e^{i\psi}-1$
$\{2\}$	$\{-\}$	1	1	0	0	0	0	0	0	0	$-e^{-i\psi}(e^{i\psi}-1)$
$\{1,2\}$	$\{+,-\}$	1	0	1	1	0	0	0	0	0	$\|\rho\|^2$

From this we see that

$$\mathrm{tr}(R_1R_2^{-1})=3+(e^{i\psi}-1)-e^{-i\psi}(e^{-i\psi}-1)+|\rho|^2=1+2\cos(\psi)+|\rho|^2.$$

Likewise, the table for $R_1R_2R_3$ is:

a_S	ϵ_S	m_-	z	p_1	n_1	p_2	n_2	p_3	n_3	w	term
$\{1\}$	$\{+\}$	0	1	0	0	0	0	0	0	0	$e^{i\psi}-1$
$\{2\}$	$\{+\}$	0	1	0	0	0	0	0	0	0	$e^{i\psi}-1$
$\{3\}$	$\{+\}$	0	1	0	0	0	0	0	0	0	$e^{i\psi}-1$
$\{1,2\}$	$\{+,+\}$	0	0	1	1	0	0	0	0	0	$-e^{i\psi}\|\rho\|^2$
$\{2,3\}$	$\{+,+\}$	0	0	0	0	1	1	0	0	0	$-e^{i\psi}\|\sigma\|^2$
$\{1,3\}$	$\{+,+\}$	0	0	0	0	0	0	1	1	0	$-e^{i\psi}\|\tau\|^2$
$\{1,2,3\}$	$\{+,+,+\}$	0	0	1	0	1	0	1	0	1	$e^{i\psi}\rho\sigma\tau$

Thus

$$\begin{aligned}\mathrm{tr}(R_1R_2R_3)&=e^{-i\psi}\big(3+e^{\psi}-1+e^{i\psi}-1+e^{i\psi}-1\\&\qquad-e^{i\psi}|\rho|^2-e^{i\psi}|\sigma|^2-e^{i\psi}|\tau|^2+e^{i\psi}\rho\sigma\tau\big)\\&=3-|\rho|^2-|\sigma|^2-|\tau|^2+\rho\sigma\tau.\end{aligned}$$

Next, we do the same thing with $R_3R_2R_1$.

a_S	ϵ_S	m_-	z	p_1	n_1	p_2	n_2	p_3	n_3	w	term
$\{1\}$	$\{+\}$	0	1	0	0	0	0	0	0	0	$e^{i\psi}-1$
$\{2\}$	$\{+\}$	0	1	0	0	0	0	0	0	0	$e^{i\psi}-1$
$\{3\}$	$\{+\}$	0	1	0	0	0	0	0	0	0	$e^{i\psi}-1$
$\{2,1\}$	$\{+,+\}$	0	0	1	1	0	0	0	0	0	$-e^{i\psi}\|\rho\|^2$
$\{3,2\}$	$\{+,+\}$	0	0	0	0	1	1	0	0	0	$-e^{i\psi}\|\sigma\|^2$
$\{3,1\}$	$\{+,+\}$	0	0	0	0	0	0	1	1	0	$-e^{i\psi}\|\tau\|^2$
$\{3,2,1\}$	$\{+,+,+\}$	0	0	0	1	0	1	0	1	-1	$(-e^{-i\psi})^{-3}(e^{i\psi})^{-1}\overline{\rho}\,\overline{\sigma}\,\overline{\tau}$

Thus

$$\begin{aligned}\operatorname{tr}(R_3R_2R_1) &= e^{-i\psi}\big(3+e^{\psi}-1+e^{i\psi}-1+e^{i\psi}-1\\ &\qquad -e^{i\psi}|\rho|^2-e^{i\psi}|\sigma|^2-e^{i\psi}|\tau|^2-e^{2i\psi}\overline{\rho}\,\overline{\sigma}\,\overline{\tau}\big)\\ &= 3-|\rho|^2-|\sigma|^2-|\tau|^2-e^{i\psi}\overline{\rho}\,\overline{\sigma}\,\overline{\tau}.\end{aligned}$$

Finally, we do the same thing for $R_1R_2R_3R_2^{-1}$.

a_S	ϵ_S	m_-	z	p_1	n_1	p_2	n_2	p_3	n_3	w	term
$\{1\}$	$\{+\}$	0	1	0	0	0	0	0	0	0	$e^{i\psi}-1$
$\{2\}$	$\{+\}$	0	1	0	0	0	0	0	0	0	$e^{i\psi}-1$
$\{3\}$	$\{+\}$	0	1	0	0	0	0	0	0	0	$e^{i\psi}-1$
$\{2\}$	$\{-\}$	1	1	0	0	0	0	0	0	0	$(-e^{-i\psi})(e^{i\psi}-1)$
$\{1,2\}$	$\{+,+\}$	0	0	1	1	0	0	0	0	0	$-e^{i\psi}\|\rho\|^2$
$\{1,3\}$	$\{+,+\}$	0	0	0	0	0	0	1	1	0	$-e^{i\psi}\|\tau\|^2$
$\{1,2\}$	$\{+,-\}$	1	0	1	1	0	0	0	0	0	$\|\rho\|^2$
$\{2,3\}$	$\{+,+\}$	0	0	0	0	1	1	0	0	0	$-e^{i\psi}\|\sigma\|^2$
$\{2,2\}$	$\{+,-\}$	1	2	0	0	0	0	0	0	0	$(-e^{-i\psi})(e^{i\psi}-1)^2$
$\{3,2\}$	$\{+,-\}$	1	0	0	0	1	1	0	0	0	$\|\sigma\|^2$
$\{1,2,3\}$	$\{+.+,+\}$	0	0	1	0	1	0	1	0	1	$e^{i\psi}\rho\sigma\tau$
$\{1,2,2\}$	$\{+,+,-\}$	1	1	1	1	0	0	0	0	0	$(e^{i\psi}-1)\|\rho\|^2$
$\{1,3,2\}$	$\{+,+,-\}$	1	0	0	1	0	1	0	1	-1	$(-e^{i\psi})^2(e^{-i\psi})\overline{\rho}\,\overline{\sigma}\,\overline{\tau}$
$\{2,3,2\}$	$\{+,+,-\}$	1	1	0	0	1	1	0	0	0	$(e^{i\psi}-1)\|\sigma\|^2$
$\{1,2,3,2\}$	$\{+,+,+,-\}$	1	0	1	1	1	1	0	0	0	$(-e^{i\psi})\|\rho\|^2\|\sigma\|^2$

Thus

$$\begin{aligned}
&\operatorname{tr}(R_1R_2R_3R_2^{-1})\\
&= e^{-2i\psi/3}\big(3 + e^{i\psi} - 1 + e^{i\psi} - 1 + e^{i\psi} - 1 + e^{-i\psi} - 1\\
&\qquad - e^{i\psi}|\rho|^2 - e^{i\psi}|\tau|^2 + |\rho|^2 - e^{i\psi}|\sigma|^2 + e^{-i\psi} - 1 + |\sigma|^2\\
&\qquad + e^{i\psi}\rho\sigma\tau + (e^{i\psi}-1)|\rho|^2 + e^{i\psi}\overline{\rho}\,\overline{\sigma}\,\overline{\tau} + (e^{i\psi}-1)|\sigma|^2 - e^{i\psi}|\rho|^2|\sigma|^2\big)\\
&= e^{i\psi/3}\big(2 - |\rho\sigma - \overline{\tau}|^2\big) + e^{-2i\psi/3}. \qquad \square
\end{aligned}$$

Acknowledgements

These notes are based on a short lecture course given at IMS of National University of Singapore in July 2010 as part of the programme *Topology and Dynamics of Character Varieties*. I would like to thank the organisers Bill Goldman, Caroline Series and Ser Peow Tan for the opportunity to deliver these lectures. I would also like to thank the referee for their very detailed and helpful report.

References

1. B.H. Bowditch; *Geometrical finiteness with variable negative curvature.* Duke Math. J. **77** (1995) 229–274.
2. S.S. Chen & L. Greenberg, *Hyperbolic spaces.* In: Contributions to Analysis, ed. L.V. Ahlfors et al., Academic Press, 1974.
3. M.W. Davis; The geometry and topology of Coxeter groups. *London Mathematical Society Monographs* **32**, Princeton University Press, 2008.
4. M. Deraux, J.R. Parker & J. Paupert, *Census of the complex hyperbolic sporadic groups.* Experimental Mathematics **20** (2011) 467–486.
5. E. Falbel; *Geometric structures associated to triangulations as fixed point sets of involutions.* Topology Appl **154** (2007) 1041–1052.
6. W. Fenchel & J. Nielsen; Discontinuous groups of isometries in the hyperbolic plane. Edited by Asmus L. Schmidt, de Gruyter 2003.
7. W.M. Goldman; *Representations of fundamental groups of surfaces.* In: Geometry and Topology, ed J. Alexander & J. Harer, *Lecture Notes in Maths.*, **1167**, Springer, 1985, 95–117.
8. W.M. Goldman; Complex Hyperbolic Geometry. Oxford Mathematical Monographs. Oxford University Press (1999).
9. W.M. Goldman; *Trace coordinates on Fricke spaces of some simple hyperbolic surfaces.* In: Handbook of Teichmüller theory. Vol. II, *IRMA Lect. Math. Theor. Phys.* **13**, Eur. Math. Soc., Zürich, 2009. 611684.
10. F. Kirwan, Complex Algebraic Curves, *London Mathematical Society Student Texts* **23**, Cambridge University Press, 1992.
11. A. Korányi & H.M. Reimann; *The complex cross-ratio on the Heisenberg group.* L'Enseign. Math. **33** (1987) 291–300.

12. C. Kourouniotis; *Complex length coordinates for quasi-Fuchsian groups.* Mathematika **41** (1994) 173–188.
13. S. Lawton; *Generators, relations and symmetries in pairs of* 3×3 *unimodular matrices.* Journal of Algebra **313** (2007) 782–801.
14. E.H. Lockwood, A Book of Curves, Cambridge University Press, 1961.
15. D.B. McReynolds; Arithmetic Lattices in $\mathrm{SU}(n, 1)$. (to appear).
16. G.D. Mostow; Strong Rigidity of Locally Symmetric Spaces; Annals of Maths. Studies **78**, Princeton University Press 1973.
17. G.D. Mostow; *On a remarkable class of polyhedra in complex hyperbolic space.* Pacific Journal of Maths. **86** (1980) 171–276.
18. Y. Okumura; *Global real analytic length parameters for Teichmüller spaces.* Hiroshima Math. J. **26** (1996) 165-179.
19. J.R. Parker; Hyperbolic Spaces. Jyväskylä Lectures in Mathematics **2**, 2008.
20. J.R. Parker; *Complex hyperbolic lattices.* Discrete Groups and Geometric Structures, ed K. Dekimpe et al., Contemporary Mathematics **501** (2009), 1–42.
21. J.R. Parker; Complex Hyperbolic Kleinian Groups; Cambridge University Press, to appear.
22. J.R. Parker & J. Paupert; *Unfaithful complex hyperbolic triangle groups II: higher order reflections.* Pacific J. Maths. **239** (2009) 357–389.
23. J.R. Parker & I.D. Platis; *Complex hyperbolic Fenchel-Nielsen coordinates.* Topology **47** (2008) 101–135.
24. J.R. Parker & I.D. Platis; *Complex hyperbolic quasi-Fuchsian groups.* Geometry of Riemann Surfaces, ed. F. Gardiner et al., London Mathematical Society Lecture Notes **368** (2010) 309–355.
25. J.R. Parker & C. Series; *Bending formulae for convex hull boundaries.* J. d'Analyse Math. **67** (1995) 165–198.
26. J. Paupert; *Unfaithful complex hyperbolic triangle groups III: arithmeticity and commensurability.* Pacific J. Maths. **245** (2010) 359–372.
27. A. Pratoussevitch; *Traces in complex hyperbolic triangle groups.* Geometriae Dedicata **111** (2005) 159–185.
28. H. Sandler; *Traces in* $\mathrm{SU}(2, 1)$ *and complex hyperbolic ideal triangle groups.* Algebras, Groups and Geometry **12** (1995) 139 156.
29. P. Schmutz; *Die Parametrisierung des Teichmüllerraumes durch geodätische Längenfunktionen.* Comment. Math. Helv. **68** (1993) 278–288.
30. S.P. Tan; *Complex Fenchel-Nielsen coordinates for quasi-Fuchsian structures.* Internat. J. Math. **5** (1994) 239–251.
31. J.M. Thompson; *Non-equilateral complex hyperbolic triangle groups.* To appear.
32. Z.-X. Wen; *Relations polynomiales entre les traces de produits de matrices* C. R. Acad. Sci. Paris **318** (1994) 99–104.
33. P. Will; *Groupes libres, groupes triangulaires et tore épointé dans* $\mathrm{PU}(2, 1)$. Ph.D. thesis University of Paris VI.
34. P. Will; *Traces, cross-ratios and 2-generator subgroups of* $\mathrm{SU}(2, 1)$. Canad. J. Math. **61** (2009) 1407–1436.

LORENTZIAN GEOMETRY

Todd A. Drumm
Department of Mathematics
Howard University
Washington, DC 20059, USA
tdrumm@howard.edu

Flat Lorentz $(3, 1)$ space is the natural home for Einstein's Special Theory of Relativity, with three "space" dimensions and one "time" dimension. The geometry naturally encodes the ideas of inertial frames, time and space dilation. Much of the mathematical terminology and research interests are infused with physical considerations, but purely mathematical questions are at the core of this treatment.

This is an introduction to Lorentzian $(n, 1)$ geometry for all $n \geq 1$, examining the constant curvature spaces: flat, de Sitter (positive curvature) and anti-de Sittter spaces. The conformal boundaries of these Lorentz spacetimes will also be constructed. The connection between Lorentz $(2, 1)$ geometry and geometry of the hyperbolic plane will also be investigated.

Keywords: Lorentzian, Einstein Universe, crooked planes

Mathematics Subject Classification 2000: 53C50, 83A05

1. Introduction

The notes are based on lectures given at the Summer School on the Geometry, Topology and Dynamics of Character Varieties at the Institute for the Mathematical Sciences, Singapore, July 2010. The notes are aimed at the graduate student level, and should not be considered an exhaustive overview of the field of Lorentzian geometry.

1.1. *Physical considerations*

Consider the flat 2-dimensional universe $\mathbb{R}^2$ and plot all the points on this plane for all times t, positive and negative. We now have the spacetime universe with points described as (x, y, t). The *spacetime path* of an (permanent) object in the universe is $\gamma : \mathbb{R} \to \mathbb{R}^3$ where $\gamma(t) = (x(t), y(t), t)$.

For example, a body at rest at the origin of the plane for all time would have the path $\gamma(t) = (0,0,t)$ and the image of the path, also referred to as its path, would be a vertical line. This path defines an *inertial frame of reference*. In particular, an observer following this path observes that, without any forces acting on them, 'bodies at rest tend to stay at rest, bodies in motion tend to stay in motion.' Another, different, inertial frame would be defined by an observer moving at a constant velocity in relation to the original observer. The spacetime path of that observer is a straight line transverse to any and all horizontal planes.

The classical Einstein example is that both the observer standing beside the train tracks and the observer on the train see that inertia (and all laws of physics) holds. Moreover, one constant which must be the same in every inertial reference frame is the speed of light.

Consider the following experiment to measure the speed of light. A stationary observer, Maxx, stands at the origin and his brother, Zakk, holds a mirror at known fixed distance d away. Maxx strikes a match and and records the time τ it takes from the moment he creates the spark until he sees the spark reflected from the mirror. The speed of light $c = 2d/\tau$. For simplicity (and for our joy and piece of mind), set $c = 1$.

Graph the experiment just performed in the spacetime universe. Maxx, the observer, has the spacetime path $(0,0,t)$ and Zakk is holding the mirror at $(0,d,t)$. Maxx creates the spark; the light from the spark goes out in all spatial directions. In every horizontal plane creating a cone of light. The cone has 'slope' 1 (or $1/c$ when $c \neq 1$) with the time dimension thought of as the vertical. The light from the spark bounces off of Zakk's mirror, at the point, $(0,d,d)$ and is redirected along a cone of the same slope and is detected by Maxx at time $\tau = 2d$.

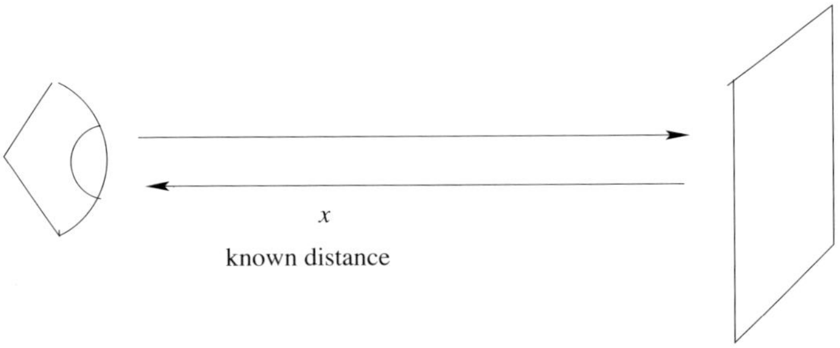

Fig. 1. Measuring the speed of light.

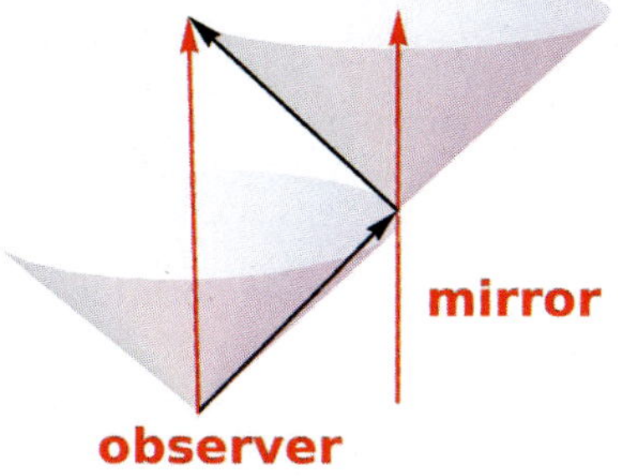

Fig. 2. Graph of measuring the speed of light.

The slope of this *(future) lightcone*, which encodes the speed of light, must be the same in every inertial reference frame. Another observer, the boys' mother Aimee, is on a train travelling at a constant speed of half the speed of light watching the proceedings. Assume spacetime path is $(1, t/2, t)$. To understand Aimee's view, we must transform the entire experiment. Straight lines (in particular, spacetime paths of objects travelling at a constant speed relative to Zakk and Maxx) are mapped to straight lines. Allowing for a change of origin, the transformations considered are of the form $\gamma(x) = g(x) + \mathsf{v}$ where $g \in \mathrm{GL}(3, \mathbb{R})$ and $\mathsf{v} \in \mathbb{R}^3$.

But not all of these transformations will preserve the slope of the light cones. Consider the points on the (entire) lightcone at the origin, $\{(x, y, t) | x^2 + y^2 - t^2 = 0\}$ (the future lightcone is the subset of points for which $t \geq 0$). The equations in the lightcone definition gives rise to the (indefinite) inner product with signature $(2, 1)$:

$$\langle \mathsf{u}, \mathsf{w} \rangle = \mathsf{u}^T \begin{bmatrix} 1 & 0 & 0 \\ 0 & 1 & 0 \\ 0 & 0 & -1 \end{bmatrix} \mathsf{w}. \tag{1.1}$$

The vector space with this inner product is denoted $\mathbb{R}^{2,1}$, and the set matrices which preserve this inner product is denoted $\mathrm{O}(2, 1)$. For the physical situation and most of this paper, we will concern ourselves with $\mathrm{SO}^0(2, 1)$, the connected component of the set of matrices determinant one inside $\mathrm{O}(2, 1)$. The matrices in $\mathrm{SO}^0(2, 1)$ preserve the nappes of the light cone and orientation.

Let $\mathrm{Isom}(\mathbb{E}^{2,1})^0(2, 1) = \mathrm{SO}^0(2, 1) \ltimes \mathbb{R}^{2,1}$. For $\gamma = (g, \mathsf{v}) \in \mathrm{Isom}(\mathbb{E}^{2,1})^0$ $(2, 1)$ the action on the affine space is defined as $\gamma(x) = g(x) + \mathsf{v}$.

2. Flat Lorentz Spaces

Denote the complete flat Lorentz $(n,1)$ space as $\mathbb{E}^{n,1}$. For two points $x, y \in \mathbb{E}^{n,1}$, the difference between two points is a vector. We write, $x - y = \mathsf{v}$ and also $x = y + \mathsf{v}$. By choosing a point $o \in \mathbb{E}^{n,1}$, the space $\mathbb{E}^{n,1}$ is identified with the vector space $\mathbb{R}^{n,1}$ via the map $\varphi : \mathbb{E}^{n,1} \to \mathbb{R}^{n,1}$ defined by $\varphi(x) = x - o$. For the rest of this paper, we will abuse notation and write $\mathbb{R}^{n,1}$ to denote $\mathbb{E}^{n,1}$ and its vector space.

The $(n+1)$-dimensional vector space $\mathbb{R}^{n,1}$ is equipped with the indefinite *Lorentzian inner product*,

$$\langle \mathsf{v}, \mathsf{w} \rangle = \mathsf{v}^T \begin{bmatrix} 1 & & & \\ & \ddots & & \\ & & 1 & \\ & & & -1 \end{bmatrix} \mathsf{w} = \mathsf{v}^T \left(\mathbb{I}_n \oplus (-\mathbb{I}_1) \right) \mathsf{w}, \tag{2.1}$$

where $\mathbb{I}_n$ denotes the $n \times n$ identity matrix.

In the literature are several different notations for these spaces. In particular, $\mathbb{R}^{n+1}_1$ is another way to write $\mathbb{R}^{n,1}$. Whereas, physicists often use $\mathbb{R}^{n+1}$; This symbol denotes essentially the same space, but with an inner product with n minus signs and one positive sign.

Borrowing much of the terminology from Special Relativity, a nonzero vector $\mathsf{v} \in \mathbb{R}^{n,1}$ is called

- *lightlike* (or *null*) if $\langle \mathsf{v}, \mathsf{v} \rangle = 0$,
- *timelike* if $\langle \mathsf{v}, \mathsf{v} \rangle < 0$, or
- *spacelike* if $\langle \mathsf{v}, \mathsf{v} \rangle > 0$.

The *light cone* or *null cone* is the set $\mathbf{N}^{n,1} = \{\mathsf{v} \in \mathbb{R}^{n,1} | \langle \mathsf{v}, \mathsf{v} \rangle = 0\}$. The superscripts will be removed when there is no confusion. The spacelike vectors are "outside" the cone and timelike vectors are "inside" the cone. In every dimension, the light cone minus the origin, $\mathbf{N} \backslash \{\mathbf{0}\}$ has two connected components, or *nappes*.

Time orientation is given by the last component (written as a t in the physical introduction). Lightlike and timelike vectors $\mathsf{v} \in \mathbb{R}^{n,1}$ are called *future pointing* if $v_{n+1} > 0$ and *past pointing* if $v_{n+1} < 0$. The set of future pointing lightlike vectors, called the *future light cone*, consists of future pointing lightlike vectors, and is denoted $\mathbf{N}^f$. Similarly, $\mathbf{N}^p$, the *past light cone*, consists of past pointing lightlike vectors. The set T of timelike vectors is similarly written as the union of T^f, the set of future pointing timelike vectors, and T^p, the set of of past pointing timelike.

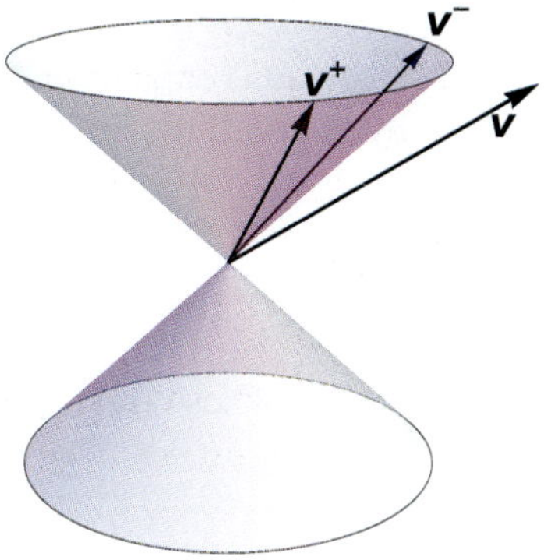

Fig. 3. The $\{\mathsf{v}^-, \mathsf{v}^+, \mathsf{v}\}$ basis for $\mathbb{R}^{2,1}$.

It is interesting to note that the Lorentzian inner product itself does not distinguish time orientation. The choice of time orientation is equivalent to choosing one (nonzero) timelike vector as future pointing. The Lorentzian inner product then does distinguish between nonorthogonal vectors with different time orientations. In particular, if a timelike vector t is chosen to be future pointing, then any other timelike or nonzero lightlike vector v is future pointing if $\langle \mathsf{v}, \mathsf{t} \rangle < 0$.

Given a nonzero vector v, the set of vectors Lorentzian orthogonal to v is written

$$\mathsf{v}^\perp = \{\mathsf{w} \in \mathbb{R}^{n,1} | \langle \mathsf{w}, \mathsf{v} \rangle = 0\rangle.$$

We have the following characterization for the Lorentzian orthogonal space.

- If u is timelike, then every vector in $\mathsf{u}^\perp$ is spacelike and the Lorentzian inner product is *positive definite* on $\mathsf{u}^\perp$.
- If w is lightike, then $\mathsf{w}^\perp$ is tangent to the lightcone at w. All vectors in $\mathsf{w}^\perp$ are spacelike except those vectors parallel to w, and the Lorentzian inner product is said to be *degenerate* on $\mathsf{w}^\perp$.
- If v is spacelike, then $\mathsf{v}^\perp$ is transverse to the lightcone and intersects the light cone along two different lines, and the Lorentzian inner product is said to be *indefinite* on $\mathsf{v}^\perp$. Define the timelike vectors $\mathsf{v}^\pm \in \mathsf{v}^\perp \in \mathbf{N}^f$, with third coordinates $= 1$, so that $\{\mathsf{v}^-, \mathsf{v}^+, \mathsf{v}\}$ is a right handed basis for $\mathbb{R}^{2,1}$ pictured in Figure 3.

3. Compactifications

We now introduce the reader to the compactifications of $\mathbb{R}^{n,1}$. See [3] for similar treatment. We start by reminding the reader about the one-point compactification of $\mathbb{R}^n$.

3.1. *Euclidean compactification*

Start with the vector space $\mathbb{R}^{n+1,1}$, whose inner product can be written as an square diagonal matrix with $n+1$ entries of $= -1$ and one $+1$. Consider the *projectivization* of the null cone:

$$\mathbf{PN}^{n+1,1} = \{[\mathsf{v}] | \mathsf{v} \in \mathbf{N}^{n+1,1} \backslash \{\mathbf{0}\} \text{ where } [\mathsf{v}] = [k\mathsf{v}] \text{ for } k \neq 0\}. \tag{3.1}$$

Let π denote the usual projection, so that $\pi(\mathsf{v}) = [\mathsf{v}]$. The coordinates of any vector $\mathsf{v} \in \mathbf{N}^{n+1,1} \backslash \{\mathbf{0}\}$ are called *homogeneous coordinates* for the point $[\mathsf{v}] \in \mathbf{PN}^{n+1,1}$. For two vectors in the same class, $[\mathsf{w}] = [\mathsf{v}]$, we write $\mathsf{w} \sim \mathsf{v}$.

For $n = 1$, the projectivization of the null cone, seen in Figure 3, is homeomorphic to a circle. The map π maps each line through the origin to a single point in $\mathbf{PN}^{2,1}$, and there are a circle's worth of directions along the null cone. Thus, $\mathbf{PN}^{2,1}$ can be thought of as S^1.

A change of basis $\mathbb{R}^{n+2}$ results in a change of form in the Lorentzian inner product, given in (2.1). For the purposes of this paper, we call this new basis the *embedding basis* as it will make the embedding of $\mathbb{R}^n$ into $\mathbf{PN}^{n+1,1}$ particularly nice. In the embedding basis, the Lorenztian inner product is written as follows:

$$\begin{aligned} \langle \mathsf{v}, \mathsf{w} \rangle &= \textstyle\sum_{i=1}^{n} v_i w_i - \frac{1}{2}\left(v_{n+1} w_{n+2} + v_{n+2} w_{n+1}\right) \\ &= \mathsf{v}^T \left(\mathbb{I}_n \oplus \left(-\tfrac{1}{2}\right) \begin{bmatrix} 0 & 1 \\ 1 & 0 \end{bmatrix} \right) \mathsf{w}. \end{aligned} \tag{3.2}$$

Exercise 1. Find the embedding basis of $\mathbb{R}^n$.

As an aside here, to simplify the equations the inner product will often be provided in different forms. The following are all equivalent ways the Lorentzian inner product can be written with respect to the embedding basis. First, the matrix in (3.2) may be used to identify the inner product:

$$\mathbb{I}_n \oplus \left(-\frac{1}{2}\right) \begin{bmatrix} 0 & 1 \\ 1 & 0 \end{bmatrix}.$$

Other times, the formula for inner product of a vector with itself is enough to define a bilinear inner product.

$$\langle \mathsf{v}, \mathsf{v} \rangle = \sum_{i=1}^{n} v_i^2 - v_{n+1} v_{n+2}.$$

Although we will not use this form, the inner product on the tangent spaces is given by the *metric element*

$$ds^2 = \sum_{i=1}^{n} dx_i^2 - dx_{n+1} dx_{n+2}.$$

There will be more on inner products when we see Lie algebras and matrix groups.

Now embed $\mathbb{R}^n$ into $\mathbf{PN}^{n+1,1}$, using the embedding basis. Define $f : \mathbb{R}^n \hookrightarrow \mathbf{N}^{n+1,1}$ so that

$$f(\mathsf{x}) = \begin{bmatrix} \mathsf{x} \\ \mathsf{x} \cdot \mathsf{x} \\ 1 \end{bmatrix} \tag{3.3}$$

where the dot product, written "·", is the standard Euclidean inner product on $\mathbb{R}^n$, given by $\mathbb{I}_n$. The desired embedding is $\pi \circ f$.

Every $\mathsf{x} \in \mathbb{R}^n$ is clearly mapped into $\mathbf{N}^{n+1,1}$ via $\pi \circ f$. Furthermore, for $\mathsf{x} \neq \mathsf{y}$ their f-images in $\mathbf{N}^{n+1,1}$, $\begin{bmatrix} \mathsf{x} \\ \mathsf{x} \cdot \mathsf{x} \\ 1 \end{bmatrix}$ and $\begin{bmatrix} \mathsf{y} \\ \mathsf{y} \cdot \mathsf{x} \\ 1 \end{bmatrix}$, do not lie on the same line through the origin. Conversely, for every point in **PN** with a nonzero last coordinate:

$$\begin{bmatrix} x_1 \\ \vdots \\ x_{n+1} \\ x_{n+2} \end{bmatrix} \sim \begin{bmatrix} x_1/x_{n+2} \\ \vdots \\ x_{n+1}/x_{n+2} \\ 1 \end{bmatrix} = f(\mathsf{x}) \text{ where } \mathsf{x} = \begin{bmatrix} x_1/x_{n+2} \\ \vdots \\ x_n/x_{n+2} \end{bmatrix}.$$

The only point in **PN** not in the image of $(\pi \circ f)(\mathbb{R}^n)$ is the point with whose last homogeneous coordinate is 0. The coordinates of the points in **N** satisfy the equation

$$\sum_{i=1}^{n} v_1^2 - v_{n+1} v_{n+2} = 0.$$

If the last coordinate is 0, then the sum of the squares of the first n coordinates must also be 0. This implies that the each of the first n

coordinates must be 0. Since the projection is not defined on $\mathbf{0}$, the next to last coordinate must be nonzero. Therefore, we define

$$\infty := \pi\left(\begin{bmatrix} 0 \\ \vdots \\ 1 \\ 0 \end{bmatrix}\right).$$

The *one-point compactification* of the real line is $S^1 = \mathbb{R} \cup \{\infty\}$. The one-point compactification of the real plane is $S^2 = \mathbb{R}^2 \cup \{\infty\}$. Adding a complex structure to S^2 we get the Riemann sphere $\widehat{\mathbb{C}}$.

The notion of angle on the one-point compactification descends from the map π. For a vector $\mathsf{v} \in \mathbf{N}^f$ (third coordinate positive), the tangent space is $\mathsf{v}^\perp$. Tangent vectors at $[\mathsf{v}]$ are images of oriented 2-planes in $\mathsf{v}^\perp$. Given vectors $\mathsf{u}, \mathsf{w} \in \mathsf{v}^\perp$, let L be the oriented plane defined by v and u and K be the oriented plane defined by v and w. For vectors $k\mathsf{v} + l\mathsf{u} \in L$ and $m\mathsf{v} + n\mathsf{w} \in K$, with all coefficients positive, their Lorentzian inner product is

$$\langle m\mathsf{v} + n\mathsf{w}, k\mathsf{v} + l\mathsf{u}\rangle = nl\langle \mathsf{u}, \mathsf{w}\rangle.$$

Dividing by the length of the two vectors, this defines the cosine of the angle between the vectors. This angle is invariant for any two vectors chosen in their respective halfplanes. This defines the angle between the planes L and K, which also defines the angles between tangent vectors at $[\mathsf{v}]$.

Exercise 2. Show that the map f is angle preserving.

The set of transformations which preserve the inner product, and thus also the null cone, is the set of matrices denoted $\mathrm{O}(n+1,1)$. The set of conformal automorphisms of the null cone, and also $S^n = \mathbf{PN}$, is

$$\mathrm{PO}(n+1,1) = \mathrm{O}(n+1,1)/\{\pm 1\}.$$

Exercise 3. For $A \in \mathrm{O}(n)$, $r \neq 0$, and $\mathsf{y} \in \mathbb{R}^n$ show the following are true in the embedding matrix:

- $\begin{bmatrix} \mathbb{I}_n & 0 & \mathsf{y} \\ 2\mathsf{y}^T & 1 & \mathsf{y}\cdot\mathsf{y} \\ 0 & 0 & 1 \end{bmatrix}$ and $\begin{bmatrix} A & 0 & 0 \\ 0 & r & 0 \\ 0 & 0 & r^{-1} \end{bmatrix}$ are elements of $\mathrm{O}(n,1)$.

- For $T(\mathsf{x}) = rA(\mathsf{x}) + \mathsf{y}$, a similarity transformation of $\mathbb{R}^n$, then

$$(f \circ T)(\mathsf{x}) \sim \begin{bmatrix} \mathbb{I}_n & 0 & \mathsf{y} \\ 2\mathsf{y}^T & 1 & \mathsf{y}\cdot\mathsf{y} \\ 0 & 0 & 1 \end{bmatrix} \begin{bmatrix} A & 0 & 0 \\ 0 & r & 0 \\ 0 & 0 & r^{-1} \end{bmatrix} [f(\mathsf{x})] .$$

Now consider another automorphism φ. (A similar automorphism will play a major part in our understanding of the compactification of Lorentz space.) The map $\varphi = \mathbb{I}_n \oplus \begin{bmatrix} 0 & 1 \\ 1 & 0 \end{bmatrix}$ is an element of $\mathrm{O}(n, 1)$ with respect to the embedding basis. It preserves the f-image of the unit sphere in $\mathbb{R}^n$, and induces an action on $\mathbf{PN}$, also denoted φ, which interchanges the image of $\mathbf{0}$ and ∞.

3.2. *The Einstein Universes*

We follow the same method described above to conformally compactify the Lorentzian spaces $\mathbb{R}^{n,1}$. As we shall see, more than one point will be added in these compactifications.

For the compactification of the Lorentz space $\mathbb{R}^{n,1}$, we consider the set of null vectors $\mathbf{N}^{n+1,2}$ inside $\mathbb{R}^{n+1,2}$. The compactification of $\mathbb{R}^{n,1}$ is called the *Einstein Universe* $\mathrm{Ein}^{n,1}$.

In the standard basis, the inner product of $\mathbb{R}^{n+1,2}$ is defined by $\mathbb{I}_{n+1} \oplus (-1)\mathbb{I}_2$,. Just as before, we define an embedding basis where the inner product has the form $\mathbb{I}_n \oplus (-1)\mathbb{I}_1 \oplus \left(-\frac{1}{2}\right) \begin{bmatrix} 0 & 1 \\ 1 & 0 \end{bmatrix}$.

The embedding $g : \mathbb{R}^{n,1} \to \mathbf{N}^{n+1,2}$ is by the map

$$f(\mathsf{x}) = \begin{bmatrix} \mathsf{x} \\ \langle \mathsf{x}, \mathsf{x} \rangle \\ 1 \end{bmatrix} ,$$

where "$\langle \, , \rangle$" here is the Lorentzian inner product (in the standard basis). This map is conformal, for similar reasons that the embedding of Euclidean space was conformal.

The coordinates on the null cone satisfy

$$\sum_{i=1}^{n} v_1^2 - v_{n+1}^2 - v_{n+2}v_{n+3} = 0.$$

and every point that is, not in the image of the embedding must have a 0 in the last coordinates. The points in $\mathrm{Ein}^{n.1}$ which are not in the image of the embedding are of three different types (seen in Figure 4):

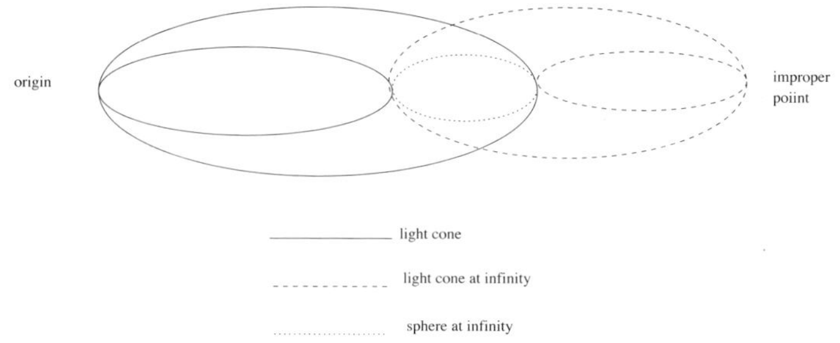

Fig. 4. Points at infinity.

- the *improper point*, denoted ∞, has homogeneous coordinates $\begin{bmatrix}\mathbf{0}\\1\\0\end{bmatrix}$ and is denoted ∞;
- generic points on the *cone at infinity* have homogeneous coordinates $\begin{bmatrix}\mathsf{x}\\1\\0\end{bmatrix}$, where $\mathsf{x}\cdot\mathsf{x}=0$ ($\mathsf{x}\neq\mathbf{0}$);
- points on the *sphere at infinity*, denoted S_∞, have homogeneous coordinates $\begin{bmatrix}\mathsf{x}\\0\\0\end{bmatrix}$, where $\mathsf{x}\cdot\mathsf{x}=0$ but $\mathsf{x}\neq\mathbf{0}$.

Homogeneous coordinates of a point on S_∞ are determined by a Lorentzian null vector x up to scalar multiplication, satisfying the equation

$$x_1^2+\cdots+x_n^2-x_{n+1}^2=0.$$

The last coordinate of x must be nonzero. Let $y_i=x_i/x_{n+1}$, so that each point in S_∞ is identified with one and exactly one solution of

$$y_1^2+\cdots+y_n^2=1.$$

The sphere at infinity in $\mathrm{Ein}^{n,1}$ is homeomorphic to an $(n-1)$-dimensional sphere.

Consider the involution

$$\varphi=\mathbb{I}_{n+1}\oplus\begin{bmatrix}0&1\\1&0\end{bmatrix}. \tag{3.4}$$

The improper point and generic points on the cone at infinity have homogeneous coordinates $\begin{bmatrix} \mathsf{x} \\ 1 \\ 0 \end{bmatrix}$ where $\mathsf{x} \cdot \mathsf{x} = 0$. Thus, $\varphi\left(\begin{bmatrix} \mathsf{x} \\ 1 \\ 0 \end{bmatrix}\right) = \begin{bmatrix} \mathsf{x} \\ 0 \\ 1 \end{bmatrix}$ which are exactly the points in the image of the null cone at the origin in Lorentz space.

3.3. *Extending Lorentzian lines in the Einstein Universe*

Lines of $\mathbb{R}^{n,1}$ are compactified via the embedding $\mathbb{R}^{n,1}$ into $\mathrm{Ein}^{n,1}$. For simplicity, consider lines through the origin. Parameterizations $\gamma : \mathbb{R} \to \mathbb{R}^{n,1}$ of lines through the origin have the form $\gamma(t) = t\mathsf{x}$ where $\mathsf{x} \neq \mathsf{O}$.

The embedding of a lightlike line through O, here $\mathsf{x} \cdot \mathsf{x} = 0$, in $\mathbb{R}^{n,1}$ is

$$\varphi(t\mathsf{x}) = \begin{bmatrix} t\mathsf{x} \\ 0 \\ 1 \end{bmatrix} \sim \begin{bmatrix} \mathsf{x} \\ 0 \\ 1/t \end{bmatrix} \xrightarrow[t\to\pm\infty]{} \begin{bmatrix} \mathsf{x} \\ 0 \\ 0 \end{bmatrix}.$$

Every lightlike line through O is compactified by the addition of a point on the sphere at infinity.

For a Lorentzian spacelike or timelike line through O, we may choose x such that $\mathsf{x} \cdot \mathsf{x} = \pm 1$.

$$\varphi(t\mathsf{x}) = \begin{bmatrix} t\mathsf{x} \\ t^2 \mathsf{x} \cdot \mathsf{x} \\ 1 \end{bmatrix} \sim \begin{bmatrix} t\mathsf{x} \\ \pm t^2 \\ 1 \end{bmatrix} \sim \begin{bmatrix} \mathsf{x}/t \\ \pm 1 \\ 1/t^2 \end{bmatrix} \xrightarrow[t\to\pm\infty]{} \begin{bmatrix} 0 \\ \pm 1 \\ 0 \end{bmatrix} \sim \begin{bmatrix} 0 \\ 1 \\ 0 \end{bmatrix} = \infty.$$

All spacelike and timelike lines through O are extended in $\mathrm{Ein}^{n,1}$ by adding the improper point.

Exercise 4. Show that all spacelike and timelike lines are extended by adding the improper point.

Exercise 5. For a generic point on the cone at infinity v, find a lightlike line $\eta(t)$ such that $\lim_{t\to\infty} \eta(t) = \mathsf{v}$.

3.4. *Covers of the Einstein Universe*

It is often viewed as unsatisfying that the same improper point compactifies the timelike and lightlike lines inside the Einstein Universe. Thus, it is interesting to consider a cover of this universe where the points which compactify timelike and lightlike curves are different. Furthermore, there are orientability issues that arise for the Einstein Universe in different dimensions which will be resolved by passing to a double cover of the Einstein Universe.

Exercise 6. Show that $\mathrm{Ein}^{n,1}$ is unorientable for n even and orientable for n odd.

Define the projection $\zeta : \mathbf{N}^{n+1,2} \setminus \{\mathsf{O}\} \to \widehat{\mathrm{Ein}^{n,1}}$ which identifies points which differ by a *positive* real number. The space $\widehat{\mathrm{Ein}^{n,1}}$ is the *double cover of the Einstein Universe.* We define the two-to-one projection

$$\delta : \widehat{\mathrm{Ein}^{n,1}} \to \mathrm{Ein}^{n,1}\,.$$

Inside of $\widehat{\mathrm{Ein}^{n,1}}$,

$$\begin{bmatrix} \mathsf{x} \\ \langle \mathsf{x}, \mathsf{x}\rangle \\ 1 \end{bmatrix} \neq \begin{bmatrix} -\mathsf{x} \\ -\langle \mathsf{x}, \mathsf{x}\rangle \\ -1 \end{bmatrix} \text{ but } \delta\left(\begin{bmatrix} \mathsf{x} \\ \langle \mathsf{x}, \mathsf{x}\rangle \\ 1 \end{bmatrix}\right) = \delta\left(\begin{bmatrix} -\mathsf{x} \\ -\langle \mathsf{x}, \mathsf{x}\rangle \\ -1 \end{bmatrix}\right).$$

There is another embedding $g : \mathbb{R}^{n,1} \to \mathbf{N}^{n+1,2}$ defined $g = (-1)f$. Their images $\zeta \circ f(\mathbb{R}^{n,1})$ and $\zeta \circ g(\mathbb{R}^{n,1})$ are completely disjoint and are called *Minkowski patches.*

As with the Einstein Universe, consider the points in $\widehat{\mathrm{Ein}^{n,1}}$:

- with respect the Minkowski patch $\zeta \circ f(\mathbb{R}^{n,1})$ the two points in $\delta^{-1}(\infty)$ are called
  - the *spatial improper point*, which has homogeneous coordinates $\begin{bmatrix} \mathbf{0} \\ 1 \\ 0 \end{bmatrix}$ is denoted ∞_{sp}, and
  - the *timelike improper point* has homogeneous coordinates $\begin{bmatrix} \mathbf{0} \\ -1 \\ 0 \end{bmatrix}$ is denoted ∞_{ti};
- the two distinct cones at infinity comprising the inverse image of the cone at infinity in the Einstein Universe, one with homogeneous coordinates $\begin{bmatrix} \mathsf{x} \\ 1 \\ 0 \end{bmatrix}$, and the other with homogeneous coordinates $\begin{bmatrix} \mathsf{x} \\ -1 \\ 0 \end{bmatrix}$ where in both cases $\mathsf{x} \cdot \mathsf{x} = 0$ ($\mathsf{x} \neq \mathbf{0}$);

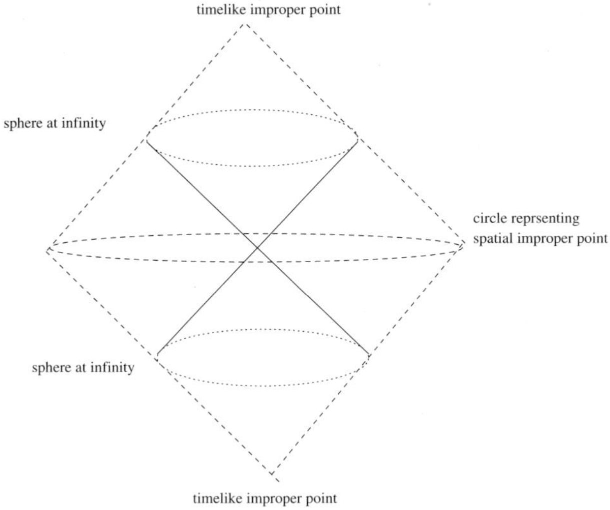

Fig. 5. Minkowski patch of the double cover of the Einstein Universe.

- there is one sphere at infinity, with homogeneous coordinates $\begin{bmatrix} \mathsf{x} \\ 0 \\ 0 \end{bmatrix}$ where $\mathsf{x} \cdot \mathsf{x} = 0$ but $\mathsf{x} \neq \mathbf{0}$, which is a double cover of the sphere at infinity in the Einstein Universe.

The point of view of one Minkowski patch is shown in Figure 5 where the timelike improper point is represented by two identified points, the spatial improper point in represented by an entire circle and the sphere at infinity is represented by the two identified circles. A point on one sphere is identified to a point on the other sphere by scalar multiplication by -1.

In the Minkowski patch $\zeta \circ f(\mathbb{R}^{n,1})$, a timelike line is compactified by the addition of the point ∞_{ti} and a spacelike line is compactified by the addition of the point ∞_{sp}. However, a timelike line in the Minkowski patch $\zeta \circ g(\mathbb{R}^{n,1})$ is compactified by the addition of the point we named ∞_{sp}. A spacelike like is compactified by the addition of the point ∞_{ti}. That is, the names (and notations) of the two improper points would be reversed as viewed from the Minkowski patch $\zeta \circ g(\mathbb{R}^{n,1})$.

Inside of $\widehat{\mathrm{Ein}^{n,1}}$, whether following a timelike line in either a future pointing direction or a past pointing direction the limit is still the one timelike improper point. In order to separate the limits of future of timelike directions and past timelike directions, we can pass to the universal cover of the Einstein Universe $\widetilde{\mathrm{Ein}}^{n,1}$.

4. Constant Curvature Spaces

We start each discussion as we did above. First, we recall the constructions for the constant curvature spherical and hyperbolic geometry. Then we apply this same method to constant curvature Lorentzian geometry.

4.1. *Positive curvature*

4.1.1. *The round sphere*

The standard round sphere S^n is defined in $\mathbb{R}^{n+1}$ as the set of points which are Euclidean unit length from the origin.

$$S^n = \left\{ \mathsf{x} \in \mathbb{R}^{n+1} \,|\mathsf{x} \cdot \mathsf{x} = 1 \right\}.$$

The distance function on S^n is the restriction of the standard Euclidean distance function. The angles are also defined via the standard dot product. The curvature at every point on S^n is identically 1. Geodesics are the arcs of great circles, the intersections of planes in $\mathbb{R}^{n+1}$ through the origin and the S^n.

4.1.2. *de Sitter space*

Define the n-dimensional de Sitter space $\mathrm{dS}^{n,1}$ to be a subset of $\mathbb{R}^{n+1,1}$ which lie a Lorentzian unit length from the origin. In particular, this model of $\mathrm{dS}^{n,1}$ is a hyperboloid of one sheet.

$$\mathrm{dS}^{n,1} = \left\{ \mathsf{x} \in \mathbb{R}^{n+1,1} \,|\langle \mathsf{x}, \mathsf{x} \rangle = 1 \right\}.$$

As before, "$\langle\, ,\, \rangle$" denotes Lorentzian $(n,1)$ inner product.

Remark 4.1. The superscripts in this notation are used to emphasize the signature of the de Sitter spacetime, but the same space is also written as dS^n.

The Lorentzian structure on $\mathrm{dS}^{n,1}$ is defined by the Lorentzian structure on the ambient space $\mathbb{R}^{n+1,1}$. In particular, the tangent space at every point is isomorphic to $\mathbb{R}^{n-1,1}$ as a vector space. As in the Euclidean case, geodesics in de Sitter space are the intersections of planes through the origin and the Lorentzian unit sphere in $\mathbb{R}^{n+1,1}$.

Fig. 6. The one sheeted hyperboloid: a model of $\mathrm{dS}^{1,1}$, and the nonconformal boundary of $\mathrm{adS}^{2,1}$.

The dimension 2 example of de Sitter space is the one-sheeted hyperboloid seen in Figure 6 defined by the equation $x_1^2 + x_2^2 - x_3^2 = 1$.

Exercise 7. Show that the one-sheeted hyperboloid model of dS^n is ruled by two transverse families of lightlike lines.

Sometimes "de Sitter space" is used to denote the universal cover of the de Sitter space defined here.

4.2. *Negative curvature: Hyperboloids*

4.2.1. *Hyperbolic space*

Recall that $T^f \subset \mathbb{R}^{n,1}$ denotes the set of future pointing timelike vectors. The set of vectors $\mathbf{H}^n = \{\mathsf{v} \in T^f \,|\, \langle \mathsf{v}, \mathsf{v} \rangle = -1\}$ is the natural model of n-dimensional hyperbolic space. The inner product on $\mathbf{H}^n$ is the Lorentzian inner product restricted to the tangent space.

Exercise 8. In $\mathbb{R}^{2,1}$, derive the metric for the Poincaré disk model by stereographically projecting $\mathbf{H}^2$ onto the unit disk from the point $\mathsf{v} = \begin{bmatrix} 0 \; 0 \; -1 \end{bmatrix}^T$.

Geodesics in $\mathbf{H}^n$ are the intersections of planes through the origin with $\mathbf{H}^n$. Totally geodesic planes in any appropriate dimension arise from the intersection of $\mathbf{H}^n$ with linear subspaces of one higher dimension. The situation for $\mathbf{H}^2$ is seen in Figure 7. The hyperboloid model of $\mathbf{H}^n$ is defined inside Lorentzian space $\mathbb{R}^{n,1}$ as the set of points which differ from the origin by a

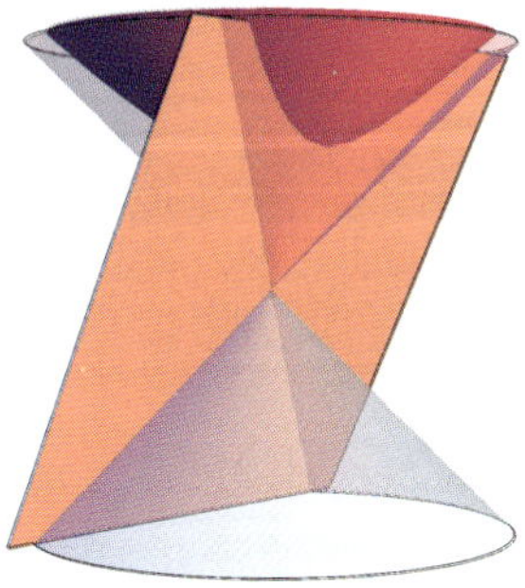

Fig. 7. A light cone, hyperboloid model of $\mathbf{H}^2$, and a plane defining a geodesic.

vector whose inner product with itself is -1 and whose third coordinate is positive.

$$\mathbf{H}^n = \left\{ \mathsf{x} \in \mathbb{R}^{n,1} \,|\langle \mathsf{x}, \mathsf{x} \rangle = -1 \text{ and } x_{n+1} > 0 \right\}.$$

The distance function is derived from the standard Lorentzian inner product restricted to $\mathbf{H}^n$. Geodesics are the intersections of planes in $\mathbb{R}^{n+1}$ through the origin with S^n.

Exercise 9. Show that all tangent vectors to $\mathbf{H}^2$ are spacelike. Integrate along a line segment to find the hyperbolic distance between the point $[0, 0.1]^T$ and $[x, 0, \sqrt{x^2+1}]^T$

The curvature at every point on $\mathbf{H}^n$ is identically -1.

In the conformal compactification, the boundary of $\mathbf{H}^n$ is the projectivization of the null cone in $\mathbb{R}^{n,1}$. That is, $\partial \mathbf{H}^n = S^{n-1}$.

Transformations of the hyperboloid model are linear maps which preserve the underlying Lorentzian structure, written $O(n,1)$. The group $O(n,1)$ has four connected components, and two of these components exchange the hyperboloid with its mirror image through the plane defined by $x_{n+1} = 0$. But points on this mirror image are identified to points on $\mathbf{H}^n$ by projection through the origin. For this reason, a more satisfying model of hyperbolic n-space may be the projective space of all timelike vectors in $\mathbb{R}^{n,1}$. The points on one sheet of the hyperboloid are in one-to-one correspondence with the points in the projective space.

Stereographic projection of the hyperboloid from the point

$$\begin{bmatrix} 0 \\ \vdots \\ 0 \\ -1 \end{bmatrix}$$

onto the plane defined by $x_{n+1} = 0$ gives the ball model of hyperbolic n-space. The boundary of the ball model is the unit sphere. That is, the one point compactification of $\mathbb{R}^n$ is the (conformal) boundary of $\mathbf{H}^n$. The geodesics in this model are the interiors of circular arcs which are perpendicular to the boundary sphere, or diameters.

The ball model of hyperbolic space can be conformally transformed by Möbius transformations to the upper half space model of hyperbolic space. Geodesics are arcs of circles perpendicular to the real line and vertical rays. In dimension 2, the transformations have a particularly nice form.

The Klein model of hyperbolic space B^n is the projection, with respect to the origin. of the hyperboloid model onto the plane

$$P = \{\mathsf{x} \in \mathbb{R}^{n,1} \,|\, x_{n+1} = 1\}.$$

The projection of the hyperboloid onto P, thought of as $\mathbb{R}^n$, is the open unit ball

$$B^n = \{\mathsf{y} \in \mathbb{R}^n \,|\, \mathsf{y} \cdot \mathsf{y} < 1\}.$$

Planes through the origin intersect the hyperboloid and B^n in Euclidean straight lines and contain the segments of the projection from $\mathbf{H}^n$ to B^n. Geodesics in Klein model appear as the interiors of Euclidean line segments. The model can be appealing for this reason, but ultimately the model is not as ubiquitous as other models, because it is not conformal and transformations in this model do not have convenient forms.

4.2.2. *Anti-de Sitter space*

The related hyperboloid model of anti-de Sitter space lies inside $\mathbb{R}^{n,2}$ and consists of projectively equivalent points which differ from the origin by vectors whose inner product with themselves is -1.

$$\mathrm{adS}^{n,1} = \left\{\mathsf{x} \in \mathbb{R}^{n+1,1} \,|\mathsf{x} * \mathsf{x} = -1\right\}.$$

The inner product here, denoted "*", is the inner product with signature $(n, 2)$.

Geodesics are the intersections of planes through the origin and $\mathrm{adS}^{n,1}$. The Lorentzian structure on $\mathrm{adS}^{n,1}$ is induced by the ambient $\mathbb{R}^{n,2}$ structure.

The Lorentzian equivalent of the Klein model of $\mathrm{adS}^{n,1}$ is the projection of $\mathrm{adS}^{n,1}$ onto $P = \{\mathsf{x} \in \mathbb{R}^{n,1} \mid x_{n+2} = 1\}$. This is a local model, as points where $x_{n+2} = 0$ are not represented in this model. Setting $y_k = \frac{x_k}{x_{n+2}}$ for $1 \leq k \leq n+1$, the equation defining the hyperboloid,

$$x_1^2 + \ldots x_n^2 - x_{n+1}^2 - x_{n+2}^2 = -1,$$

turns into

$$y_1^2 + \ldots y_n^2 - y_{n+1}^2 = 1 - \frac{1}{x_{n+2}^2}.$$

This Klein model of $\mathrm{adS}^{2,1}$ is therefore the interior of the one-sheeted hyperboloid seen in Figure 6.

This is not a conformal model of $\mathrm{adS}^{2,1}$.

Exercise 10. Describe the light cones in the Klein model of $\mathrm{adS}^{2,1}$. Show that (unlike $\mathbb{R}^{2,1}$) there are points in $\mathrm{adS}^{2,1}$ whose futures do not intersect.

5. Three Dimensional Flat Lorentz Manifolds

Now, we take an in depth look at complete three dimensional flat Lorentz manifolds. This emerging theory is of a different nature from the preceding sections. Many of the techniques apply only to dimension three in the field of Lorentzian geometry, but are have applications in exploring noncompact hyperbolic surfaces and higher dimensional flat manifolds with indefinite inner products.

Complete three dimensional Lorentz manifolds are of particular interest as counterexamples to a question posed by Milnor. Milnor [13] proved that any *virtually polycyclic group* could arise as the fundamental group of some complete affinely flat manifold and then asked whether the converse was true. (Tits [14] showed that the fundamental group of any complete affine manifold is virtually polycyclic or contains a free group of rank 2.) Fried and Goldman [6] then showed that Lorentzian manifolds were the only possible counterexamples to Milnor's question in dimension 3. Margulis, [10] and [11], constructed free groups of rank 2 (free groups are not virtually polycyclic) which act freely and properly on $\mathbb{R}^{2,1}$. In an influential preprint, published with notes in [12], Mess was the first to show that complete flat Lorentz manifolds cannot have the fundamental group isomorphic to fundamental group of a closed surface group. Classification of flat Lorentz manifolds with free fundamental groups will complete the classification of all affinely flat three-manifolds.

5.1. *Group actions*

In the most general setting, a group action of isometries G for a geometric space X is called *proper* if the mapping $G \times X \to X \times X$ that sends $(g, x) \mapsto (gx, x)$ is a proper map, that is, the preimage of any compact set is compact. If G is discrete and proper then X/G is a geometric manifold or orbifold modelled on X. Presently, we will be interested in exactly which discrete subgroups of act properly and freely on a space X with no fixed points, that is, when X/G is complete. The use of "discrete and proper" is equivalent to the more often used "properly discontinuous": G acts *properly discontinuously* on X if for any compact set $K \subset X$, the set $\{g \in G \,|\, g(K) \cap K \neq \emptyset\}$ is finite.

5.2. *Transformations*

We start with a discussion of affine transformations of $\mathbb{R}^{2,1}$. We shall continue to exploit the deep connection between the affine geometry of $\mathbb{R}^{2,1}$ and hyperbolic geometry of surfaces.

The linear maps of interest to us here are $\mathrm{SO}^0(2,1)$, the identity component of $\mathrm{O}(2,1)$. Each element of $\mathrm{SO}^0(2,1)$ preserves $\mathbf{H}^2$ and its orientation, the future pointing lightcone and thus a time orientation.

Nonidentity elements of $\mathrm{SO}^0(2,1)$ are classified by their trace. An element $g \in \mathrm{SO}^0(2,1)$ is called

- *hyperbolic* if $\mathrm{tr}(g) > 3$,
- *parabolic* if $\mathrm{tr}(g) = 3$, and
- *elliptic* if $0 < \mathrm{tr}(g) < 3$.

A hyperbolic element of $\mathrm{SO}^0(2,1)$ has three positive real eigenvalues $\lambda < 1 < \lambda^{-1}$. A nonzero eigenvector corresponding to the nonidentity eigenvalue must be lightlike (null). In relationship to the hyperbolic plane, these lightlike fixed eigenvectors correspond to fixed points on $\partial\mathbf{H}^2$. A fixed eigenvector v corresponding to the eigenvalue 1 is spacelike and $\mathsf{v}^\perp$ contains all the nonfixed eigenvectors. The plane $\mathsf{v}^\perp$ intersects $\mathbf{H}^2$ in a geodesic called the *axis of* g whose endpoints on $\partial\mathbf{H}^2$ are the fixed points.

A parabolic element of $\mathrm{SO}^0(2,1)$ has exactly one eigenvalue equal to 1. All nonzero fixed eigenvectors are lightlike and correspond to a single fixed point on $\partial\mathbf{H}^2$.

An elliptic element of $SO^0(2,1)$ has one real and two complex eigenvalues. All nonzero fixed eigenvectors are timelike, corresponding to a fixed point in $\mathbf{H}^2$.

The group of affine isometries of $\mathbb{R}^{2,1}$ is the following semidirect product:

$$\mathrm{Isom}(\mathbb{E}^{2,1}) = \mathrm{O}(2,1) \ltimes \mathbb{R}^{2,1}.$$

Unless otherwise noted, we consider the identity component. Each element can be written

$$\gamma(x) = g(x) + \mathsf{v}$$

when $g \in SO^0(2,1)$ and $\mathsf{v} \in \mathbb{R}^{2,1}$. Given the projection $\mathbb{L} : \mathrm{Isom}(\mathbb{E}^{2,1}) \to \mathrm{O}(2,1)$, we say that $g = \mathbb{L}(\gamma)$ is the *linear part* of the affine map γ. The *translational part* of γ is $u(\gamma)$, where $u : \mathrm{Isom}(\mathbb{E}^{2,1}) \to \mathbb{R}^{2,1}$ is a projection on to the vector space $\mathbb{R}^{2,1}$.

Every element of which is not a pure translation will be called *hyperbolic*, *parabolic*, or *elliptic* depending whether its linear part is hyperbolic, parabolic, or elliptic.

5.3. *Margulis invariant*

One important tool used in investigating discrete group actions on $\mathbb{E}$ is the *Margulis invariant* $\alpha(\gamma)$ of a hyperbolic or parabolic Lorentzian transformation $\gamma \in SO^0(2,1)$.

Suppose that γ is hyperbolic. The linear part g has three positive eigenvalues $\lambda < 1 < \lambda^{-1}$. The corresponding eigenvectors are called the *repelling*, *fixed*, and *attracting* and denoted x_g^-, x_g^0, and x_g^+, respectively. The lengths of $\mathsf{x}_g^\pm$ are not identifiable via the Lorentzian structure, but we (arbitrarily) impose the condition that the third coordinates are 1 and that they are both future pointing vectors. Then x_g^0 is defined to be the unique eigenvector with unit Lorentzian length such that $\{\mathsf{x}_g^-, \mathsf{x}_g^+, \mathsf{x}_g^0\}$ is a right-handed basis for $\mathbb{R}^{2,1}$.

Exercise 11. Show that the repelling and attracting eigenvectors $\mathsf{x}_g^\pm$ are lightlike and the fixed eigenvector x_g^0 is spacelike. Furthermore, the $\mathsf{x}_g^\pm$ are Lorentzian perpendicular to x_g^0.

The vectors in Figure 3 represent these vectors, where $\mathsf{v} = \mathsf{x}_g^0$ and $\mathsf{v}^\pm = \mathsf{x}_g^\pm$.

For a hyperbolic element γ there exists a unique fixed line C_γ which is parallel to the vector x_g^0.

Exercise 12. Show that for any hyperbolic element γ there exists a unique fixed line C_γ which is parallel to the vector x_g^0.

Following Margulis in [10] and [11] we define the real valued function α defined on hyperbolic elements in $\mathrm{Isom}(\mathbb{E}^{2,1})$ as follows

$$\alpha(\gamma) = \langle \gamma(\mathsf{x}) - \mathsf{x}, \mathsf{x}_g^0 \rangle \text{ where } \mathsf{x} \in C_\gamma. \tag{5.1}$$

Assuming that γ has no fixed points, $\mathbb{R}^{2,1}/\langle\gamma\rangle$ is a complete Lorentzian manifold. The number $\alpha(\gamma)$ measures the Lorentzian length of the unique closed geodesic in $\mathbb{R}^{2,1}/\langle\gamma\rangle$. As x_g^0 defines an orientation on C_γ, note that $\alpha(\gamma)$ can be negative. As we shall see shortly, this property will have important repercussions in our investigation.

The following facts are not hard to prove:

Theorem 1. *For the function α:*

- *the definition* (5.1) *is independent of choice of $\mathsf{x} \in \mathbb{R}^{2,1}$, that is, $\alpha(\gamma) = \langle \gamma(\mathsf{x}) - \mathsf{x}, \mathsf{x}_g^0 \rangle$ for any $\mathsf{x} \in \mathbb{R}^{2,1}$;*
- *$\alpha(\gamma) = 0$ if and only if γ has a fixed point;*
- *$\alpha(\gamma) = \alpha\left(\eta\gamma\eta^{-1}\right)$ for any $\eta \in \mathrm{Isom}(\mathbb{E}^{2,1})$, that is, α is a* class function*;*
- *$\alpha\left(\gamma^n\right) = |n|\alpha(\gamma)$, and in particular $\alpha\left(\gamma^{-1}\right) = \alpha(\gamma)$.*

The following result, called *Margulis's Opposite Sign Lemma*, is still one of the only ways that one can tell if a group of Lorentzian transformations whose underlying linear group is discrete does not act properly on $\mathbb{R}^{2,1}$.

Lemma 5.1. *For hyperbolic $\gamma_1, \gamma_2 \in \mathrm{Isom}(\mathbb{E}^{2,1})$, if the signs of $\alpha(\gamma_1)$ and $\alpha(\gamma_2)$ differ then the group $\langle\gamma_1, \gamma_2\rangle$ does not act properly on $\mathbb{E}$.*

The sign of α cannot be extended in any meaningful way to elliptic elements. However, the sign of α has been extended to parabolic transformations (see [4]). Versions of Theorem 1 and Margulis' Opposite Sign Lemma are presented in [1], [7], and [8].

Briefly consider extending α to higher dimensional affine geometries.

Exercise 13. Show that for $A \in GL(n, \mathbb{R})$ and $\mathsf{x}, \mathsf{v} \in \mathbb{R}^n$, if A does not have 1 as an eigenvalue, then the map $A(\mathsf{x}) + \mathsf{v}$ has a fixed point.

Generic linear isometries of $\mathbb{R}^{n,1}$ for $n \geq 3$ do not have 1 as an eigenvalue. For example, let us restrict our attention elements in $\mathrm{SO}(3,1)$. While it is true that elements conjugate to $\{1\} \times g$ where $g \in \mathrm{SO}(2,1)$ do have 1 as an eigenvalue, these elements have measure 0 inside the larger Lie group $\mathrm{SO}(3,1)$. That is, generic Lorentzian isometries in dimension 4 (or higher) have a fixed point. The Lorentzian length of closed geodesics in these dimensions does not appear to be a useful invariant.

Alternatively, the function α does have an meaningful extension to higher dimensional complete flat manifolds whose inner product has signature $(n+1,n)$ [2]. Generic elements in $\mathrm{SO}^0(n+1,n)$ do have 1 as an eigenvalue. Margulis' Sign Lemma is true in these dimensions.

In $\mathbb{R}^{3,2}$, a generic element of $g \in \mathrm{SO}^0(3,2)$ is conjugate to

$$g = \begin{bmatrix} \lambda & & & & \\ & \frac{1}{\lambda} & & & \\ & & \mu & & \\ & & & \frac{1}{\mu} & \\ & & & & 1 \end{bmatrix},$$

for $\lambda, \mu > 1$. The direction of the eigenvector, x_g^0 corresponding to the unit eigenvalue is well defined as in $\mathbb{R}^{2,1}$. But in stark contrast to $\mathbb{R}^{2,1}$, the well-defined direction of this fixed eigenvector does not change sign when passing to the inverse, that is, $\mathsf{x}_{g^{-1}}^0 = \mathsf{x}_g^0$. For the affine transformation $\gamma(\mathsf{x}) = g(\mathsf{x}) + \mathsf{v}$ we have $\alpha(\gamma) = -\alpha\left(\gamma^{-1}\right)$. Linear subgroups of $\mathrm{SO}^0(3,2)$ contain generic elements if they are "Zariski dense" in $\mathrm{SO}^0(3,2)$. Thus, any noncyclic free group $\Gamma \subset \mathrm{Isom}^+(\mathbb{E}^{3,2})$ whose whose underlying linear group is Zariski dense in $\mathrm{SO}^0(3,2)$ does not act properly on $\mathbb{E}^{3,2}$.

The general theorem is as follows:

Theorem 5.2. *Consider noncyclic groups* $\Gamma \subset \mathrm{Isom}^+(\mathbb{E}^{n,n-1})$ $(n \geq 2)$ *whose underlying linear group is Zariski dense in* $\mathrm{SO}^0(n,n-1)$.

- *For n even, there exist Γ which act properly and without fixed points on $\mathbb{E}^{n,n-1}$.*
- *For n odd there are no Γ which act properly and without fixed points on $\mathbb{E}^{n,n-1}$.*

5.4. *Deformations of hyperbolic surfaces*

Back in $\mathbb{R}^{2,1}$, another viewpoint of the function α was proposed and used by Goldman and Margulis [9]. This interpretation of α relies on the following amazing facts:

$$\mathrm{SO}^0(2,1) \cong \mathrm{PSL}(2,\mathbb{R}) \text{ and } \mathfrak{sl}(2,\mathbb{R}) \cong \mathbb{R}^{2,1}.$$

The group $\mathfrak{sl}(2,\mathbb{R})$ is the Lie algebra of $\mathrm{PSL}(2,\mathbb{R})$, that it is its tangent space.

We start by establishing the vector space isomorphism $\mathfrak{sl}(2,\mathbb{R}) \cong \mathbb{R}^{2,1}$. The Lie algebra $\mathfrak{sl}(2,\mathbb{R})$ consists of traceless 2×2 real matrices. The vector space is generated by the following matrices:

$$X = \begin{bmatrix} 1 & 0 \\ 0 & -1 \end{bmatrix}, \; Y = \begin{bmatrix} 0 & 1 \\ 1 & 0 \end{bmatrix}, \; Z = \begin{bmatrix} 0 & 1 \\ -1 & 0 \end{bmatrix}.$$

The inner product is given by

$$\mathbb{B}(U,V) = \frac{1}{2}\operatorname{tr}(UV)$$

Exercise 14. Show that the inner product on $\mathfrak{sl}(2,\mathbb{R})$ is indeed equivalent to the Lorentzian inner product for $\mathbb{R}^{2,1}$.

Define the isometry $\psi : \mathfrak{sl}(2,\mathbb{R}) \to \mathbb{R}^{2,1}$ as

$$\psi\left(\begin{bmatrix} a & b \\ c & -a \end{bmatrix}\right) = \begin{bmatrix} a \\ (b+c)/2 \\ (-b+c)/2 \end{bmatrix}$$

The action of $\mathrm{SL}(2,\mathbb{R})$ on $\mathfrak{sl}(2,\mathbb{R})$ is by conjugation.

Exercise 15. Find the isomorphism between $\mathrm{PSL}(2,\mathbb{R})$ and $\mathrm{SO}^0(2,1)$ by looking at the action of $\mathrm{PSL}(2,\mathbb{R})$ on the Lie algebra $\mathfrak{sl}(2,\mathbb{R})$.

We now interpret an affine transformation $\gamma \in \mathrm{SO}^0(2,1) \ltimes \mathbb{R}^{2,1}$ as an element of the isometry group of the hyperbolic plane along with a tangent direction. Since $\mathrm{SO}^0(2,1) \ltimes \mathbb{R}^{2,1} \cong \mathrm{PSL}(2,\mathbb{R}) \times \mathfrak{sl}(2,\mathbb{R})$, we write $\gamma = (\mathsf{g}, V)$ where $\mathsf{g} \in \mathrm{PSL}(2,\mathbb{R})$ corresponding to $g \in \mathrm{SO}^0(2,1)$ and $V \in \mathfrak{sl}(2,\mathbb{R})$ corresponding to $\mathsf{v} \in \mathbb{R}^{2,1}$.

Let g have positive eigenvalues $\mu < 1 < \mu^{-1}$. Up to conjugation

$$\mathsf{g} = \begin{bmatrix} \mu^{-1} & \\ & \mu \end{bmatrix},$$

which we consider to lie in $\mathrm{SL}(2,\mathbb{R})$ without any repercussions. Let Π be the projection

$$\Pi : \mathfrak{gl}(2,\mathbb{R}) \to \mathfrak{sl}(2,\mathbb{R})$$

defined by

$$\Pi(\mathsf{g}) = \mathsf{g} - \frac{\operatorname{tr}(\mathsf{g})}{2}\mathbb{I}.$$

The element of the Lie algebra $\Pi(\mathsf{g})$ is certainly invariant under the action (conjugation) of g. Thus, $\Pi(\mathsf{g})$ is parallel to $\mathsf{x}^0_{\mathsf{g}}$. Without proof (see [9]) we claim

$$\mathsf{x}^0_{\mathsf{g}} = \frac{\Pi(\mathsf{g})}{\operatorname{sgn}(\operatorname{tr}(\mathsf{g}))\sqrt{-\det(\Pi(\mathsf{g}))}}.$$

Here, $\operatorname{sgn}(\operatorname{tr}(\mathsf{g}))$ denotes the sign of the trace of g.

Consider a differentiable path g_t of elements in the Lie group $\mathrm{PSL}(2,\mathbb{R})$ whose images are all hyperbolic elements within a neighborhood of 0. for each value t there is an associated noncompact hyperbolic surface $\Sigma_{\mathsf{g}_t} = \mathbf{H}^2/\langle \mathsf{g}_t\rangle$. Each of these surfaces has a unique closed geodesic g, and the length of the closed geodesic $\ell(g)$ is determined by the absolute value of the trace of g.

$$|\operatorname{tr}(\mathsf{g})| = 2\cosh\left(\ell(g)/2\right).$$

From this equation, we see that $|\operatorname{tr}(\mathsf{g})|$ increases and decreases with $\ell(g)$.

The path g_t of elements in $\mathrm{PSL}(2,\mathbb{R})$ corresponds to the path of *deformations* of surface Σ_{g}. The local geometry of the surfaces does not change, because they all have constant curvature -1, but the global geometry is changed, or deformed.

Writing g_0 as g and Σ_{g}, the *infinitesimal deformation* of the surface is determined by the tangent vector to the path in $\mathrm{PSL}(2,\mathbb{R})$. The tangent vector lies in the Lie algebra $\mathfrak{sl}(2,\mathbb{R})$ which is the tangent space of the Lie group. Specifically, denote the vector tangent to the path at g as $V \in \mathfrak{sl}(2,\mathbb{R})$. The path is then paramaterized as follows:

$$\mathsf{g}_t = \mathsf{g}\left(\exp\left(tV + O(t^2)\right)\right). \tag{5.2}$$

Examine the the derivative of the absolute value of the trace, related to the length of the closed geodesic on the surface, along the path in $\mathrm{PSL}(2,\mathbb{R})$ described in (5.2).

$$\begin{aligned}
\tfrac{d}{dt}|_{t=0}|\operatorname{tr}\mathsf{g}_t| &= \pm\tfrac{d}{dt}|_{t=0}\operatorname{tr}\left(g\left(\exp\left(tV + O(t^2)\right)\right)\right)\\
&= \pm\tfrac{d}{dt}|_{t=0}\operatorname{tr}\left(g\mathbb{I}g^{-1} + tgVg^{-1} + gO(t^2)g^{-1}\right)\\
&= \pm\operatorname{tr}\left(gVg^{-1}\right)\\
&= \tfrac{\sqrt{\operatorname{tr}(\mathsf{g})^2-4}}{2}\,\alpha(\mathsf{g}).
\end{aligned}$$

The resulting correspondence,

$$\left\{ \begin{array}{l} \text{affine map } \gamma = (g, \mathsf{v}): \\ g \in \mathrm{SO}^0(2,1) \text{ and} \\ \mathsf{v} \in \mathbb{R}^{2,1} \end{array} \right\} \leftrightarrow \left\{ \begin{array}{l} \text{path tangent } (\mathsf{g}, V): \\ \mathsf{g} \in \mathrm{PSL}(2,\mathbb{R}) \text{ and} \\ V \in \mathfrak{sl}(2,\mathbb{R}) \end{array} \right\}$$

provides the relationship between the two interpretations of positive Margulis invariant:

$$\left\{ \begin{array}{l} \text{positive Lorentzian} \\ \text{length of closed} \\ \text{geodesic in } \mathbb{R}^{2,1}/\langle\gamma\rangle \end{array} \right\} \leftrightarrow \left\{ \begin{array}{l} \text{infinitesimal increase} \\ \text{of } \ell(\mathsf{g}) \text{ on } \Sigma_{\mathsf{g}_t} \end{array} \right\}.$$

5.5. *Crooked planes*

We now introduce a more geometric method of finding discrete groups of Lorentz transformations which act properly and without fixed points on $\mathbb{E}^{2,1}$. We remind the reader about Dirichlet domains, review some basic examples and then describe their analogs in Lorentz geometry.

5.5.1. *Domains in Euclidean and hyperbolic spaces*

Suppose that X in an n-dimensional manifold with a distance d and that G is a discrete group of isometries of X. For two points $x, y \in X$, define the (closed) half-space of points in X which are closer to x than to y.

$$H(x,y) = \{z \in X \,|\, d(z,x) \leq d(z,y)\}$$

If G is a discrete group of isometries which acts properly on X then the *Dirichlet domain* of G, a particular type of fundamental domain, in X centered at a point x_0 is the set of points closer to x_0, called *the base point*, than to any other image of x_0 under the action of G, that is,

$$\begin{aligned} \Delta_G(x_0) &= \{y \in X \,\big|\, \forall g \in G, d(y,x_0) \leq d(y, g(x_0))\} \\ &= \cap_{g\in G} H\left(x_0, g(x_0)\right) \\ &= X \setminus \cup_{g\in G} \operatorname{int}\left(H\left(g(x_0), x_0\right)\right). \end{aligned}$$

(Here, "int" refers to the interior.) For any $x_0 \in X$, the Dirichlet domain $\Delta_G(x_0)$ is a fundamental domain for the action of G on X. The $(n-1)$-dimensional faces on the boundary of the Dirichlet domain are subsets of *equidistant surfaces* between the base point and its images.

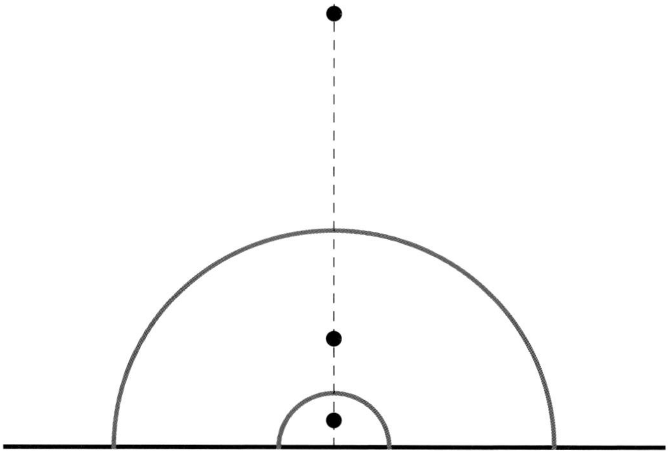

Fig. 8. Dirichlet for a cyclic group of $\mathrm{PSL}(2,\mathbb{R})$ acting on $\mathbf{H}^2$.

Consider the situation where X is the hyperbolic plane $\mathbf{H}^2$. We will use the upper half-plane model of the hyperbolic plane

$$\mathbf{H}^2 = \{z \in \hat{\mathbb{C}} \,|\, \mathrm{Im}(z) > 0\}.$$

The boundary of $\mathbf{H}^2$ is $\partial\mathbf{H}^2 = \mathbb{R} \cup \{\infty\}$. Geodesics of this model are vertical rays or semicircles whose centers lie on the real axis. The group $\mathrm{PSL}(2,\mathbb{R})$ is the set of isometries acting by fractional linear transformations;

$$\begin{bmatrix} a & b \\ c & d \end{bmatrix}(z) = \frac{az+b}{cz+d}.$$

The element $g = \begin{bmatrix} 2 & 0 \\ 0 & 1/2 \end{bmatrix}$ has as an invariant geodesic, called the *axis*, the vertical ray along the imaginary axis. For the discrete group $\langle g \rangle$, choose i as the basepoint for the Dirichlet domain, so $\Delta_{\langle g \rangle}(i)$ is the region bounded by the the geodesic with endpoints $\pm\frac{1}{2}$ and the geodesic with endpoints ± 2, seen in Figure 8.

5.5.2. *Crooked half-spaces*

Returning to our Lorentzian world, let Γ be a discrete group of Lorentz transformations acting on $\mathbb{E}$. If we can construct a fundamental domain for the action of Γ then we would be able to deduce that Γ acts properly. Moreover, this fundamental domain with its identifications may allow us to

Fig. 9. A crooked plane.

determine the topological type of $\mathbb{R}^{2,1}/\Gamma$ fundamental domain Since there exists no metric compatible with the Lorentz structure (the inner product is indefinite after all), there is no chance to find a Dirichlet domain. However, we now define *crooked planes*, which are the Lorentzian analogs of equidistant surfaces in a metric space.

Recall that if u is a lightlike vector, $\mathsf{u}^\perp$ is a plane which is tangent to the light cone. The plane can be divided by the linear subspace defined by u. These half-spaces can be defined by spacelike vectors in $\mathsf{u}^\perp$.

$$\mathcal{W}(\mathsf{v}) := \{\mathsf{w} \in \mathbb{R}^{2,1} | \langle \mathsf{w}, \mathsf{v} \rangle \geq 0 \text{ and } \langle \mathsf{w}, \mathsf{x}_\mathsf{v}^+ \rangle = 0\}.$$

Given a spacelike vector $\mathsf{v} \in \mathsf{u}^\perp$, we have $\mathsf{u}^\perp = \mathcal{W}(\mathsf{v}) \cup \mathcal{W}(-\mathsf{v})$

A *crooked half-space* is determined by its vertex $p \in \mathbb{E}$ and a spacelike *spinal vector* $\mathsf{v} \in \mathbb{R}^{2,1}$. The crooked half-space $\mathcal{H}(p, \mathsf{v})$ is defined as follows:

$$\mathcal{H}(p, \mathsf{v}) := \left\{ q \in \mathbb{E} \,\middle|\, \begin{matrix} \langle (q-p), \mathsf{x}_\mathsf{v}^+ \rangle \leq 0 & \text{if } \langle (q-p), \mathsf{v} \rangle \geq 0 \\ \langle (q-p), \mathsf{x}_\mathsf{v}^- \rangle \geq 0 & \text{if } \langle (q-p), \mathsf{v} \rangle \leq 0 \end{matrix} \right\}.$$

The boundary of a crooked half-space is a *crooked plane*. The crooked plane $\mathcal{C}(p, \mathsf{v})$ is defined as follows:

$$\mathcal{C}(p, \mathsf{v}) := \left\{ q \in \mathbb{E} \,\middle|\, \begin{matrix} \langle (q-p), \mathsf{x}_\mathsf{v}^+ \rangle = 0 & \text{if } \langle (q-p), \mathsf{v} \rangle \geq 0 \\ (\langle (q-p), \mathsf{x}_\mathsf{v}^+ \rangle) \, (\langle (q-p), \mathsf{x}_\mathsf{v}^- \rangle) \geq 0 & \text{if } \langle (q-p), \mathsf{v} \rangle = 0 \\ \langle (q-p), \mathsf{x}_\mathsf{v}^- \rangle = 0 & \text{if } \langle (q-p), \mathsf{v} \rangle \leq 0 \end{matrix} \right\}.$$

The first and third lines in the definition of the crooked plane, the points $p+\mathcal{W}(\mathsf{v})$ and $p+\mathcal{W}(-\mathsf{v})$, define the *wings* of the crooked plane. The wings are half-planes tangent to the light cone through p, where $p+k\mathsf{x}^+(\mathsf{v})$ is the boundary of $p+\mathcal{W}(\mathsf{v})$ and the other wing $p+\mathcal{W}(-\mathsf{v})$ has boundary $p+\mathsf{x}^-(\mathsf{v})$. All vectors from the vertex p to any other point in $p+\mathcal{W}(\mathsf{v})$ are Lorentzian perpendicular to the lightlike vector x_v^+.

Exercise 16. Show that any two spacelike vectors from the same vertex lying in a wing define that wing, that is, show that if $\mathsf{w} \in \mathcal{W}(\mathsf{v})$ then $\mathcal{W}(\mathsf{w}) = \mathcal{W}(\mathsf{v})$. Furthermore, show any two different wings with the same vertex p only share the point p,

Wings consist entirely of points outside the light cone at the vertex p and the set of wings with a common vertex p *foliate* the the exterior of the light cone at p. Of particular interest for a single crooked plane, the two wings of any crooked planes intersect only at the vertex.

The *spine* of the crooked plane, or equivalently of one of the associated crooked half-space, is the line $p+\mathbb{R}\mathsf{v}$.

Exercise 17. Show that the spine of a crooked plane is the only spacelike line that lies completely on the crooked plane.

The second line in the definition of the crooked plane is called the *stem* of the crooked plane. The stem consists of two infinite triangles bounded by the lines $p+\mathsf{x}_\mathsf{v}^+$ and $p+\mathsf{x}_\mathsf{v}^-$ through the origin. All vectors from the vertex p to points on the stem are timelike, that is, the stem entirely consists of points inside the light cone at p.

Emphasizing the connection between crooked planes and Dirichlet domains, consider crooked planes whose vertex is the origin, $\mathcal{C}(o,\mathsf{v})$. The crooked plane, or more specifically its stem, intersects the hyperboloid model of $\mathbf{H}^2$ along geodesics of $\mathbf{H}^2$. Geodesics in $\mathbf{H}^2$ are indeed the equidistant "surfaces" of $\mathbf{H}^2$, that is, geodesic arcs are the faces of Dirichlet domains in $\mathbf{H}^2$.

Any two nonzero parallel spacelike vectors determine the same geodesic in $\mathbf{H}^2$, and in particular the spacelike v and $-\mathsf{v}$ determine the same hyperbolic geodesic. Alternatively, any two nonzero spacelike vectors which are not parallel determine different geodesics in $\mathbf{H}^2$. That is, we have the following correspondence

$$\{\text{spacelike lines in } \mathbb{E} \text{ through the origin}\} \leftrightarrow \left\{\text{geodesics in } \mathbf{H}^2\right\}.$$

Fig. 10. Wings of a crooked plane with spine.

The endpoints of the geodesic corresponding to v themselves correspond the $\mathsf{x}_{\mathsf{v}}^{\pm}$.

Each crooked plane is the boundary for two crooked half-spaces. For a fixed crooked plane, every timelike and lightlike line through the origin either lies entirely in one crooked halfspace or entirely in the other. Alternatively, except for the singular example of the spine of the crooked plane, the two rays emanating from the origin for any other spacelike line through the origin lie in different crooked halfspaces.

To get a better feel for how crooked planes and halfspaces are used in the theory of discrete group actions in Lorentzian geometry, consider the action of a single linear element of $g \in \mathrm{SO}^0(2,1)$ on a single crooked plane whose vertex is the origin o. Let g be a hyperbolic element; it fixes a unique geodesic $\ell \in \mathbf{H}^2$. Choose a vector v which corresponds to a geodesic in $\mathbf{H}^2$ transverse to ℓ.

Let g denote the affine transformation whose linear part is g and whose translational part is just the zero vector. The affine map g fixes the origin and the hyperboloid model of the hyperbolic plane. Consider $g\left(\mathcal{C}(o,\mathsf{v})\right)$. The action of g restricts to a hyperbolic isometry which maps the geodesic

Fig. 11. Stem of a crooked plane with spine and hyperboloid model of $\mathbf{H}^2$.

corresponding to the vector v to a the geodesic corresponding to $g(\mathsf{v})$ which do not intersect because of the transversality condition above. Furthermore, g generates a discrete group which acts properly on $\mathbf{H}^2$ and a fundamental domain for the action is the region bounded by the geodesics corresponding to v and $g(\mathsf{v})$. This situation is drawn in the disk model in Figure 12.

The transformation g maps the stem of the crooked plane $\mathcal{C}(o, \mathsf{v})$ to the stem of $\mathcal{C}(o, g(\mathsf{v}))$. The lightlike lines bounding the stem of $\mathcal{C}(o, \mathsf{v})$ are mapped to the lightlike lines bounding the stem of $\mathcal{C}(o, g(\mathsf{v}))$. That is $g(\mathsf{x}_{\mathsf{v}}^{\pm})$ are parallel to $\mathsf{x}_{g(\mathsf{v})}^{\pm}$. We again emphasize that $g \in \mathrm{SO}^0(2,1)$ preserves the Lorentzian inner product, and thus maps Lorentzian orthogonal vectors to Lorentzian orthogonal vectors. The wings of the crooked plane $\mathcal{C}(o, \mathsf{v})$, which are halfplanes Lorentzian perpendicular to $\mathsf{x}_{\mathsf{v}}^{\pm}$, are mapped to the wings of the crooked plane $\mathcal{C}(o, g(\mathsf{v}))$. which are halfplane Lorentzian perpendicular to $\mathsf{x}_{(g(\mathsf{v})}^{\pm}$. We have shown the following:

$$g\left(\mathcal{C}(o, \mathsf{v})\right) = \mathcal{C}(o, g(\mathsf{v})) = \mathcal{C}(o, -g(\mathsf{v})).$$

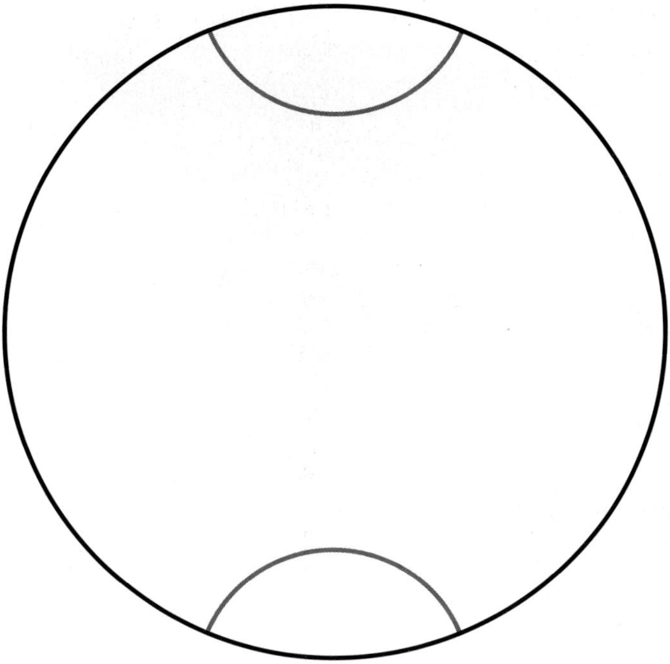

Fig. 12. A fundamental domain for a cyclic group in the disk model.

All of the four wings of $\mathcal{C}(o, \mathsf{v})$ and $\mathcal{C}(o, g(\mathsf{v}))$ are pairwise disjoint away from the origin. The crooked planes $\mathcal{C}(o, \mathsf{v})$ and $\mathcal{C}(o, g(\mathsf{v}))$ bound a region in $\mathbb{R}^{2,1}$ which is a fundamental domain for the action of the group generated by g on $\mathbb{R}^{2,1}$ minus two halfplanes at o which are Lorentzian tangent to $\mathsf{x}_g^{\pm}$.

The language of crooked halfspaces leads to more concise language. If we add the condition that $\mathsf{v} \cdot \mathsf{x}\ (g) > 0$, then the fundamental domain is the closure of the set

$$\mathbb{R}^{2,1} \setminus (\mathcal{H}(o, \mathsf{v}) \cup \mathcal{H}(o, -g(\mathsf{v}))) \,.$$

Now, the discussion moves from the linear to the affine. For an affine transformation $\gamma = (g, \mathsf{u})$, note that

$$\gamma\,(\mathcal{C}(p, \mathsf{v})) = \mathcal{C}(p + \mathsf{u}, g(\mathsf{v})).$$

The effect of adding a translational part to the linear transformation is to "move" the crooked plane. To this end, we consider when two crooked planes are disjoint.

Fig. 13. A fundamental domain for a linear cyclic group in the $\mathbb{R}^{2,1}$.

The set of points in the interior of the crooked half-space on the same plane as the stem of the crooked plane bounding the crooked half-space is called the *stem-quadrant.*

$$\{q \in \mathbb{E} | \langle (q-p), \mathsf{v}\rangle = 0, \langle (q-p), \mathsf{x}_{\mathsf{v}}^{+}\rangle < 0, \text{ and } \langle (q-p), \mathsf{x}_{\mathsf{v}}^{-}\rangle > 0\}.$$

Exercise 18. Suppose that v and w are nonparallel spacelike vectors such that interiors of $\mathcal{H}(o, \mathsf{v})$ and $\mathcal{H}(o, \mathsf{w})$ are disjoint. (Caution: they may share an lightlike line.) Show that if u is in the stem-quadrant of v and y is in the stem-quadrant w then

$$\mathcal{H}(p + \mathsf{u}, \mathsf{v}) \cap \mathcal{H}(p + \mathsf{y}, g(\mathsf{y})) = \emptyset.$$

This leads us to a very useful theorem.

Theorem 2. *Suppose there are $2n$ disjoint crooked halfspaces $\mathcal{H}(p_i, \mathsf{v}_i)$, $i = \pm 1, \ldots, n$, and n Lorentz transformations γ_i, $i = 1, \ldots, n$ such that $\gamma_i(\mathcal{H}(p_{-i}, \mathsf{v}_i))$ is the closure of the complement of $\mathcal{H}(p_{-i}, \mathsf{v}_i)$, then $\langle \gamma_1, \ldots, \gamma_n\rangle$ is a discrete group which acts properly and without fixed points on $\mathbb{R}^{2,1}$.*

See [5]. This theorem can be employed to show that any free discrete group in $\mathrm{SO}(2,1)$ can occur as the linear part of an affine group which acts properly and without fixed points on $\mathbb{R}^{2,1}$. (The groups may include elements not in the identity component.) It may also be used to completely characterize the groups of Lorentz transformations which act properly on $\mathbb{R}^{2,1}$ when the linear part corresponds to a three-holed sphere and a one-holed torus. Furthermore, if Γ is a group which satisfies the conditions of Theorem 2 then the topological type of $\mathbb{R}^{2,1}/\Gamma$ is determined. It is solid handlebody and thus tame: A manifold is *tame* if it is homeomorphic to a compact manifold with a closed subset of the boundary removed.

Finally, we state the *Crooked Plane Conjecture* which would imply that Lorentz manifolds with free fundamental group are tame.

Conjecture 1. *Every free group of rank $n \geq 2$ of Lorentzian transformations that acts properly on $\mathbb{R}^{2,1}$ has a fundamental domain bounded by $2n$ disjoint crooked planes.*

References

1. Abels, H., *Properly discontinuous groups of affine transformations, A survey,* Geom. Ded. **87** (2001) 309–333.
2. Abels, H., Margulis, G., and Soifer, G., *Properly discontinuous groups of affine transformations with orthogonal linear part,* C. R. Acad. Sci. Paris Sér. I Math. **324** No. 3, (1997), 253–258.
3. Barbot, T., Charette, V., Drumm, T., Goldman, W., and Melnick, K., *A Primer on the Einstein Universe,* in "Recent developments in pseudo-Riemannian Geometry, D. Alekseevsky and H. Baum (eds.), Erwin Schrödinger Institute Lectures in Mathematics and Physics, Eur. Math. Soc. (2008), 179 – 221.
4. Charette, V., and Drumm, T., *The Margulis Invariant for Parabolic Transformations,* Proc. A.M.S. **133** (2005), pp. 2439–2447.
5. Drumm, T., *Fundamental polyhedra for Margulis space-times,* Topology **31** (4) (1992), 677–683.
6. Fried, D. and Goldman, W., *Three-dimensional affine crystallographic groups,* Adv. Math. **47** (1983), 1–49.
7. Goldman, W., *The Margulis Invariant of Isometric Actions on Minkowski (2+1)-Space,* in "Ergodic Theory, Geometric Rigidity and Number Theory," Springer-Verlag (2002), 149–164.
8. Goldman, W., Labourie, F. and Margulis, G., *Proper affine actions and geodesic flows of hyperbolic surfaces,* Ann. Math. **170** (2009), 1051–1083. `math.DG/0406247`.
9. Goldman, W. and Margulis, G., *Flat Lorentz 3-manifolds and cocompact Fuchsian groups,* in "Crystallographic Groups and their Generalizations," Contemp. Math. **262** (2000), 135–146.
10. Margulis, G., *Free properly discontinuous groups of affine transformations,* Dokl. Akad. Nauk SSSR **272** (1983), 937–940.
11. Marguls, G., *Complete affine locally flat manifolds with a free fundamental group,* J. Soviet Math. **134** (1987), 129–134.
12. Mess, G., *Lorentz spacetimes of constant curvature,* Geom. Ded. **126**, No. 1, (2007), 3–45.

13. Milnor, J., *On fundamental groups of complete affinely flat manifolds,* Adv. Math. **25** (1977), 178–187.
14. Tits, J., *Free subgroups in linear groups,* J. Algebra **20** (1972), 250–270.

CONNECTED COMPONENTS OF PGL(2, $\mathbb{R}$)-REPRESENTATION SPACES OF NON-ORIENTABLE SURFACES

Frédéric Palesi

LATP, Aix-Marseille Université
Avenue de l'escadrille Normandie Niemen, Case A
13397 Marseille Cedex 20, France
frederic.palesi@univ-cezanne.fr

The Teichmüller space of a surface naturally embeds as a connected component in the moduli space of representations from the fundamental group of the surface into the group of isometries of the hyperbolic plane. We present invariants that distinguish all the connected components of the space of representations. This allows us to compute the number of connected components of these spaces both in the orientable and in the non-orientable case.

Keywords: Representations of surface groups, characteristic classes, non-orientable surfaces

Mathematics Subject Classification 2000: 57M05, 57R20

1. Introduction

Let G be a Lie group and S be a closed surface (orientable or not) with Euler characteristic $\chi(S) < 0$. Its fundamental group is denoted by $\pi = \pi_1(S)$ with respect to a base point. The *representation space* $\mathrm{Hom}(\pi, G)$ is the set of all homomorphisms $\phi : \pi \to G$. The group G acts by conjugation on the representation space, and the quotient of $\mathrm{Hom}(\pi, G)$ by this action is called the *moduli space of representations*

$$\mathrm{Hom}(\pi_1(S), G)/G = \mathcal{X}(S, G).$$

Moduli spaces of representations appear naturally in many contexts, depending on the group G. For example we can identify the moduli space as the set of isomorphism classes of flat G-bundles on the surface. These spaces admit a natural action of the mapping class group exhibiting very interesting dynamics ([8, 14]).

When G is non-compact, the moduli space $\mathcal{X}(S, G)$ contains contractible components which correspond to locally homogeneous geometric structures on the surface (see [9] for a survey on such structures). In particular, when

$G = \mathrm{Isom}(\mathbb{H}^2)$ is the group of isometries of the hyperbolic plane, the related geometric structures on the surface will be hyperbolic. Other geometric structures can be recovered from other groups G such as complex projective structures when $G = \mathrm{PSL}(2, \mathbb{C})$ (see [5]), or convex projective structures when $G = \mathrm{PSL}(3, \mathbb{R})$ (see [4]).

The Teichmüller space $\mathcal{T}(S)$ is equivalent by the uniformization theorem to the set of equivalence classes of complete hyperbolic structures on S. There is an isomorphism between $\mathrm{Isom}(\mathbb{H}^2)$ and the Lie group $\mathrm{PGL}(2, \mathbb{R})$ through the identification of $\mathbb{H}^2$ with the upper-half plane. When the surface S is given a hyperbolic structure, the holonomy map gives a representation of π into $\mathrm{PGL}(2, \mathbb{R})$. This representation is only well defined up to conjugation by an element of $\mathrm{PGL}(2, \mathbb{R})$ (corresponding to a choice of a base point for $\pi_1(S)$). Hence we have a well defined map

$$hol : \mathcal{T}(S) \longrightarrow \mathcal{X}(S, \mathrm{PGL}(2, \mathbb{R})).$$

This map is a proper embedding, and its image is exactly the set of $\mathrm{PGL}(2, \mathbb{R})$-equivalence classes of discrete and faithful representations. The subset of such representations is well-understood inside the moduli space as shown in the following theorem (see [7]):

Theorem 1.1. *Let S be a closed surface (orientable or not). The set of $\mathrm{PGL}(2, \mathbb{R})$-equivalence classes of discrete and faithful representations is a connected component of the moduli space $\mathcal{X}(S, \mathrm{PGL}(2, \mathbb{R}))$.*

Obviously, representation spaces and moduli spaces contain more representations than the discrete and faithful ones, so these spaces have more than one connected component. A natural question is to find some topological invariants that allow one to distinguish the connected components and hence compute the number of components.

For a representation $\phi : \pi_1(S) \to \mathrm{PGL}(2, \mathbb{R})$, a classical invariant is the so-called *Euler class* that is associated to the circle bundle over the surface defined by ϕ. This invariant can be defined in several ways, and the main objective of this note is to give a simple construction valid both in the orientable and the non-orientable case. And we give some insight into how this invariant distinguishes the connected components of the representation space.

In the classical case of orientable surfaces, the holonomy of a hyperbolic structure lies in the group of orientation-preserving isometries $\mathrm{Isom}^+(\mathbb{H}^2)$, which is identified with the group $\mathrm{PSL}(2, \mathbb{R})$.

In Section 2, we will give some basic properties of the group $\mathrm{PSL}(2,\mathbb{R})$ and its universal cover. In particular, we will study the behavior of the map corresponding to the relation of the fundamental group of the surface. This is the key ingredient to define a topological invariant using a simple algorithm.

In Section 3, we first define this invariant for closed orientable surfaces, and then generalize it to orientable surfaces with boundary. This invariant is all that we need to distinguish the connected components of the representation space $\mathrm{Hom}(\pi_1(\Sigma),\mathrm{PSL}(2,\mathbb{R}))$, and give a characterization for the discrete and faithful representations. In [7], Goldman proves the following:

Theorem 1.2. *Let Σ be a closed orientable surface with $\chi(\Sigma)<0$. The representation space $\mathrm{Hom}(\pi,\mathrm{PSL}(2,\mathbb{R}))$ has $2|\chi(\Sigma)|+1$ connected components, indexed by the Euler class $\chi(\Sigma)\leq e\leq -\chi(\Sigma)$.*

Moreover, a representation is discrete and faithful if and only if it is in the maximal components satisfying $|e|=|\chi(\Sigma)|$.

We give only some elements of the proof of this result and refer to Goldman's original paper for a complete proof. This question has also been generalized and solved for a variety of other Lie groups, for example when $G=\mathrm{PSL}(n,\mathbb{R})$ (see [10]), $G=\mathrm{Sp}(2n,\mathbb{R})$ (see [2, 6]), or when the symmetric space associated to G is Hermitian (see [17]).

In Section 4, we extend the construction of the topological invariant defined in Section 3 to the case of a non-orientable surface N. In this case, the invariant is only well defined in $\mathbb{Z}/2\mathbb{Z}$ and the representation space $\mathrm{Hom}(\pi_1(N),\mathrm{PSL}(2,\mathbb{R}))$ has only 2 connected components. These components do not contain any holonomy representations for hyperbolic structures (see [15]).

In this case, it is natural to consider not only orientation-preserving isometries, but also orientation reversing isometries. Hence, the interesting representation space is $\mathrm{Hom}(\pi,\mathrm{PGL}(2,\mathbb{R}))$. For representation into $\mathrm{PGL}(2,\mathbb{R})$, we have to define another invariant that takes values in $(\mathbb{Z}/2\mathbb{Z})^k$, and then understand the set of representations for each value of this invariant. The main result of [16] is the following:

Theorem 1.3. *Let N be a closed non-orientable surface with $\chi(N)<0$, then the representation space $\mathrm{Hom}(\pi,\mathrm{PGL}(2,\mathbb{R}))$ has $2^{3-\chi(n)}-2\chi(N)-1$ connected components indexed by two obstruction classes.*

2. The Group PSL(2, $\mathbb{R}$)

The fundamental group of a surface is a finitely presented group with p generators and q relations. So for any Lie group G the representation space $\mathrm{Hom}(\pi, G)$ can be identified with the subset of G^p defined by q equations in words in G, corresponding to the q relations. So to understand the representation space, we study the properties of G and of the map

$$\begin{aligned} G^p &\longrightarrow G^q \\ (g_1, \ldots, g_p) &\longmapsto (R_1(g_1, \ldots, g_p), \ldots, R_q(g_1, \ldots, g_p)) \end{aligned}$$

where the R_i are the q relations, each in p G-valued unknowns.

In the following section, we shall note $\widehat{G} = \mathrm{SL}(2, \mathbb{R})$ and $G = \mathrm{PSL}(2, \mathbb{R}) = \widehat{G}/\{\pm I\}$. We see that $\widehat{G}$ is a double cover of G.

2.1. *Universal cover*

The standard polar decomposition of $\mathrm{SL}(2, \mathbb{R})$ allows one to decompose any $A \in \widehat{G}$ in a unique way into $A = R \cdot P$, with $R \in \mathrm{SO}(2)$ and $P \in \mathrm{Sym}$, the group of symmetric positive definite matrices of determinant 1. The group Sym being contractible, we have a natural deformation retraction of G onto the circle $\mathrm{SO}(2) \simeq \mathbb{S}^1 \simeq \mathbb{R}/\mathbb{Z}$.

So, the fundamental group of G is $\pi_1(G) \simeq \pi_1(\mathbb{S}^1) \simeq \mathbb{Z}$. Let $\widetilde{G}$ be the universal cover of G and $p : \widetilde{G} \to G$, the covering map. The center of $\widetilde{G}$, $Z(\widetilde{G}) = \ker(p)$ is isomorphic to $\mathbb{Z}$ and $G \simeq \widetilde{G}/\mathbb{Z}$. We denote by z a generator of $Z(\widetilde{G})$ so that $Z(\widetilde{G}) = \langle z \rangle$.

Two lifts of the same element will differ by an element of $Z(\widetilde{G})$. In other words, if $A, B \in \widetilde{G}$ satisfy $p(A) = p(B)$, then there exists $n \in \mathbb{Z}$ such that $A = z^n B$.

The group $\widehat{G}$ is a connected 2-fold cover of G, and thus $\widetilde{G}$ is also the universal cover of $\widehat{G}$. We have a canonical identification $G = \widetilde{G}/\langle z^2 \rangle$.

2.2. *Classification of elements*

The orientation-preserving isometries can be classified through their number of fixed points on the boundary. An element f in $\mathrm{Isom}^+(\mathbb{H})^2 \setminus \{Id\}$, is said to be *elliptic* (resp. *parabolic* and *hyperbolic*) when f has no fixed points on the boundary $\partial\mathbb{H}^2$ (resp. one and two fixed points).

This can be interpreted in terms of the trace of corresponding elements of $\mathrm{PSL}(2, \mathbb{R})$. Notice that only the absolute value of the trace is well defined in G. An element A in $G \setminus \{I\}$, is elliptic (resp. parabolic and hyperbolic) when $|\operatorname{tr}(A)| < 2$ (resp. when $|\operatorname{tr}(A)| = 2$ or $|\operatorname{tr}(A)| > 2$.

Lifts of elliptic (resp. parabolic, hyperbolic) elements in the universal cover $\widetilde{G}$ will also be called elliptic (resp. parabolic, hyperbolic) and we denote by $\mathcal{E}$, $\mathcal{P}$ and $\mathcal{H}$ the corresponding subsets of $\widetilde{G}$. These subsets decompose $\widetilde{G}$ into infinitely many disjoint subsets, indexed by integers $\mathbb{Z}$ corresponding to different copies of G inside $\widetilde{G}$, according to Figure 1.

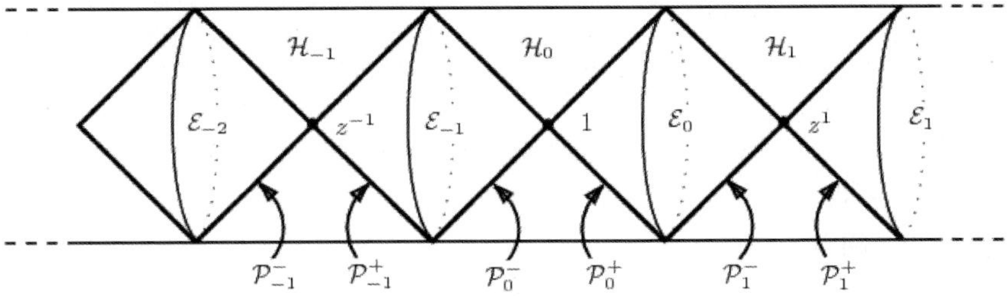

Fig. 1. Domains of $\widetilde{G}$.

We can distinguish these regions using a simple invariant of $A \in \widetilde{G}$. The action of G on $\mathbb{H}^2$ extends to an action on the circle at infinity $\partial\mathbb{H}^2 \simeq \mathbb{S}^1$. This action lifts to an action of $\widetilde{G}$ on $\widetilde{\mathbb{S}^1} \simeq \mathbb{R}$. Hence for any element in $\widetilde{G}$ we can define

$$\underline{m}A = \min\{A \cdot x - x \mid x \in \mathbb{R}\},$$
$$\overline{m}A = \max\{A \cdot x - x \mid x \in \mathbb{R}\}.$$

Using these invariants, we can give the following characterizations [11]:

$$A \in \mathcal{E}_i \Longleftrightarrow [\underline{m}A, \overline{m}A] \subset]i, i+1[,$$
$$A \in \mathcal{H}_i \Longleftrightarrow i \in]\underline{m}A, \overline{m}A[,$$
$$A \in \mathcal{P}_i^+ \Longleftrightarrow \underline{m}A = i < \overline{m}A,$$
$$A \in \mathcal{P}_i^- \Longleftrightarrow \underline{m}A < i = \overline{m}A.$$

There are some direct consequences of these formulas, that are important to understand the case of surfaces with Euler characteristic -1. For example, we have the following proposition (see [7]):

Proposition 2.1. *If $A, B \in \mathcal{H}_0$ and $AB \in \mathcal{H}_i$ then $i \in \{-1, 0, 1\}$.*

We will see in Section 3 how this will be useful to understand the case of the pair-of-pants.

2.3. *The commutator map*

The fundamental group of an orientable surface of genus g is given by:

$$\pi_1(\Sigma_g) = \left\langle X_1, Y_1, \dots, X_g, Y_g \mid \prod_1^g [X_i, Y_i] \right\rangle .$$

Let $A_1, B_1, \dots, A_g, B_g$ be elements of G and choose arbitrary lifts $\widetilde{A_1}, \dots, \widetilde{B_g}$ in $\widetilde{G}$. The element defined by:

$$\widetilde{R_g}(A_1, \dots, B_g) = [\widetilde{A_1}, \widetilde{B_1}] \cdots [\widetilde{A_g}, \widetilde{B_g}],$$

is independent of the chosen lifts. Indeed two different lifts differ by a central element of $Z(\widetilde{G})$ which will cancel out in $\widetilde{R}$. For example, assume that we have two different lifts $\widetilde{A_1}' = g\widetilde{A_1}$ where $g \in Z(\widetilde{G})$ then

$$\begin{aligned}\widetilde{R_g}' &= [\widetilde{A_1}', \widetilde{B_1}] \cdots [\widetilde{A_g}, \widetilde{B_g}] = (g\widetilde{A_1}) \cdot \widetilde{B_1} \cdot (\widetilde{A_1}^{-1} g^{-1}) \cdot \widetilde{B_1}^{-1} \cdots [\widetilde{A_g}, \widetilde{B_g}] \\ &= gg^{-1} \cdot [\widetilde{A_1}, \widetilde{B_1}] \cdots [\widetilde{A_g}, \widetilde{B_g}] = \widetilde{R_g}(A_1, \dots, B_g).\end{aligned}$$

So we can define the map

$$\widetilde{R_g} : G^{2g} \longrightarrow \widetilde{G}. \tag{2.1}$$

When $g = 1$, the image of the commutator map $\widetilde{R_1}$ can be explicitely described using the following proposition (see [7]).

Proposition 2.2. *If $\widetilde{R_1}(A, B)$ is in $\mathcal{H}_i$ then $i \in \{-1, 0, 1\}$.*

We will see in Section 3 how this will be useful to understand the case of the one-holed torus.

2.4. *The square map*

The fundamental group of a non-orientable surface of genus k is given by

$$\pi_1(N_k) = \left\langle X_1, \dots, X_k \mid X_1^2 \cdots X_k^2 \right\rangle .$$

Let $A_1, \dots, A_k$ be elements of G and choose arbitrary lifts $\widetilde{A_1}, \dots, \widetilde{A_k}$ in $\widetilde{G}$. The element defined by

$$Q_k(A_1, \dots, A_k) = \left(\widetilde{A_1}\right)^2 \cdots \left(\widetilde{A_k}\right)^2,$$

is not independent of the chosen lifts. For example, assume that we have two different lifts $\widetilde{A_1}' = g\widetilde{A_1}$ where $g \in Z(\widetilde{G})$ then

$$Q'_k = \left(\widetilde{A_1}'\right)^2 \cdots \left(\widetilde{A_k}\right)^2 = g^2 \left(\widetilde{A_1}\right)^2 \cdots \left(\widetilde{A_k}\right)^2 = g^2 Q_k.$$

However, we see that this gives a well defined element of $\widetilde{G}/\langle z^2 \rangle = \widehat{G}$, hence this defines a continuous map:

$$Q_k : G \longrightarrow \widehat{G}. \tag{2.2}$$

Using this, we can define the corresponding lifted square map $\widetilde{Q_k} : \widetilde{G}^k \longrightarrow \widetilde{G}$.

When $k = 1$, the image of Q_1 is easily computed (see [15]) to be:

$$\mathrm{Im}(Q_1) = J = \{K \in \mathrm{SL}(2, \mathbb{R}) | \operatorname{tr}(K) > -2\} \cup \{I\}.$$

Hence, the image of the lifted square map is:

$$\mathrm{Im}(\widetilde{Q_1}) = \mathcal{J} = \widetilde{G} \setminus \left(\bigcup_{k \in \mathbb{Z}} \mathcal{H}_{2k+1} \cup \mathcal{P}_{2k+1} \right). \tag{2.3}$$

We will see in Section 4 how this will be useful to understand the case of one-holed projective plane.

3. Components for Orientable Surfaces

In this section, we give a simple algorithm to compute the Euler class of a representation $\phi : \pi \to G$ when the surface is orientable and $G = \mathrm{PSL}(2, \mathbb{R})$.

3.1. *A topological invariant*

Let $\pi = \pi_1(\Sigma_g)$ be the fundamental group of the closed orientable surface of genus g and let ϕ be an element of $\mathrm{Hom}(\pi, G)$. The relation of the fundamental group implies that the element $\widetilde{R}(\phi(X_1), \ldots, \phi(Y_g))$ defined by (2.3) is an element $\ker(p) = \mathbb{Z}$, *i.e.* there exists $n \in \mathbb{Z}$ such that

$$\widetilde{R}(\phi(X_1), \ldots, \phi(Y_g)) = z^n.$$

We define $o_2(\phi) = n \in \mathbb{Z}$ and call this number the *second obstruction class* of the representation ϕ. (We shall see the first obstruction class in the last section.) This defines a map:

$$o_2 : \mathrm{Hom}(\pi, \mathrm{PSL}(2, \mathbb{R})) \longrightarrow \mathbb{Z}.$$

This map is continuous and takes values in a discrete set, so o_2 is constant on connected components of the representation space.

We can relate this invariant with other classical invariants, such as the characteristic class $w_2(E_\phi)$ of the flat G-bundle $E_\phi \to \Sigma$, the Euler class $e(\phi)$ of the associated circle bundle over Σ, or also the Toledo invariant $T(\phi)$ of the representation. All these invariants coincide when $G = \mathrm{PSL}(2,\mathbb{R})$. However, in higher dimension they give different generalizations. For precise definitions and relations between these invariants, we refer to [3].

3.2. *Milnor-Wood inequality and a result of Goldman*

A circle bundle over Σ is simply a fiber bundle $E \to \Sigma$ whose fibers are isomorphic to $\mathbb{S}^1$. A representation $\rho : \pi(S) \to \mathrm{PSL}(2,\mathbb{R})$ defines a circle bundle over Σ through the embedding $G \hookrightarrow \mathrm{Homeo}(\mathbb{S}^1)$ defined by the action of G on $\partial\mathbb{H}^2$.

Consider the free product $P = \widetilde{\Sigma} \times \mathbb{S}^1$ and the action of $\pi_1(\Sigma)$ on P given by:

$$\begin{aligned} \pi_1(\Sigma) \times (\widetilde{\Sigma} \times \mathbb{S}^1) &\longrightarrow \widetilde{\Sigma} \times \mathbb{S}^1 \\ \gamma \cdot (x, \theta) &\longmapsto (\gamma \cdot x, \rho(\gamma) \cdot \theta) \end{aligned}$$

where the action of $\pi_1(\Sigma)$ on $\widetilde{\Sigma}$ is by deck transformations. The quotient of P by this action defines a circle bundle over the surface S:

$$E_\rho = (\widetilde{\Sigma} \times \mathbb{S}^1)/\pi_1(\Sigma),$$

since the action on the first factor is free. The Euler class of the representation, denoted $e(\rho) \in \mathbb{Z}$ is defined to be the evaluation of the characteristic class $e(E_\rho)$ of the circle bundle on the fundamental cycle $[\Sigma]$. We have $e(\rho) = o_2(\rho)$.

Euler classes of circles bundles were studied by Milnor and Wood [12, 13] who established the so-called Milnor-Wood inequality, stating that the Euler class of an orientable circle bundle E over Σ is bounded by the Euler characteristic of S, namely

$$|e(E)| \leq |\chi(\Sigma)|.$$

Hence the Euler class $e(\rho)$ of a representation can only take $2|\chi(\Sigma)| + 1$ values, corresponding to the integers contained in the interval $[\chi(\Sigma), -\chi(\Sigma)]$. The Theorem 1.2 establishes the converse result and we can reformulate it as:

Theorem 3.1. *Let Σ be an orientable closed surface with $\chi(\Sigma) < 0$, for all $n \in \mathbb{Z}$, if $|n| \leq |\chi(\Sigma)|$ then $e^{-1}(n)$ is non-empty and connected.*

The proof of the theorem goes by induction on smaller surfaces that are necessarily surfaces with boundary. So next, we define the invariant in this case and generalize Theorem 1.2 for surfaces with boundary.

3.3. *Surfaces with boundary*

For surfaces with boundary, we need to be a little more precise in the definition of the obstruction class. Let $\Sigma = \Sigma_{(g,m)}$, the genus g surface with $m \geq 1$ boundary components denoted $\partial_1\Sigma, \dots, \partial_m\Sigma$. Its fundamental group has the following presentation:

$$\pi_1(\Sigma) = \left\langle X_1, Y_1, \dots, X_g, Y_g, C_1, \dots, C_m \middle| \prod_1^g [X_i, Y_i] \cdot C_1 \cdots C_m \right\rangle,$$

where C_i corresponds to the boundary circle $\partial_i\Sigma$. It is clear that $\pi_1(\Sigma)$ is isomorphic to the free group F_N in $N = 2g + m - 1$ generators, because we can express one of the C_i as a word in the other generators.

In this case, the representation space $\mathrm{Hom}(F_N, G)$ is simply the product group G^N, which is naturally connected as G is itself connected. Indeed when one tries to apply Milnor's algorithm to the presentation above, we get that the element of $\pi_1(G)$ obtained, depends on the choice of the lifts of the image of the m generators $C_1, \dots, C_m$. So in order to have a non-trivial topological invariant, we need to be able to choose canonical lifts for the image of these generators.

An element g in G will have a canonical lift in $\widetilde{G}$ if it has a fixed point on $\partial\mathbb{H} \simeq \mathbb{S}^1$. The canonical lift, denoted $\widetilde{g}^0$, is the unique lift of g in $\widetilde{G}$ that also has fixed points on $\widetilde{\mathbb{S}^1} \simeq \mathbb{R}$. By the classification of hyperbolic isometries, an isometry has a fixed point if and only if it is not elliptic. The canonical lift of a non-elliptic element will belong to $\{I\} \cup \mathcal{H}_0 \cup \mathcal{P}_0$.

Now, we can define an invariant on the set

$$W(\Sigma) = \{\phi \in \mathrm{Hom}(\pi_1(\Sigma), G) \,|\, \forall C_i \subset \partial\Sigma,\, \phi(C_i) \text{ is not elliptic}\}.$$

Let ϕ be an element of $W(\Sigma)$. Choose arbitrary lifts $\widetilde{\phi(A_1)}, \dots, \widetilde{\phi(B_g)}$ of the images of the generators into $\widetilde{G}$. The relation of the fundamental group implies that the element

$$\widetilde{R} = \left[\widetilde{\phi(X_1)}, \widetilde{\phi(Y_1)}\right] \dots \left[\widetilde{\phi(X_g)}, \widetilde{\phi(Y_g)}\right] \cdot \widetilde{\phi(C_1)}^0 \cdots \widetilde{\phi(C_m)}^0,$$

is a lift of the identity element of G into $\widetilde{G}$ that is independent of the chosen lifts. So there exists $n \in \mathbb{Z}$ such that $\widetilde{R} = z^n$. We define $o_2(\phi) = n$ and call it the *relative second obstruction class* of the representation ϕ. This defines the map $o_2 : W(\Sigma) \longrightarrow \mathbb{Z}$

The following theorem is the generalization of Theorem 1.2.

Theorem 3.2. *Let S be an orientable surface with $\chi(S) < 0$. For all $n \in \mathbb{Z}$, if $|n| \leq |\chi(S)|$ then $o_2^{-1}(n)$ is non-empty and connected.*

As a consequence, the space $W(\Sigma)$ has $2|\chi(\Sigma)| + 1$ connected components. We can see that Theorem 1.2 is just a particular case of Theorem 3.2, when Σ has no boundaries.

3.4. *Summary of the proof*

The proof of this theorem goes by induction. First, we prove the theorem for surface of Euler characteristic -1, namely the pair-of-pants $\Sigma_{(0,3)}$ and the one-holed torus $\Sigma_{(1,1)}$. This is mostly a consequence of Propositions 2.1 and 2.2, that can describe the set $W(\Sigma)$.

Next, we decompose any orientable surface into surfaces of characteristic -1. We want the graph dual to this decomposition to be a tree so that the induction process works. For the surface $\Sigma_{(g,m)}$, such a decomposition consists of g one-holed tori and $g + m - 2$ pair-of-pants. One can find a continuous path from any representation to a representation such that the curves defining the decomposition are sent to hyperbolic elements. So it is possible to compute the relative Euler class on each subsurface of the decomposition. Then, using the result on $\Sigma_{(0,3)}$ and $\Sigma_{(1,1)}$, one can show that if two representations have their relative Euler class coincide on each subsurface, then they are in the same connected component.

Then it suffices to show that if two representations have the same Euler class, we can find paths to representations such that the relative Euler classes coincide on each subsurface. First, one can prove this fact for each surface obtained by gluing two surfaces of Euler characteristic -1, namely $\Sigma_2, \Sigma_{(1,2)}$ and $\Sigma_{(0,4)}$. Using this, one can prove the result on a general surface by induction along the tree corresponding to the dual graph of the decomposition (see [7] for a complete proof of these arguments).

4. Non-Orientable Surfaces

In this section, we extend the definition of the obstruction invariant to the representation of the fundamental group of a non-orientable surface into $\mathrm{PSL}(2, \mathbb{R})$, and then in $\mathrm{PGL}(2, \mathbb{R})$. We denote $G_+ = \mathrm{PSL}(2, \mathbb{R})$ and

$G_\pm = \mathrm{PGL}(2,\mathbb{R})$. So the group $G_\pm$ is the disjoint union $G_+ \sqcup G_-$, where G_- is the set of orientation-reversing isometries of $\mathbb{H}^2$.

4.1. *Non-orientable surface group*

Recall that a closed non-orientable surface N_k is homeomorphic to the connected sum of k projective planes. The number k is called the non-orientable genus of the surface and the Euler characteristic is $\chi(N_k) = 2-k$.

The fundamental group is given by the following presentation:

$$\pi_1(N_k) = \langle A_1, \ldots, A_k | A_1^2 \cdots A_k^2 \rangle.$$

In this presentation, the generators are represented by one-sided curves, which are curves whose neighborhoods are Mobiüs strips. In particular, when N_k is given an hyperbolic structure on N_k , the holonomy of a one-sided curve is sent to an orientation-reversing isometry.

4.2. *Second obstruction class*

Let us first understand the connected components of $\mathrm{Hom}(\pi, G_+)$, where $\pi = \pi_1(N_k)$. Let $\phi \in \mathrm{Hom}(\pi, G_+)$. The relation defining the fundamental group implies that the element $Q_k(A_1, \ldots, A_k)$ defined in Section 2.2 is a lift of the identity in G_+ to $\widehat{G}$. This defines a map into the group $Z(\widehat{G}) = \{\pm I\} \simeq \mathbb{Z}/2\mathbb{Z}$. So we have the following map:

$$o_2 : \mathrm{Hom}(\pi, G_+) \to \mathbb{Z}/2\mathbb{Z}.$$

As the map Q_k is continuous, the map o_2 is also continuous and takes values in a discrete set, so we can infer that o_2 is constant on connected components of the representation space.

We can also generalize the construction to non-orientable surfaces with boundaries, denoted $N_{(k,m)}$. As in the orientable case, we have to impose the condition that boundary components are sent to non-elliptic elements, so we define

$$W(N_{(k,m)}) = \left\{\phi \in \mathrm{Hom}(\pi, G_+) \,|\, \forall C \in \partial N_{(k,m)},\, \phi(C) \text{ is not elliptic}\right\}.$$

The map $o_2 : W(N_{(k,m)}) \to \mathbb{Z}/2\mathbb{Z}$ is defined as in the case of orientable surfaces. The characteristic class o_2 is sufficient to distinguish the connected components of $W(N)$, and we can state the following:

Theorem 4.1. *Let $N_{(k,m)}$ be a non-orientable surface with $\chi(N_{(k,m)}) < 0$. For $[n] \in \mathbb{Z}/2\mathbb{Z}$, the set $o_2^{-1}([n]) \subset W(N_{(k,m)})$ is non-empty and connected.*

This establishes that the representation space $\mathrm{Hom}(\pi_1(N_k), G_+)$ has exactly 2 connected components.

It may seem surprising that in the non-orientable case, the number of connected components does not depend on the non-orientable genus. In fact, representations in G_+ do not parametrize hyperbolic structures on the surfaces or oriented circle bundles over the surface. Indeed, if a representation $\phi : \mathrm{Hom}(\pi_1(N), G_\pm)$ defines an oriented circle bundle over N, then necessarily if $A \in \pi_1(N)$ is represented by a one-sided curve, then $\phi(A) \in G_-$. Hence we have to study representations in $G_\pm$.

4.3. *Non-orientable Euler class*

The group $G_\pm$ is not connected anymore as $G_\pm = G_+ \sqcup G_-$, hence $\pi_0(G_\pm) \simeq \mathbb{Z}/2\mathbb{Z}$. When studying the representation into $G_\pm$, one has to use another topological invariant for the representation, namely the *first obstruction class* denoted o_1 and defined as follows.

The obstruction class $o_1(\phi)$ of a representation $\phi \in \mathrm{Hom}(\pi, G_\pm)$ is an element of the set $\mathrm{Hom}(\pi, \pi_0(G_\pm))$ and is obtained by composition of the representation $\phi : \pi \to G_\pm$ with the epimorphism $G_\pm \to \pi_0(G_\pm)$. As π is finitely generated by $\{A_1, \ldots, A_k\}$, the image of the generators by the map $o_1(\phi)$ completely determines $o_1(\phi)$. So we can see $\mathrm{Hom}(\pi, \pi_0(G_\pm))$ as a subset of $(\pi_0(G_\pm))^k \simeq (\mathbb{Z}/2\mathbb{Z})^k$. We have the following notation for the first obstruction:

$$o_1 : \mathrm{Hom}(\pi_1(N_k), G_\pm) \to (\mathbb{Z}/2\mathbb{Z})^k.$$

For example the representation space $\mathrm{Hom}(\pi, G_+)$ is exactly the set of representations $\phi \in \mathrm{Hom}(\pi, G_\pm)$ such that $o_1(\phi) = (0, \ldots, 0)$.

On the other hand, given a hyperbolic structure ϕ on N, the holonomy along a 1-sided curve is necessarily an orientation-reversing isometry as the curve reverse the orientation on the surface. Hence, the image of the 1-sided curves are all orientation-reversing isometries. The generators in the standard presentation are represented by 1-sided curves, so any holonomy representation of a hyperbolic structure satisfies $o_1(\phi) = (1, \ldots, 1)$.

4.4. *Square map*

To understand the representations inside the non-zero components $o_1^{-1}(\varepsilon_1, \ldots, \varepsilon_k)$, we need to understand squares of orientation-reversing isometries. First, we notice that if $A \in G_-$, then $A^2 \in G_+$. Moreover, all orientation-reversing isometries are glide-reflections, i.e. the product of a reflection and a hyperbolic translation along the same axis. Hence, the square A^2 of a glide-reflection is an hyperbolic element or the identity element. Simple matrix calculation (see [15]) also shows that the square map:

$$\begin{aligned} G_- &\longrightarrow \mathcal{H} \cup \{I\} \\ A &\longmapsto A^2 \end{aligned}$$

is continuous, surjective with connected fibers.

Let $\varepsilon = (\varepsilon_1, \ldots, \varepsilon_k) \in (\mathbb{Z}/2\mathbb{Z})^k$ and define l as the number of ε_i that are zero. The fundamental group $\pi_1(N_{(l,k-l)})$ can be seen as a subgroup of $\pi_1(N_k)$ by cutting the surface along the curve A_i^2 whenever $\varepsilon_i \neq 0$.

Using this property we can define a map:

$$\begin{aligned} T_\varepsilon : o_1^{-1}(\varepsilon) &\longrightarrow W(N_{l,n}) \\ \phi &\longmapsto T_\varepsilon(\phi), \end{aligned}$$

where $T_\varepsilon(\phi)(A_i) = \phi(A_i)$ if $\varepsilon_i = 0$ and else $T_\varepsilon(\phi)(A_i) = \phi(A_i)^2$. This map is continuous, surjective with connected fibers because the square map defined above is. We infer that $o_1^{-1}(\varepsilon_1, \ldots, \varepsilon_k)$ and $W(N_{(l,n)})$ have the same number of connected components.

For all ε such that $l > 0$, Theorem 4.1 asserts that $W(N_{(l,n)})$ has two connected components indexed by $o_2 \in \mathbb{Z}/2\mathbb{Z}$. However, when $l = 0$, we identify $N_{(0,k)} = \Sigma_{(0,k)}$ as the n-holed sphere. In this case, Theorem 3.2 tells us that $W(\Sigma_{(0,k)})$ has $2k - 3$ connected components indexed by the relative obstruction class $o_2 \in \mathbb{Z}$.

We can now state the following theorem on the total number of connected component of $\mathrm{Hom}(\pi, G_\pm)$.

Theorem 4.2. *The representation space* $\mathrm{Hom}(\pi_1(N_k), \mathrm{PGL}(2, \mathbb{R}))$ *has* $2^{k+1} + 2k - 5$ *connected components.*

The set of representations such that $l = 0$ is the fiber $o_1^{-1}(1, \ldots, 1)$. Through the map T_ε we can define the obstruction of a representation $\phi \in o_1^{-1}(1, \ldots, 1)$ as the relative obstruction class of $T_\varepsilon(\phi) \in W(\Sigma_{(0,k)})$. It can be shown that this invariant coincides with the Euler class of the oriented circle bundle over N_k defined by ϕ. The components of $\mathrm{Hom}(\pi, G_\pm)$ consisting of discrete and faithful representations are exactly the ones with $\varepsilon = (1, \ldots, 1)$ and maximal Euler class.

4.5. *General formula*

In [18], Xia proves that for an orientable surface of genus g, the space $\mathrm{Hom}(\pi_1(\Sigma_g), G_\pm)$ has $2^{2g+1} + 4g - 5$ connected components. When translating the formula in terms of the Euler characteristic of the surface, one obtain the following general result.

Theorem 4.3. *Let S be a closed surface orientable or not, with Euler characteristic $\chi(S) < 0$ and $G_\pm = \mathrm{PGL}(2, \mathbb{R})$, then the number $N(S)$ of connected components of the space $\mathrm{Hom}(\pi_1(S), G_\pm)$ is given by:*

$$N(S) = 2^{3-\chi(S)} - 2\chi(S) - 1.$$

Acknowledgements

The author would like to thank all the organizers for the very interesting and enriching program. This paper was written while the author was a CIRGET postdoctoral fellow at the Université de Sherbrooke.

References

1. Bradlow, S.B., Garca-Prada, O., Gothen, P.B.: *Maximal surface group representations in isometry groups of classical Hermitian symmetric spaces*, Geom. Dedicata **122** 2006, 185–213
2. Burger, M., Iozzi, A., Labourie, F., Wienhard, A.: *Maximal representations of surface groups: Symplectic Anosov structures*, Pure Appl. Math. Q. 1 **3** (2005), 543–590
3. Burger, M., Iozzi, A., Wienhard, A.: *Higher Teichmüller spaces: From* $\mathrm{SL}(2, \mathbb{R})$ *to other Lie groups*, Preprint, arXiv:1004.2894v2
4. Choi, S., Goldman W.: *The classification of real projective structures on compact surfaces*, Bull. Amer. Math. Soc. **34** (1997), no. 2, 161-171.
5. Dumas, D. *Complex projective structures*, Handbook of Teichmller theory. Vol. II, 455–508.
6. Garcia-Prada, O., Gothen, P.B., Mundet i Riera, I.: *Representations of surface groups in Sp(2n,R)*, arXiv:0809.0576 (September 2008).
7. Goldman, W.: *Topological components of spaces of representations*, Invent. Math. **93** (1988), 557–607.
8. Goldman W., *Mapping class group dynamics on surface group representation* in "Problems on Mapping Class Groups and Related Topics", B. Farb, ed. Proc. Symp. in Pure Math. **74**, American Mathematical Society (2006), 189–214.
9. Goldman W., *Locally Homogeneous Geometric Structures* Proceedings of the 2010 International Congress of Mathematicians, Hyderabad, India (2010), 717–744.
10. Hitchin, N.J.: *Lie groups and Teichmüller space*, Topology **31** (1992), 449–473.

11. Jankins, M., Neumann, W.: *Homomorphisms of fuchsian groups to* $\mathrm{PSL}(2, \mathbb{R})$, Comment. Math. Helvetici, **60** (1985), 480-495.
12. Milnor, J.: *On the existence of a connection with zero curvature*, Comment. Math. Helv. **32** (1958), 215–223.
13. Wood, J.: *Bundles with totally disconnected structure group*, Comment. Math. Helv. **46** (1971), 257–273.
14. Palesi, F.: *Ergodic action of the mapping class group on moduli spaces of representation of non-orientable surfaces*, Geometriae Dedicata **151** (2011), 107-140.
15. Palesi, F.: *Connected components of representation spaces of non-orientable surfaces*, Communications in Analysis and Geometry, **18** (2010), 195-21.
16. Palesi, F.: *Non-Teichmüller components for non-orientable surfaces*, in preparation.
17. Toledo, D.: *Representations of surface groups in complex hyperbolic space*, J. Differential Geom. **29** (1989), 125–133.
18. Xia, E.Z.: *Components of* $\mathrm{Hom}(\pi_1, \mathrm{PGL}(2, \mathbb{R}))$, Topology **36** (1997), 481–499.

RIGIDITY AND FLEXIBILITY OF SURFACE GROUPS IN SEMISIMPLE LIE GROUPS

Inkang Kim
School of Mathematics
Korea Institute for Advanced Study
Hoegiro 85, Dongdaemun-gu, 130-722, Seoul, Korea
inkang@kias.re.kr

We survey recent results on surface group representations in semisimple Lie groups and local rigidity phenomena of lattices in semisimple Lie groups. We review basics about the character varieties of surface groups, its application and structures. We show that a surface group contained in a reductive real algebraic group can be deformed to become Zariski dense, unless one of the noncompact factors of the symmetric space of its Zariski closure is Hermitian of tube type. This is a kind of converse to rigidity results of Burger, Iozzi and Wienhard.

Keywords: Flexible representation, quadratic model, local rigidity, tube type, Hermitian symmetric space, Toledo invariant

Mathematics Subject Classification 2000: 51M10, 57S25

1. Introduction

Recently there has been a burst of research related to surface group representations in semisimple Lie groups. Probably the most well-known examples are Fuchsian groups, discrete subgroups of $SL(2,\mathbb{R})$ which uniformize hyperbolic surfaces. There is a rich deformation theory for hyperbolic structures on a fixed closed surface S of genus $g \geq 2$. Indeed this theory of Riemann surfaces dates back to Bernhard Riemann in 1800, followed by Felix Klein and Adolf Hurwitz, who inspired the study of moduli spaces and character varieties.

Let S be a closed surface of negative Euler number and $\tilde{S} = D$, a unit disk, over which the deck group $\pi_1(S) = \Gamma \subset \mathrm{Diff}^+(D)$ acts. Let $\pi : D \to S$ be the projection. The following diagram is well-known:

$$\begin{array}{ccccccc} \{e\} \longrightarrow \Gamma \longrightarrow & N^+ & \longrightarrow \mathrm{Diff}^+(S) \longrightarrow \{e\} \\ & \big\downarrow A & \\ & Aut(\pi_1(S)) & \end{array}$$

where N^+ is the normalizer of Γ and the map A is induced by the conjugation action of N^+ on $\Gamma = \pi_1(S)$. Let N^+_{inn} be the set of elements in N^+ inducing inner automorphisms of $\pi_1(S)$. Then for any $\phi \in N^+_{inn}$, there exists $\eta \in \pi_1(S)$ so that $\phi\gamma\phi^{-1} = \eta\gamma\eta^{-1}$ and hence

$$\eta^{-1}\phi\gamma = \eta^{-1}\eta\gamma\eta^{-1}\phi = \gamma\eta^{-1}\phi,\ \forall\gamma \in \pi_1(S)$$

i.e., $\eta^{-1}\phi$ commutes with the action of Γ, hence $\pi(\eta^{-1}\phi) = \pi\phi$ is homotopic to the identity on S. Now we have the diagram:

$$\begin{array}{ccccccc} \{e\} \longrightarrow \Gamma \longrightarrow & N^+/N^+_{inn} & \longrightarrow MCG(S) = \mathrm{Diff}^+(S)/\mathrm{Diff}^+_0(S) \longrightarrow \{e\} \\ & \big\downarrow A & \\ & Out(\pi_1(S)) & \end{array}$$

where $\mathrm{Diff}^+_0(S)$ is the set of diffeomorphisms isotopic to the identity and $Aut(\pi_1(S))/Inn(\pi_1(S)) = Out(\pi_1(S))$. Indeed $MCG(S)$ is isomorphic to $Out^+(\pi_1(S))$ by Dehn-Nielsen-Baer. See [5] for some discussions.

If S admits a hyperbolic metric h, denote the induced metric on D by the same notation h by abusing notation. Fix a isometry

$$f_h : (D, (o, v)) \to (H^2_{\mathbb{R}}, (0, w))$$

where o is a fixed reference point in D with a fixed base vector $v \in T_oD$, and 0 is the origin of the Poincare disk $H^2_{\mathbb{R}}$ with a fixed base vector $w \in T_0H^2_{\mathbb{R}}$ so that $df_h(v) \in \mathbb{R}^+w$. Then for any $\phi \in \mathrm{Diff}^+(D)$ and the push-forward metric ϕ^*h, we have

$$c(\phi, h) = f_{\phi^*h}\phi f_h^{-1} : H^2_{\mathbb{R}} \to H^2_{\mathbb{R}}$$

is an isometry. Hence

$$c : \mathrm{Diff}^+(D) \times Hyp(D) \to PSL(2, \mathbb{R})$$

is a cocycle i.e., $c(\phi_1\phi_2, h) = c(\phi_1, h)c(\phi_2, \phi_1^*h)$. When $\phi \in \Gamma$, we have a holonomy representation

$$\rho_h = c(\cdot, h) : Hyp(S) \to Hom(\pi_1(S), PSL(2, \mathbb{R})).$$

If $\phi \in \mathrm{Diff}^+(S)$ is isotopic to identity, we would like to consider two hyperbolic metrics h and ϕ^*h equal. Indeed if $\phi \in N^+_{inn}$, then $A_\phi(\gamma) = \alpha^{-1}\gamma\alpha,\ \forall\gamma \in \pi_1(S)$ for some α, hence

$$\begin{aligned} \rho_{\phi^*h}(\gamma) &= c(\phi, h)^{-1}\rho_h(A_\phi\gamma)c(\phi, h) \\ &= c(\phi, h)^{-1}\rho_h(\alpha^{-1}\gamma\alpha)c(\phi, h) \\ &= (\rho_h(\alpha)c(\phi, h))^{-1}\rho_h(\gamma)(\rho_h(\alpha)c(\phi, h)). \end{aligned}$$

This induces the following holonomy map

$$\delta : \mathcal{T}(S) = Hyp(S)/(\mathrm{Diff}_0^+(S)) \to Hom(\pi_1(S), G)//G = \mathcal{R}(\pi_1(S), G),$$

where $\mathcal{T}(S)$ is Teichmüller space, and $\mathcal{R}(\pi_1(S), G)$ is the character variety with $G = PSL(2, \mathbb{R})$. We remark that one can obtain the holonomy map using the notion of developing maps as well. It is easy to see that the above argument goes through with an arbitrary reasonable Lie group G, for example if one can define geometric structures on S using the action of G. If an abstract group Γ has a finite generating set F, and if G is a real algebraic group, then from

$$Hom(\Gamma, G) \subset G^F,$$

it is easy to see that $Hom(\Gamma, G)$ is also a real algebraic set with finitely many connected components, and each component is a real semi-algebraic set.

If we denote $Hom_{d,f}(\Gamma, G)$ the subset of discrete and faithful representations, one can show [8] that, it is a closed subset of $Hom(\Gamma, G)$ if Γ has a property that every normal nilpotent subgroup is finite. Furthermore if $\Gamma \subset G$ is a lattice in a connected reductive Lie group G, then

$$Hom_0(\Gamma, G) = \{\rho : \Gamma \to G | \rho \text{ discrete, faithful, } \rho(\Gamma)\backslash G \text{ compact}\}$$

is an open set of $Hom(\Gamma, G)$, indeed equal to the set $Hom_{d,f}(\Gamma, G)$, and hence a union of finitely many connected components [19]. Indeed if G is a semisimple Lie group of noncompact type, not equal to $SL(2, \mathbb{R})$, then Mostow [16] rigidity implies that $Hom_{d,f}(\Gamma, G)$ has a unique component.

2. Geometric Invariants

Let G be a semisimple Lie group associated to a Hermitian symmetric space $(X = G/K, \omega)$ of noncompact type where ω is a Kähler form. Given a representation $\rho : \pi_1(S) \to G$, one can form a flat fiber bundle

$$E = \tilde{S} \times_\rho X$$

over S with fiber X, where the action of $\pi_1(S)$ on $\tilde{S}$ is via deck group, and on X is via ρ. Since the fiber X is contractible, one can find a smooth section $s : S \to E$, equivalently a ρ-equivariant map $f_\rho : \tilde{S} \to X$. Then $f_\rho^*\omega$ is ρ-equivariant, hence descends to S, denoted by the same notation $f_\rho^*\omega$. The Toledo invariant is defined as:

$$T(\rho) = \frac{1}{2\pi}\int_S f_\rho^*\omega.$$

It is known that there exists $\ell_X \in \mathbb{Q}$ such that $T(\rho) \in \ell_X \mathbb{Z}$. This geometric quantity is independent of the choice of f_ρ, and constant over each component of the character variety $\mathcal{R}(\pi_1(S), G)$.

In our example of $\mathcal{R}(\pi_1(S), PSL(2, \mathbb{R}))$, Goldman [6] (see also [15]) showed that

$$|T(\rho)| \leq |\chi(S)|$$

and $T(\rho) = 2g - 2$ iff ρ is a hyperbolic structure on the oriented surface S. $T(\rho) = 2 - 2g$ corresponds to the hyperbolic structure on S with orientation reversed. This shows that the Toledo invariant distinguishes different components of $\mathcal{R}(\pi_1(S), PSL(2, \mathbb{R}))$.

In this special case $G = PSL(2, \mathbb{R})$, one can view the Toledo invariant as the Euler class as follows. Note that an element of $PSL(2, \mathbb{R})$ gives rise to an orientation preserving homeomorphism of $S^1 = \partial H^2_{\mathbb{R}}$. Let $T(x) = x + 1$ on $\mathbb{R}$ and identify S^1 with $\mathbb{R}/\mathbb{Z}$. There exists a central extension

$$0 \to \mathbb{Z} \to H^+_{\mathbb{Z}}(\mathbb{R}) \xrightarrow{p} Homeo_+(S^1) \to 0$$

where $H^+_{\mathbb{Z}}(\mathbb{R}) = \{f : \mathbb{R} \to \mathbb{R} :$ increasing homeomorphisms commuting with $T\}$. One can define a section s of p as follows. For $f \in Homeo_+(S^1)$, find a unique lift $\tilde{f} : \mathbb{R} \to \mathbb{R}$ with $0 \leq \tilde{f}(0) < 1$. This section s associating $\tilde{f}$ to f is not in general homomorphism, and there exists 2-cocycle ϵ measuring nonhomomorphism as

$$\tilde{f} \circ \tilde{g} = \widetilde{f \circ g} \circ T^{\epsilon(f,g)}.$$

Let $e \in H^2(Homeo_+(S^1), \mathbb{Z})$ be the Euler class defined by ϵ.

Then the Euler number $e(\rho)$ of a representation

$$\rho : \pi_1(S) \to PSL(2, \mathbb{R}) \subset Homeo_+(S^1)$$

is defined by

$$\langle \rho^*(e), [S] \rangle$$

via the isomorphism

$$H^2(S, \mathbb{Z}) \to H^2(\pi_1(S), \mathbb{Z}).$$

By integrating the Kähler form ω on the Poincare disk over geodesic triangles

$$c(g_1, g_2) = \frac{1}{2\pi} \int_{\triangle(0, g_1 0, g_1 g_2 0)} \omega,$$

one obtain a bounded cohomology class κ, and this κ corresponds to the Euler class e. Hence

$$e(\rho) = \langle \rho^*(e), [S] \rangle = \frac{1}{2\pi} \int_S f_\rho^* \omega = T(\rho).$$

A natural question arises what happens if we enlarge the target group G. For example in $\mathcal{R}(\pi_1(S), PSL(2,\mathbb{R})) \subset \mathcal{R}(\pi_1(S), PSL(3,\mathbb{R}))$, $\delta(\mathcal{T}(S))$ is contained in a component, called the Hitchin component, which constitutes the holonomy representations of strictly convex real projective structures on S [4].

3. Real Zariski Tangent Space of Character Variety and Local Rigidity

The real Zariski tangent space of the character variety $\mathcal{R}(\Gamma, G)$ at $[\rho]$ can be identified with $H^1(\Gamma, \mathfrak{g}_{Ad\rho})$ where $\mathfrak{g}_{Ad\rho}$ is a G-module via the adjoint action of $\rho(G)$. Geometrically, any nonzero element $\mu \in H^1(\Gamma, \mathfrak{g}_{Ad\rho})$ could possibly represent an infinitesimal deformation of $[\rho]$, and indeed a necessary condition [17] is

$$0 = [\mu, \mu] \in H^2(\Gamma, \mathfrak{g}_{Ad\rho}).$$

Due to Margulis superrigidity [13] if Γ is a lattice in a higher rank semisimple Lie group, there is no local deformation for any $\rho : \Gamma \to G$ where G is any simple Lie group. The same is true for any lattice in $Sp(n,1)$ and F_4 due to Corlette [3].

In the example $\mathcal{R}(\pi_1(S), PSL(2,\mathbb{R})) \subset \mathcal{R}(\pi_1(S), PSL(3,\mathbb{R}))$, recall that $\delta(\mathcal{T}(S))$ is contained in a component, called the Hitchin component, which constitutes the holonomy representations of strictly convex real projective structures on S. This means that if we enlarge the group to $PSL(3,\mathbb{R})$, then there are more deformations which do not appear in a smaller group. Also in $PSL(2,\mathbb{R}) \subset PSL(2,\mathbb{C})$ case, Teichmüller space is strictly contained in quasi-fuchsian space. Hence one might expect that if we enlarge the target group H to G, there might be a local deformation even if the original representation $\rho : \Gamma \to H$ is locally rigid in $\mathcal{R}(\Gamma, H)$.

This is not always possible due to an example by Goldman-Millson [8]. They showed that a lattice Γ in $SU(n,1)$

$$\Gamma \subset SU(n,1) \subset SU(m,1)$$

is locally rigid in $SU(m,1), m > n$.

Recently Kim-Klingler-Pansu [12] extended this result to bigger groups

$$\Gamma \subset SU(n,1) \subset SU(m,1) \subset Sp(m,1) \subset SU(2m,2) \subset SO(4m,4), \;\; n>1$$

so-called a quaternionic series. The associted symmetric spaces corresponding to latter three groups are quaternionic Kähler. Basically what they proved is that every local deformation is a deformation of a Hodge structure, which implies that a harmonic one form representing a possible deformation in quaternionic space lies in

$$H^1(M,\mathbb{R}) \oplus H^0(M, S^3T^*M \otimes L_{\chi^2})$$

where $M = H^n_{\mathbb{C}}/\Gamma$, $H^1(M,\mathbb{R})$ is the deformation in the centralizer of $SU(n,1)$ in $U(n,1)$, and $\chi : U(n) \to \mathbb{C}$ is the determinant character, whence L_{χ^2} is the automorphic line bundle associated to χ^2. In more plain language, a method of Matsushima-Murakami [14] gives that any such a harmonic 1-form α should have values in $S^3\mathfrak{p}^* \otimes \chi^2$ where S denotes the symmetric tensor.

Furthermore there is a form $\lambda : \mathfrak{sp}(n,1) \to \mathbb{R}$ such that

$$\lambda \circ [\alpha,\alpha] \wedge \omega_M^{n-1} = c|\alpha|^2\omega_M^n$$

for some constant c where ω_M is a natural Kähler form on M. In summary, this approach concludes that any such nonzero α should be

$$[\alpha,\alpha] \neq 0,$$

which proves the local rigidity in quaternionic space. The other cases are similar.

Indeed, after A. Weil's local rigidity theorem [24] for lattices in semisimple Lie groups, Raghunathan [18] showed that

Theorem 3.1. *Let G be a connected semisimple Lie group. $\Gamma \subset G$ be a uniform irreducible lattice and $\rho : G \to Aut(E)$ a simple nontrivial linear representation. Then $H^1(\Gamma, E) = 0$ except possibly when $\mathfrak{g} = \mathfrak{so}(n+1,1)$ (resp. $\mathfrak{g} = \mathfrak{su}(n,1)$) and the highest weight of ρ is a multiple of the highest weight of the standard representation of $\mathfrak{so}(n+1,1)$ (resp. of the standard representation of $\mathfrak{su}(n,1)$) or of its contragredient representation).*

Due to this theorem, there is a local rigidity of a uniform lattice $\Gamma \subset SO(3,1), SO(4,1)$ inside $Sp(n,1)$ [11]. Indeed,

$$\mathfrak{sp}(n,1) = \mathfrak{sp}(1) \oplus \mathbb{H}^n \oplus \mathfrak{sp}(n-1,1)$$

and the first group cohomology splits accordingly. Here $\mathbb{H}^n$ is the n-tuples of quaternions viewed as a right module over quaternions $\mathbb{H}$. The first factor corresponds to the deformation stabilizing a quaternionic line, hence not interesting, and the second factor has a trivial deformation due to Raghunathan's criterion. The corresponding representation turns out to be the spin representation. We remark here that for non-uniform lattices, to apply Weil-Raghunathan-Matsushima-Murakami theory, one has to use L^2-cohomology. If a parabolic element deforms to a hyperbolic element, it is difficult to have L^2-cohomology. In that matter, if we assume that throughout the deformation the parabolicity is preserved, one can still shows local rigidity. Note that one can deform a cusped hyperbolic 3-manifold into a closed hyperbolic 3-manifold via Dehn surgery [22]. As far as I know, this area is wide open.

4. Surface Group Representation

In this section we concentrate on surface group representations in real reductive algebraic groups. This theory is particularly well-developed due to the rich structure present in the character variety. Harmonic theory, Higgs bundle theory, representation theory are well balanced in this domain. In contrast to local rigidity phenomena, there are rich deformation theories. We have seen already in Teichmüller theory and in convex real projective structures, quasi-fuchsian deformation theory corresponding to Lie groups $PSL(2,\mathbb{R}), PSL(3,\mathbb{R})$ and $PSL(2,\mathbb{C})$ respectively.

Toledo invariant theory is particularly useful in this case. Recently Burger-Iozzi-Wienhard [2] showed that every maximal Toledo invariant representation from a closed surface group into a Hermitian group is a representation into a tube type Hermitian Lie group, and the representation is necessarily discrete and faithful. Here a tube type Hermitian Lie group is a semisimple Lie group associated to the tube type Hermitian space. A Hermitian symmetric space is of tube type if it can be realized as a domain in $\mathbb{C}^n$ of the form $\mathbb{R}^n + iC$ where $C \subset \mathbb{R}^n$ is a proper open cone.

Siegel's upper half spaces with isometry groups $Sp(2n,\mathbb{R})/\{\pm I\}$ and Grassmannians with isometry groups $PO(2,q)$ are of tube type. The Grassmannian $\mathcal{D}_{p,q}$, $p \leq q$, with isometry group $PU(p,q)$ is of tube type iff $p = q$. The Grassmannian $\mathcal{G}_n$ with isometry group $SO^*(2n)$ is of tube type iff n is even.

The exceptional Hermitian symmetric space of dimension 27 with isometry group $E_7(-25)$ is of tube type, the other one (of dimension 16) is not.

Products of tube type spaces are of tube type, so polydisks are of tube type.

We remark here that the Toledo invariant can be defined whenever one can use an almost complex structure, hence first Chern class. This motivates us to consider Toledo invariant technique for general reductive Lie groups which are not necessarily Hermitian. The method is to decompose the Lie algebra into complex root spaces using the center c of the centralizer of the image $\rho(\Gamma)$ in G [9]

$$\mathfrak{g}_{\mathbb{C}} = \mathfrak{g} \otimes_{\mathbb{R}} \mathbb{C} = \oplus \mathfrak{g}_\alpha,$$

$$\mathfrak{g} = \oplus \mathfrak{g}_{\alpha,\mathbb{R}}, \ \mathfrak{g}_{\alpha,\mathbb{R}} = (\mathfrak{g}_\alpha \oplus \mathfrak{g}_{-\alpha} \oplus \mathfrak{g}_{\bar{\alpha}} \oplus \mathfrak{g}_{-\bar{\alpha}}) \cap \mathfrak{g}.$$

This decomposition is Γ-invariant, and using the Killing form of G, each $\mathfrak{g}_{\alpha,\mathbb{R}}$ admits a symplectic structure ω_α. Then using a compatible almost complex structure, one can define Toledo invariant for each induced representation

$$\rho_\alpha : \Gamma \to G \stackrel{\text{Adj}}{\longrightarrow} Aut(\mathfrak{g}) \to Sp(\mathfrak{g}_{\alpha,\mathbb{R}}, \omega_\alpha).$$

For each α, on the associated flat bundle

$$E_\alpha = (\tilde{S} \times_{\rho_\alpha} \mathfrak{g}_{\alpha,\mathbb{R}}, J_\alpha),$$

Meyer's signature formula tells us that the quadratic form on $H^1(S, E_\alpha)$ has signature $4c_1$ where c_1 is the first Chern class of the flat bundle using the almost complex structure J_α compatible with ω_α, i.e., $\omega_\alpha(v, J_\alpha v) > 0$ for any nonzero v. We define the Toledo invariant of ρ_α to be $c_1([S])$.

Note that

$$H^2(\Gamma, \mathfrak{g}) = H^0(\Gamma, \mathfrak{g}) = \{x \in \mathfrak{g} | [\rho(\gamma), x] = 0, \forall \gamma \in \Gamma\} = \text{Lie}(Z) = \mathfrak{c} \oplus \mathfrak{s}$$

where Z is the centralizer of $\rho(\Gamma)$ in G, and it decomposes into the center c and semisimple part s. The cup product on $H^1(\Gamma, \mathfrak{g})$ followed by the projection to $\mathfrak{c}$ gives rise to a quadratic form Q

$$Q : H^1(\Gamma, \mathfrak{g}) \times H^1(\Gamma, \mathfrak{g}) \to \mathfrak{c},$$

which can be written as

$$Q = \bigoplus Re(Q_\lambda t_\lambda)$$

where Q_λ is complex valued and $t_\lambda \in \mathfrak{c}$.

Goldman-Millson [7] and Simpson [21] analyzed the representation space $Hom(\Gamma, G)$ where Γ is the fundamental group of a compact Kähler manifold. They showed that if $\rho : \Gamma \to G$ is reductive, the neighborhood of ρ in $Hom(\Gamma, G)$ is analytically isomorphic to the quadratic model Q_ρ

$$Q_\rho = \{u \in Z^1(\Gamma, \mathfrak{g}_{Ad\rho}) | [u, u] = 0 \in H^2(\Gamma, \mathfrak{g})\}.$$

Using this quadratic model, we can show [9] that the following are equivalent.

(1) Q_ρ has a dense set of smooth points.
(2) 0 is contained in the interior of the convex hull of P + Span (N) where P is the set of pure imaginary roots whose Toledo invariants are positive and maximal, and N is the complement of $\pm P$ in root space.

We call the center c of the centralizer of the image balanced with respect to ρ if the second condition is satisfied. A simple corollary of this is:

Theorem 4.1. *Let G be a semisimple real algebraic group and Γ the fundamental group of a closed surface of genus $\geq 2dim(G)^2$. Let $\rho : \Gamma \to G$ be a homomorphism with reductive Zariski closure H. Then ρ can be deformed to a Zariski dense representation iff c is balanced with respect to ρ.*

As an application we get

Theorem 4.2. *Let G be a reductive real algebraic group and Γ, ρ as above. If ρ cannot be deformed to a Zariski dense representation, then H admits a tube type Hermitian Levi factor.*

One can test this criterion for rank one semisimple Lie groups and prove flexibility for surface groups in real rank one semisimple Lie groups except Toledo's rigidity [23], i.e., the surface group stabilizes a complex line in complex n-hyperbolic space,

$$\Gamma \subset SU(1,1) \subset SU(n,1).$$

5. Classical Simple Lie Groups and Flexibility

For $SL(n,\mathbb{C})$, our method proves flexibility of surface group except Bradlow-Garcia-Prada-Gothen's case [1]

$$\Gamma \subset SU(1,1) \subset SU(p,p) \subset SU(p,q).$$

We can push our method to classical simple Lie groups. Classical simple real Lie groups are special linear groups of division rings R with center $\mathbb{R}$ and special unitary groups of nondegenerate binary forms over R, see the appendix in [20].

There are only 3 such division rings: $\mathbb{R}$, $\mathbb{C}$ and $\mathbb{H}$, leading to 3 special linear groups, $SL(n,\mathbb{R})$, $SL(n,\mathbb{C})$, and $SL(n,\mathbb{H})$.

Let ι be a continuous (anti)-automorphism of R: ι can be identity, complex conjugation or quaternionic conjugation. Let $\epsilon = \pm 1$. By a (ι, ϵ)-symmetric binary form on a right R-vectorspace V, we mean a R-valued $\mathbb{R}$-bilinear map $h : V \times V \to R$ such that

(1) for all v, $v' \in V$ and all $q \in R$, $h(v, v'q) = h(v, v')q$;
(2) for all v, $v' \in V$, $h(v', v) = \epsilon\iota(h(v, v'))$.

Note that from (2), it follows that $h(vq, v') = \bar{q}h(v, v')$. Say h is *nondegenerate* if the only $v \in V$ such that $h(v, v') = 0$ for all $v' \in V$ is 0.

The group $U(V, h)$ consists of right R-linear self maps of V which preserve h. We are interested in the special unitary group $SU(V, h) = U(V, h) \cap SL(V, R)$.

When $R = \mathbb{R}$, the only choice for ι is identity, leading to

- if $\epsilon = 1$, real special orthogonal groups, indexed by dimension and signature and denoted by $SO(p, q)$, where $p + q = \dim_{\mathbb{R}}(V)$;
- if $\epsilon = -1$, real symplectic groups, indexed by dimension only and denoted by $Sp(n, \mathbb{R})$, where $n = \dim_{\mathbb{R}}(V)$ is even.

When $R = \mathbb{C}$ there are two choice for ι. If ι is identity, this leads to complex special orthogonal (if $\epsilon = 1$) and symplectic (if $\epsilon = -1$) groups, indexed by dimension and denoted respectively by $SO(n, \mathbb{C})$ and $Sp(n, \mathbb{C})$. If ι is complex conjugation, $\epsilon = \pm 1$ lead to groups called special unitary groups, indexed by dimension and signature and denoted by $SU(p, q)$, where $p + q = \dim_{\mathbb{C}}(V)$.

When $R = \mathbb{H}$ there are two choices for ι. If ι is identity, no nonzero (ι, ϵ)-symmetric forms exist. If ι is quaternionic conjugation, this leads to

- if $\epsilon = 1$, quaternionic unitary groups, indexed by dimension and signature and denoted by $Sp(p, q)$, $p + q = \dim_{\mathbb{H}}(V)$;
- if $\epsilon = -1$, quaternionic skew-unitary groups, indexed by dimension only and denoted by $SO^*(2n)$, where $n = \dim_{\mathbb{H}}(V)$ is even.

The following is proved in [10].

Theorem 5.1. *Let G be a classical real Lie group, i.e. a real form of $SL(n, \mathbb{C})$, $O(n, \mathbb{C})$ or $Sp(n, \mathbb{C})$. Let Γ be the fundamental group of a closed surface of genus $\geq 2\dim(G)^2$. A homomorphism $\phi : \Gamma \to G$ can be approximated by Zariski dense representations, unless the symmetric space X of G is Hermitian and not of tube type, and ϕ is maximal.*

In other words, the exceptions are $G = SU(p,q)$, $q > p$ and $\phi(\Gamma)$ is contained in a conjugate of $S(U(p,p) \times U(q-p)) \subset SU(p,q)$, or $G = SO^*(2n)$, n odd, and $\phi(\Gamma)$ is contained in a conjugate of $SO^*(2n-2) \times SO(2) \subset SO^*(2n)$.

Acknowledgments

The content of these notes are based on lectures given at IMS of National University of Singapore during the summer school of Geometry, Topology and Dynamics of Character varieties program from 28 June to 16 July 2010. The author thanks IMS for providing such a nice opportunity. This work was also partially supported by KRF grant 0409-20060066.

References

1. S. Bradlow, O. García-Prada and P. Gothen, Surface group representations and $U(p,q)$-Higgs bundles, JDG 64 (2003), 111-170.
2. M. Burger, A. Iozzi and A. Wienhard, Ann. Math. 172 (2010), no. 1, 517-566.
3. K. Corlette, Archimedean superrigidity and hyperbolic geometry, Ann. Math. (2) 135 (1992), no. 1, 165-182.
4. N. Hitchin, Lie groups and Teichmüller space, Topology 31 (1992), no. 3, 449-473.
5. B. Farb and D. Margalit, A primer on mapping class groups, 2008.
6. W. Goldman, Discontinuous groups and the Euler class, Thesis, Univ. Cal at Berkeley (1980).
7. William Goldman and John Millson, The deformation theory of representations of fundamental groups of compact Kähler manifolds, Publ. Math. I.H.É.S. 67, (1988), 43-96.
8. W. Goldman and J. Millson, Local rigidity of discrete groups acting on complex hyperbolic space, Inv. Math. 88 (1987), no. 3, 495-520.
9. I. Kim and P. Pansu, Density of Zariski dense representations, preprint arXiv:1009.2258.
10. I. Kim and P. Pansu, Flexibility of surface groups in classical simple Lie groups, preprint arXiv:1101.1159.
11. I. Kim and P. Pansu, Local rigidity in quaternionic hyperbolic space, JEMS 11 (2009), no. 6, 1141-1164.
12. I. Kim, B. Klingler and P. Pansu, Local quaternionic rigidity for complex hyperbolic lattices, J. Inst. Math. Jussieu 11 (2012), no. 1, 133-159.
13. G. Margulis, Discrete subgroups of semisimple Lie groups, Ergebnisse der Mathematik ihrer Grengebiete 17, Springer-Verlag (1991).
14. Y. Matsushima and S. Murakami, On vector bundle valued harmonic forms and automorphic forms on symmetric spaces, Ann. Math. 78 (1963), 365-416.
15. J. Milnor, On the existence of a connection with curvature zero, Comment. Math. Helv. 32 (1958), 215-223.

16. G. D. Mostow, Strong rigidity of locally symmetric spaces, Annals of Mathematics studies, no. 78, Princeton University Press, Princeton (1973).
17. A. Nijenhuis and R. Richardson, Deformations of homomorphisms of Lie groups and Lie algebras, Bull. AMS. 73 (1967), 175-179.
18. M. S. Raghunathan, On the first cohomology of discrete subgroups of semisimple Lie groups, Amer. J. Math. 87 (1965), 103-139.
19. M. S. Raghunathan, Discrete subgroups of Lie groups of Lie groups, Ergebnisse der Mathematik und ihrer Grenzgebiete, Band 68, Springer-Verlag (1972).
20. Ichirô Satake, Algebraic structures of symmetric domains. Kanô Memorial Lectures, **4**. Iwanami Shoten, Tokyo; Princeton University Press, Princeton, N.J., (1980).
21. Carlos Simpson, Higgs bundles and local systems. Publ. Math. I.H.É.S. 75, (1992), 5-95.
22. W. P. Thurston, The Geometry and Topology of 3-manifolds, Lecture notes, Princeton (1983).
23. D. Toledo, Representations of surface groups on complex hyperbolic space, JDG. 29 (1989), 125-133.
24. A. Weil, Discrete subgroups of Lie groups, II, Ann. Math. 75 (1962), 97-123.

ABELIAN AND NON-ABELIAN COHOMOLOGY

Eugene Z. Xia
National Center for Theoretical Sciences
Department of Mathematics
National Cheng Kung University
Tainan 70101, Taiwan
ezxia@ncku.edu.tw

We place the representation variety in the broader context of abelian and nonabelian cohomology. We outline the equivalent constructions of the moduli spaces of flat bundles, of smooth integrable connections, and of holomorphic integrable connections over a compact Kähler manifold. In addition, we describe the moduli space of Higgs bundles and how it relates to the representation variety. We attempt to avoid abstraction, but strive to present and clarify the unifying ideas underlying the theory.

Keywords: Moduli space, representation variety, local systems, flat connections, holomorphic flat connections, Higgs bundles

Mathematics Subject Classification 2000: 14J10, 14F10, 32L05, 32L10, 57M05.

1. Prelude

At first glance, the representation variety is a strange object. From the perspective of group theory, the fundamental group of a surface does not seem to have anything to do with Lie groups such as $\mathrm{SL}(2,\mathbb{R})$. Yet there is a deep connection from uniformization: the group $\mathrm{SL}(2,\mathbb{R})$ acts on the symmetric space $\mathrm{SL}(2,\mathbb{R})/\mathrm{U}(1)$ which is identified with the upper half of the complex plane $\mathbb{C}$. Given a discrete subgroup $\Gamma < \mathrm{SL}(2,\mathbb{R})$, the quotient $X = \Gamma\backslash \mathrm{SL}(2,\mathbb{R})/\mathrm{U}(1)$ is a Riemann surface with fundamental group isomorphic to Γ. Thus a discrete representation ρ of the fundamental group of X gives rise to a Riemann (hyperbolic) surface. Hence representation varieties are related to Teichmüler spaces. This is the main perspective and focus of the summer school.

There is another and related perspective of the representation variety that may also be traced back to Riemann and which is just as important.

One is interested in the holomorphic linear ordinary differential equation

$$\frac{\partial^2 f}{\partial z^2} + a(z)\frac{\partial f}{\partial z} + b(z)f = 0,$$

where a and b are rational functions and together with at most three distinct poles on the Riemann sphere $\mathbb{P}^1(\mathbb{C})$. Then the solutions are the classical Gauss hypergeometric functions. These are complicated functions with multiple indices that can make any non-analyst dizzy and satisfy still more complicated Kummer relations.

However under mild conditions and by a standard elementary transformation, one may rewrite this second order equation as a system of first order equations

$$\frac{\partial F}{\partial z} + \frac{A_0 F}{z} + \frac{A_1 F}{z-1} = 0, \quad A_i \in \mathfrak{gl}(2,\mathbb{C}),\ i = 0, 1, \tag{1.1}$$

where $\mathfrak{gl}(2,\mathbb{C})$ consists of 2×2 complex matrices and is the Lie algebra of $\mathrm{GL}(2,\mathbb{C})$. Let $A_\infty = -A_0 - A_1$. If $A_0 = A_1 = 0$, then F is a meromorphic function on $\mathbb{P}^1(\mathbb{C})$ with poles at $\{0, 1, \infty\}$ such that $\frac{\partial F}{\partial z} = 0$. This implies that F is a constant function. In general, such solution F only exists locally. Globally, F exists as a multivalued function with monodromy at $\{0, 1, \infty\}$. This gives rise to the concept of a local system (system of locally constant functions) and a monodromy representation

$$\rho : \pi_1(\mathbb{P}^1(\mathbb{C})\backslash\{0, 1, \infty\}) \longrightarrow \mathrm{GL}(2,\mathbb{C}).$$

Riemann pointed out (in our modern language) that the $\mathrm{GL}(2,\mathbb{C})$-representation variety of a 3-holed sphere is a point. In the language of hyperbolic geometry, the hyperbolic structure on a pair of pants is uniquely determined by the lengths of its three geodesic boundary components when $\mathrm{im}(\rho) < \mathrm{SL}(2,\mathbb{R})$. In other words, the solutions to the equation are determined by the eigenvalues of A_1, A_2 and A_∞ only. These eigenvalues are easy to compute and together they determine the monodromy matrices C_0, C_1 and C_∞ uniquely up to simultaneous conjugation equivalence, i.e. C_i is equal to a conjugate of $\exp(A_i)$ and $C_0 C_1 C_\infty = \mathsf{id}$. This set of C_i's determines ρ. Riemann then concluded that the hypergeometric functions corresponding to the solutions of the equation are uniquely determined by the corresponding monodromy representation ρ. In this way, he read off the Kummer relations among the hypergeometric functions without much calculation [12, 16]. The representation varieties into other Lie groups have similar interpretations in terms of holomorphic differential equations.

It is then of great interest to obtain as much information as possible concerning the representation varieties. To understand a space means to discover natural and canonical topological and geometric structures on that space. Let X be a manifold and G a Lie group and $\mathcal{M}_B$ the resulting representation variety (the Betti moduli space) of X. A priori, $\mathcal{M}_B$ is a set, one point for each representation up to conjugation equivalence. However the algebraic/analytic structure on G gives $\mathcal{M}_B$ a variety structure, making it a representation variety. If G is reductive, then the Poincaré duality and a nondegenerate symmetric bi-invariant form on G give a symplectic structure on $\mathcal{M}_B$, making it a symplectic manifold or orbifold.

One method to discover topological and geometric structures is to give alternative constructions that give rise to new topological and geometric structures. A representation corresponds to a flat G-vector bundle on X. A flat vector bundle is one where (locally) constant sections make sense. In order for this to happen, the transition maps over the overlapping covering open sets of X must be constant. These constant transition maps then give rise to a representation. Hence the representation variety corresponds to the moduli space of flat G-vector bundles. This gives rise to the Čech moduli space $\check{\mathcal{M}}$ and is the point of view of local systems.

A function f is (locally) constant if and only if its partial derivatives vanish everywhere. Differentiation is $\mathbb{R}$-linear and satisfies the Leibniz rule when applied to products of functions. Differentiation is an example of more general first order linear differential operators, namely, vector fields that also satisfy $\mathbb{R}$-linearity and the Leibniz rule. This allows one to say that f is *constant* if and only if $Vf = 0$ for each vector field V. The set of vector fields $\mathcal{T}$ forms an (infinite dimensional) vector space over $\mathbb{R}$. The very fact that vector fields operate on functions implies that $\mathcal{T}$, considered as an (additive) abelian group, acts on the space of functions, making the latter a $\mathcal{T}$-module. One then considers the family of $\mathcal{T}$-modules where the $\mathcal{T}$-action satisfies the $\mathbb{R}$-linearity condition and the Leibniz rule. This leads to the concept of *connection*. In such a module, one may then ask whether there is a reasonable notion of constant e defined as $Ve = 0$ for all $V \in \mathcal{T}$. This further leads to the idea of *integrability* and $\mathcal{D}$-modules. On a smooth manifold X, vector fields are well-defined; hence, the notion of integrable connections may be extended to smooth manifolds. Hence the Čech moduli space has an equivalent construction as the moduli space of sheaves of $\mathcal{D}$-modules or moduli space of flat G-connections. This is the smooth de Rham moduli space $\mathcal{M}_d$.

Suppose X is a complex manifold. Then a function f on X is constant if it is holomorphic and all its holomorphic partial derivatives vanish. This gives rise to the holomorphic de Rham moduli space, parameterizing holomorpic structures with integrable holomorphic connections. This is the perspective of holomorphic differential equations (see (1.1) above). The holomorphic connections are further generalized to holomorphic λ-connections for $\lambda \in \mathbb{C}$. This is an algebraic/complex analytic construction and the resulting moduli space $\mathcal{M}_d^\lambda$ has a natural variety structure. If $\lambda \in \mathbb{C}^\times$, then all the moduli spaces paramaterized by λ have isomorphic underlying variety structure. When X is compact, $\mathcal{M}_d^0$ acquires a *different* variety structure while maintaining the same underlying topological/differentiable structure as that of $\mathcal{M}_d^1$. This $\mathcal{M}_d^0$ is the Dolbeault moduli space. This new variety structure is compatible with that on the Betti moduli space. These two variety structures together with the symplectic structure give rise to a hyperkähler structure on the underlying space. To see applications of this structure in the study of geometric structures on X, see the survey paper [5].

Section 2 constructs the representation variety, the Betti moduli space $\mathcal{M}_B$, for a manifold and gives two simple examples and points out the subtle issues of the construction. Section 3 is a quick and informal introduction to groupoids. It first serves an accounting purpose to keep track of the various objects we are dealing with. More importantly, with the possible exception of the Betti moduli space, the constructions of the other moduli spaces as varieties are technically daunting. Rather, this set of introductory notes only describes the groupoids of the objects these moduli spaces parameterize and the relations between the objects among these groupoids.

Sections 4 and 5 deal with local constructions of the smooth and holomorphic de Rham spaces on the unit ball. Section 6 gives some examples of connections and especially important is the example that shows that a connection must be integrable for it to define constant functions.

Section 7 is a brief outline of the concepts of smooth and complex manifolds in preparation for global constructions on manifolds. Section 8 describes the smooth, flat and holomorphic bundles on a manifold. Sections 9 and 10 run in parallel to Sections 4 and 5 and construct the flat connections, holomorphic structures and flat holomorphic λ-connections respectively.

Section 11 describes the relations between the Čech, Betti and de Rham moduli spaces and how they relate to the Dolbeault moduli space in the case of compact Kähler manifolds. As we are leaving out the actual constructions of these moduli spaces, we state the equivalence and give examples of these equivalences in the rank-1 trivial cases in Section 12. The final Section 13 hints on the complex structure arising from the Dolbeault moduli space and how this complex structure and the one from the Betti construction and the symplectic structure give rise to a hyperkähler structure on the smooth part of the representation variety.

The subject is structure rich. These notes attempt to cover only the most basic ideas — the leitmotif is the simple observation that a smooth function is (locally) constant if and only if all its partial derivatives vanish. They are also designed to complement and supplement [8] which focuses on the rank-1 sheaves on Riemann surfaces. These notes attempt to expand the scope to high rank sheaves over Kähler manifolds, but focus more on detailed local descriptions. While [8] is succinct and rigorous, these notes are informal and avoid abstraction. The emphasis is on presenting the basic unifying ideas underlying the theory.

2. The Representation Variety

Let X be a manifold and G the general linear group $\mathrm{GL}(r, \mathbb{C})$. Let $x \in X$ and $\pi = \pi_1(X, x)$ be the topological fundamental group of X. We assume that π is finitely generated with N generators and M relations. Then the space of representations $\mathrm{Hom}(\pi, G)$ is a subvariety of G^N defined by the M relations. The group G acts on $\mathrm{Hom}(\pi, G)$ by conjugation:

$$G \times \mathrm{Hom}(\pi, G) \longrightarrow \mathrm{Hom}(\pi, G), \quad (g, \rho) \mapsto g\rho g^{-1}.$$

One would like to form a quotient of this G-action. However the orbits of the G-conjugation action may not be closed in $\mathrm{Hom}(\pi, G)$. This implies that the natural geometric quotient (orbit space) of the G-conjugation action may not be Hausdorff. To resolve this and obtain a Hausdorff quotient, one must either remove the non-closed orbits or identify them with their closure.

Since G is the rank-r linear group, each representation $\rho \in \mathrm{Hom}(\pi, G)$ induces a rank-r representation of π by composition. We say ρ is reductive if every π-invariant subspace has a π-invariant complement in $\mathbb{C}^{\oplus r}$. Denote by $\mathrm{Hom}(\pi, G)^+$ the subspace of reductive representations. The G-conjugation action preserves $\mathrm{Hom}(\pi, G)^+$ and with closed orbits. Denote by $\mathcal{M}_B$ the quotient $\mathrm{Hom}(\pi, G)^+/G$. Then $\mathcal{M}_B$ parameterizes the equivalence classes of reductive representations.

Alternatively, $\mathrm{Hom}(\pi, G)$ is an affine subvariety of G^N, defined by the M equations with its resulting coordinate ring R. Hence the G-action on $\mathrm{Hom}(\pi_1, G)$ induces a G-action on R. Denote by R^G the invariant subring. Then $\mathcal{M}_B$ is the affine variety defined by the ring R^G.

Example. Let X be the compact orientable surface of genus g and $G = \mathrm{GL}(1, \mathbb{C}) = \mathbb{C}^\times$. Set $[A, B] = ABA^{-1}B^{-1}$. Then

$$\pi = \left\langle A_i, B_i, 1 \le i \le g \mid \prod_{i=1}^{g} [A_i, B_i] \right\rangle.$$

Since G is abelian, $\mathrm{Hom}(\pi_1, \mathbb{C}^\times) \cong (\mathbb{C}^\times)^{2g}$ and the G-action is trivial. Hence $\mathcal{M}_B \cong (\mathbb{C}^\times)^{2g}$.

Example. Let X be the compact orientable surface of genus g and $G = \mathrm{GL}(2, \mathbb{C})$. Then

$$\mathrm{Hom}(\pi_1, G) \cong \left\{ a_i, b_i \in G : 1 \le i \le g, \prod_{i=1}^{g} [a_i, b_i] = e \in G \right\}.$$

Consider the representation ρ corresponding to

$$a_i, b_i = \begin{bmatrix} 1 & 1 \\ 0 & 1 \end{bmatrix}, \quad 1 \le i \le g.$$

The G-orbit of ρ is not closed because it does not contain the trivial representation. Hence the corresponding geometric quotient space is not Hausdorff.

3. A Reluctant Tour of Category and Groupoid

The basic problem is to classify a family of objects up to isomorphism and impose a natural and universal geometry over the isomorphism classes. This problem is often cast in the language of category and groupoid. We will take a somewhat informal approach and refer to (Section 1, [8]) for a careful introduction of these concepts. For us,

Definition 3.1. A **groupoid** $\mathcal{C}$ is a category consisting of

(1) A set of objects $\mathrm{Obj}(\mathcal{C})$.
(2) A set (quite often a group) $\mathcal{G}$ of equivalence relations on the objects.

We often use $\mathcal{C}$ to denote both the groupoid and its set of objects $\mathrm{Obj}(\mathcal{C})$ as well. The equivalence relation is often described by a group action $\mathcal{G} \times \mathcal{C} \longrightarrow \mathcal{C}$. From the groupoid, one obtains a set $\mathrm{Iso}(\mathcal{C})$ of isomorphism classes and then attempts to construct a natural geometric structure on $\mathrm{Iso}(\mathcal{C})$ in a universal way. This is the concept of moduli space.

In our specific construction,

Definition 3.2. The **Betti groupoid** $\mathrm{Hom}(\pi, G)$ consists of

(1) The set of objects: $\mathrm{Hom}(\pi, G)$.
(2) The equivalence group G-action:

$$G \times \mathrm{Hom}(\pi, G) \longrightarrow \mathrm{Hom}(\pi, G), \quad (g, \rho) \mapsto g\rho g^{-1}.$$

From these two ingredients, one obtains the orbit family $\mathrm{Iso}(\mathrm{Hom}(\pi, G))$. The set $\mathrm{Hom}(\pi, G)$ acquires a variety structure from G and $\mathcal{M}_B$ parameterizes the reductive representations in $\mathrm{Iso}(\mathrm{Hom}(\pi, G))$.

The Betti groupoid $\mathrm{Hom}(\pi, G)$ is our concern. To understand this space better, we present a few other related groupoids, namely the Čech, de Rham and Dolbeault groupoids. See Definitions 8.3, 8.4, 9.4 and 10.8 for their constructions.

4. Local Functions and Their Derivatives over $\mathbb{R}^n$

4.1. *Functions*

Let X be the unit ball in $\mathbb{R}^n$ with coordinates $x = (x_1, \cdots, x_n)$ and suppose that $f : X \longrightarrow \mathbb{C}$ is a smooth (infinitely differentiable) function. A simple observation is that f is a constant function if and only if

$$\frac{\partial f}{\partial x_i} := \frac{\partial\,\mathrm{Re}(f)}{\partial x_i} + \mathbf{i}\frac{\partial\,\mathrm{Im}(f)}{\partial x_i} = 0$$

for all $1 \leq i \leq n$; here $\mathbf{i} \in \mathbb{C}$ with $\mathbf{i}^2 = -1$. This simple observation is the kernel of the theory. Denote by $\mathcal{O}$ the set of smooth $\mathbb{C}$-valued functions on X and by $\mathcal{O}_c$ the subset of constant functions. These two sets form commutative rings with $\mathcal{O}_c \cong \mathbb{C}$ and $\mathcal{O}_c \hookrightarrow \mathcal{O}$ a ring homomorphism.

4.2. *Exterior differentiation*

The linear first order differential operator $\frac{\partial}{\partial x_i}$ is an example of vector fields on X. Denote by $\mathcal{T}$ the space of all such vector fields. If $f \in \mathcal{O}$ is a function and $V \in \mathcal{T}$ is a vector field, then fV is again a vector field. This means that the space $\mathcal{T}$ is a left $\mathcal{O}$-module. A function $f \in \mathcal{O}$ is constant if and only if $V(f) = 0$ for each vector field $V \in \mathcal{T}$. This motivates the definition of the exterior differentiation. Let Ω^1 be the space of 1-forms (covector fields) on X. For each smooth function $f : X \to \mathbb{C}$, df is a 1-form:

$$d : \mathcal{O} \longrightarrow \Omega^1 \quad : \quad df(V) = V(f), \quad \forall V \in \mathcal{T},$$

i.e. df takes a vector field V and returns a function: $V(f)$. Then $f \in \mathcal{O}_c$ if and only if $df(V) = 0$ for each $V \in \mathcal{T}$, or equivalently, $df = 0 \in \Omega^1$ as a covector field. In other words, the ring $\mathcal{O}_c$ may be defined as

$$\mathcal{O}^d = \{f \in \mathcal{O} : df = 0\}. \tag{4.1}$$

From this, one constructs the $\mathcal{O}$-modules of i-forms $\Omega^i = \wedge_{j=1}^i \Omega^1$ and generalizes d to obtain the standard de Rham complex

$$(\Omega^\bullet, d) : \Omega^0 \xrightarrow{d} \Omega^1 \cdots \Omega^i \xrightarrow{d} \Omega^{i+1} \cdots,$$

such that $d^2 = 0$ and satisfying the (generalized) Leibniz rule

$$d(a \wedge b) = da \wedge b + (-1)^i a \wedge db, \quad a \in \Omega^i.$$

4.3. *Connections*

Let $\mathcal{E} = \{f : X \longrightarrow \mathbb{C}^{\oplus r}\}$. Then $\mathcal{E}$ is a free $\mathcal{O}$-module isomorphic to $\mathcal{O}^{\oplus r}$. Let $\mathcal{E}_c \subseteq \mathcal{E}$ be the $\mathcal{O}_c$-module isomorphic to $\mathcal{O}_c^{\oplus r} \subseteq \mathcal{O}^{\oplus r}$. The exterior differentiation is now generalized to an operator

$$D : \mathcal{E} \longrightarrow \Omega^1(\mathcal{E}) := \Omega^1 \otimes \mathcal{E}, \quad D(f) = D(\oplus_{i=1}^r f_i) = \oplus_{i=1}^r df_i.$$

Now we further generalize the notion of exterior differentiation by enforcing linearity and the Leibniz rule.

Definition 4.1. Let $\Omega^i(\mathcal{E}) := \Omega^i \otimes \mathcal{E}$. An $\mathcal{O}_c$-linear operator

$$\nabla : \mathcal{E} \longrightarrow \Omega^1(\mathcal{E})$$

is a **connection** if $\nabla(fe) = df \otimes e + f\nabla(e)$ for $f \in \mathcal{O}$ and $e \in \mathcal{E}$. Denote by $\mathcal{C}_d$ the set of all connections on $\mathcal{E}$.

For each i, a connection ∇ extends to an $\mathcal{O}_c$-linear map

$$\nabla : \Omega^i(\mathcal{E}) \longrightarrow \Omega^{i+1}(\mathcal{E})$$

satisfying the generalized Leibniz rule

$$\nabla(\eta \otimes e) = d\eta \otimes e + (-1)^i \eta \wedge \nabla(e).$$

Remark 4.2. A connection ∇ together with a vector field $V \in \mathcal{T}$ induce a map

$$\nabla_V : \mathcal{E} \xrightarrow{\nabla} \Omega^1(\mathcal{E}) \xrightarrow{V} \mathcal{E}.$$

Hence ∇ turns $\mathcal{E}$ into a $\mathcal{T}$-module, satisfying $\mathcal{O}_c$-linearity and the Leibniz rule.

Notice that $D \in \mathcal{C}_d$. Suppose ∇ and ∇_0 are two connections. Then for $f \in \mathcal{O}, e \in \mathcal{E}$,

$$(\nabla - \nabla_0)(fe) = df \otimes e + f\nabla e - df \otimes e - f\nabla_0 e = f(\nabla - \nabla_0)(e).$$

Hence $\eta = \nabla - \nabla_0$ is $\mathcal{O}$-linear and a connection is of the form $\nabla = D + A$, where

$$A \in \Omega^1(\mathrm{End}(\mathcal{E})) := \Omega^1 \otimes \mathrm{End}(\mathcal{E})$$

and the set of connections identifies with $\Omega^1(\mathrm{End}(\mathcal{E}))$. One may think of A as an $r \times r$ matrix with 1-forms for entries. Then for $f \in \mathcal{O}, e \in \mathcal{E}$,

$$\nabla^2(fe) = \nabla(df \otimes e + f\nabla(e)) = d^2 f \otimes e - df \wedge \nabla(e) + df \wedge \nabla(e) + f\nabla^2(e) = f\nabla^2(e).$$

This implies that the operator ∇^2 is $\mathcal{O}$-linear, hence, $\nabla^2 \in \Omega^2(\mathrm{End}(\mathcal{E}))$. Again one may think of elements in $\Omega^2(\mathrm{End}(\mathcal{E}))$ as matrices with scalar 2-form entries.

Definition 4.3. The **curvature operator** is a map

$$\mathsf{F} : \mathcal{C}_d \longrightarrow \Omega^2(\mathrm{End}(\mathcal{E})), \quad \mathsf{F}(\nabla) = \nabla^2.$$

For $\nabla \in \mathcal{C}_d$, $\mathsf{F}(\nabla)$ is the **curvature** of ∇. A connection ∇ is **integrable** if $\mathsf{F}(\nabla) = 0$. If ∇ is integrable, then $\mathcal{E}^\nabla = \{e \in \mathcal{E} : \nabla e = 0\}$ is the $\mathcal{O}_c$-module of constant functions with respect to ∇. Denote by $\mathcal{F}_d \subseteq \mathcal{C}_d$ the subspace of integrable connections.

Remark 4.4. Notice that $D \in \mathcal{F}_d$ and $\mathcal{E}^D = \mathcal{E}_c$. If we consider a connection ∇ to be a generalized D and use ∇ to define $\mathcal{O}_c$-modules of constant functions, then it is not enough that ∇ is $\mathcal{O}_c$-linear and satisfies the Leibniz rule. ∇ must satisfy the integrability condition in addition.

Identifying $\mathcal{C}_d$ with $\Omega^1(\mathrm{End}(\mathcal{E}))$, there is an exact sequence of maps

$$\mathcal{F}_d \hookrightarrow \Omega^1(\mathrm{End}(\mathcal{E})) \xrightarrow{\mathsf{F}} \Omega^2(\mathrm{End}(\mathcal{E})).$$

Notice that $\mathcal{F}_d$ is not necessarily a vector space because $\nabla = D + A_1 + A_2$ is not necessarily integrable even if $\nabla_1 = D + A_1$ and $\nabla_2 = D + A_2$ are. Another way of saying this is that F is not a linear operator in general.

For an integrable $\nabla \in \mathcal{F}_d$, the ∇-de Rham complex is

$$(\Omega^\bullet(\mathcal{E}), \nabla) : \Omega^0(\mathcal{E}) \xrightarrow{\nabla} \Omega^1(\mathcal{E}) \cdots \Omega^i(\mathcal{E}) \xrightarrow{\nabla} \Omega^{i+1}(\mathcal{E}) \cdots .$$

Remark 4.5. Vector fields in $\mathcal{T}$ are first order differential operators on $\mathcal{O}$. By composing these operators, $\mathcal{T}$ generates $\mathcal{D}$, the ring of differential operators. Integrability of ∇ is precisely the needed condition for $\mathcal{E}$ to inherit a $\mathcal{D}$-module structure from its $\mathcal{T}$-module structure [1, 3, 17].

4.4. *The gauge group and its action*

The group G acts on $\mathbb{C}^{\oplus r}$ by matrix-vector multiplication. Define the G-gauge group

$$\mathcal{G} = \Omega^0(G) := \mathcal{O}(G) = \{g : X \longrightarrow G\}.$$

Then $\mathcal{G}$ acts on $\mathcal{E}$ by $\mathcal{O}$-module automorphism

$$\mathcal{G} \times \mathcal{E} \longrightarrow \mathcal{E}, \quad (g, e) \mapsto ge, \quad ge(x) = g(x).e(x),$$

where $g(x).e(x)$ means matrix-vector multiplication over a point $x \in X$. Suppose $g \in \mathcal{G}$ and $\nabla \in \mathcal{C}_d$. Then $g : \mathcal{E} \stackrel{\cong}{\longrightarrow} \mathcal{E}$ and ∇ pulls back to another connection $g^*\nabla$ on $\mathcal{E}$ as follows:

$$g^*\nabla : \mathcal{E} \stackrel{g}{\longrightarrow} \mathcal{E} \stackrel{\nabla}{\longrightarrow} \Omega^1(\mathcal{E}) \stackrel{g^{-1}}{\longrightarrow} \Omega^1(\mathcal{E}).$$

This gives a $\mathcal{G}$-action on $\mathcal{C}_d$.

Remark 4.6. The $\mathcal{O}$-module $\mathrm{End}(\mathcal{E})$ should be thought of as the Lie algebra valued functions

$$\mathcal{O}(\mathfrak{g}) = \Omega^0(\mathfrak{g}) = \{v : X \longrightarrow \mathfrak{g}\}$$

with the bracket as the commutator, where $\mathfrak{g}$ is the Lie algebra of G.

Identify $\mathcal{C}_d$ with $\Omega^1(\mathrm{End}(\mathcal{E})) \cong \Omega^1(\mathfrak{g})$ and let $\nabla = D + A$ for $A \in \Omega^1(\mathrm{End}(\mathcal{E}))$ and $g \in \mathcal{G}$. Then

$$g^{-1}\nabla g = D + g^{-1}(Dg) + g^{-1}Ag,$$

$$\mathsf{F}(\nabla) = \nabla^2 = DA + A \wedge A, \quad \mathsf{F}(g^{-1}\nabla g) = g^{-1}\nabla^2 g,$$

where Dg and DA mean entry-wise exterior differentiation and $A \wedge A$ means matrix multiplication with entry-wise wedge product. Hence $\nabla^2 = 0$ if and only if $g^{-1}\nabla^2 g = 0$, i.e. the gauge group action preserves and restricts to the integrable connections $\mathcal{G} \times \mathcal{F}_d \longrightarrow \mathcal{F}_d$. Notice that $\mathfrak{g}$ may not be commutative; hence, $A \wedge A$ (more canonically written as $\frac{1}{2}[A, A]$ in the language of differential graded Lie algebra [7]) is not necessarily zero.

Definition 4.7. The orbit space $\mathcal{M}_d = \mathcal{F}_d/\mathcal{G}$ is called the **de Rham moduli space** of X.

The connection D defines a map

$$\mathsf{D} : \mathcal{G} \longrightarrow \mathcal{C}_d, \quad \mathsf{D}(g) = g^{-1}Dg.$$

Identifying $\mathcal{C}_d$ with $\Omega^1(\mathrm{End}(\mathcal{E}))$, the following sequence of maps is exact

$$\Omega^0(G) \xrightarrow{\mathsf{D}} \Omega^1(\mathrm{End}(\mathcal{E})) \xrightarrow{\mathsf{F}} \Omega^2(\mathrm{End}(\mathcal{E})).$$

Hence the space $\mathcal{M}_d$ is a point. This is the same as saying that on the unit ball, all integrable connections are $\mathcal{G}$-gauge equivalent. The general principle is that the curvature is the only local obstruction to a connection being trivial, i.e. isomorphic to D.

Remark 4.8. An integrable connection ∇ produces an $\mathcal{O}_c$-module $\mathcal{E}^\nabla \cong \mathcal{E}_c \subseteq \mathcal{E}$. The reverse operation is $\mathcal{E} = \mathcal{O} \otimes \mathcal{E}^\nabla$.

5. Local Functions and Their Derivatives over $\mathbb{C}^n$

Let X be the unit ball in $\mathbb{C}^n$ with coordinates $z = (z_1, \ldots, z_n)$ and suppose that $f : X \longrightarrow \mathbb{C}$ is a smooth function.

5.1. *Holomorphic functions and forms*

As X has a complex structure, one may reformulate the leitmotif as: f is a constant function if and only if $\frac{\partial f}{\partial \bar{z}_i} = 0$ and $\frac{\partial f}{\partial z_i} = 0$ for all $1 \leq i \leq n$. In other words, one may define intermediate objects, halfway between the very flexible smooth functions and the very rigid constant functions:

Definition 5.1. A function $f \in \mathcal{O}$ is **holomorphic** if it satisfies the Cauchy-Riemann equation $\frac{\partial f}{\partial \bar{z}_i} = 0$ for all $1 \leq i \leq n$. Denote by $\mathcal{O}_h$ the ring of holomorphic functions on X.

Then f is constant if and only if f is holomorphic and, in addition, all its holomorphic derivatives are zero, i.e. a constant function f is a holomorphic function that satisfies the additional equations $\frac{\partial f}{\partial z_i} = 0$ for all i. Again notice that $\mathcal{O}_c \hookrightarrow \mathcal{O}_h \hookrightarrow \mathcal{O}$ are ring homomorphisms.

The complex structure on the unit ball X in $\mathbb{C}^n$ gives a decomposition $\mathcal{T} = \mathcal{T}^{1,0} \oplus \mathcal{T}^{0,1}$ as $\mathcal{O}$-modules, where $\mathcal{T}^{1,0}$ is generated by $\{\frac{\partial}{\partial z_i} : 1 \leq i \leq n\}$ and $\mathcal{T}^{0,1}$ is generated by $\{\frac{\partial}{\partial \bar{z}_i} : 1 \leq i \leq n\}$. This induces a decomposition on the dual space $\Omega^1 = \Omega^{1,0} \oplus \Omega^{0,1}$. More generally, by taking exterior products of $\Omega^{1,0}$ and $\Omega^{0,1}$, one obtains decompositions and projections

$$\Omega^m = \bigoplus_{q+p=m} \Omega^{q,p}, \quad P_{q,p} : \Omega^m \longrightarrow \Omega^{q,p}$$

and the exterior differential operator $d : \Omega^m \longrightarrow \Omega^{m+1}$ decomposes as

$$d = \partial + \overline{\partial} \text{ with } \partial = P_{q+1,p} \circ d \text{ and } \overline{\partial} = P_{q,p+1} \circ d.$$

Moreover, for fixed p and q, the de Rham complex decomposes as

$$(\Omega^{q,\bullet},\bar\partial) : \Omega^{q,0} \xrightarrow{\bar\partial} \Omega^{q,1} \cdots \Omega^{q,i} \xrightarrow{\bar\partial} \Omega^{q,i+1} \cdots$$

$$(\Omega^{\bullet,p},\partial) : \Omega^{0,p} \xrightarrow{\partial} \Omega^{1,p} \cdots \Omega^{i,p} \xrightarrow{\partial} \Omega^{i+1,p} \cdots ,$$

and

$$\mathcal{O}_h := \mathcal{O}^{\bar\partial} = \{f \in \mathcal{O} : \bar\partial f = 0\}. \tag{5.1}$$

5.2. *The holomorphic structures*

The operator D also decomposes according to types $D = D' + D''$:

$$D'' : \mathcal{E} \longrightarrow \Omega^{0,1}(\mathcal{E}) = \Omega^{0,1} \otimes \mathcal{E}, \quad D''(f) = D(\oplus_{i=1}^r f_i) = \oplus_{i=1}^r \bar\partial f_i,$$

$$D' : \mathcal{E} \longrightarrow \Omega^{1,0}(\mathcal{E}) = \Omega^{1,0} \otimes \mathcal{E}, \quad D'(f) = D'(\oplus_{i=1}^r f_i) = \oplus_{i=1}^r \partial f_i.$$

Definition 5.2. Let $\Omega^{q,p}(\mathcal{E}) := \Omega^{q,p} \otimes \mathcal{E}$. An almost holomorphic structure on $\mathcal{E}$ is an operator

$$\bar\nabla : \mathcal{E} \longrightarrow \Omega^{0,1}(\mathcal{E}), \quad \text{with } \bar\nabla(fe) = \bar\partial f \otimes e + f\bar\nabla(e)$$

for $f \in \mathcal{O}$ and $e \in \mathcal{E}$. Denote by $\mathcal{C}_h$ the space of almost holomorphic structures on $\mathcal{E}$.

For $p + q = i$, $\bar\nabla$ extends to an $\mathcal{O}_c$-linear map

$$\bar\nabla : \Omega^{p,q}(\mathcal{E}) \longrightarrow \Omega^{p,q+1}(\mathcal{E})$$

satisfying the generalized Leibniz rule

$$\bar\nabla(\eta \otimes e) = \bar\partial\eta \otimes e + (-1)^i \eta \wedge \bar\nabla(e).$$

Define map

$$\mathsf{F} : \mathcal{C}_h \longrightarrow \Omega^{0,2}(\mathrm{End}(\mathcal{E})), \quad \mathsf{F}(\bar\nabla) = \bar\nabla^2.$$

Definition 5.3. A **holomorphic structure** $\bar\nabla$ on $\mathcal{E}$ is an almost holomorphic structure satisfying the integrability condition $\mathsf{F}(\bar\nabla) = 0$. Denote by $\mathcal{F}_h$ the set of holomorphic structures on $\mathcal{E}$.

Suppose that $\bar\nabla$ and $\bar\nabla_0$ are two almost holomorphic structures. Then

$$(\bar\nabla - \bar\nabla_0)(fe) = \bar\partial f \otimes e + f\bar\nabla e - \bar\partial f \otimes e - f\bar\nabla_0 e = f(\bar\nabla - \bar\nabla_0)(e).$$

Hence $\bar\nabla - \bar\nabla_0 \in \Omega^{0,1}(\mathrm{End}(\mathcal{E}))$ and an almost holomorphic structure is of the form $\bar\nabla = D'' + A$, where $A \in \Omega^{0,1}(\mathrm{End}(\mathcal{E}))$. Hence the space of almost holomorphic structures $\mathcal{C}_h$ identifies with $\Omega^{0,1}(\mathrm{End}(\mathcal{E}))$.

When the integrability condition is satisfied, define the Dolbeault complex

$$(\Omega^{q,\bullet}(\mathcal{E}), \bar{\nabla}) : \Omega^{q,0}(\mathcal{E}) \xrightarrow{\bar{\nabla}} \Omega^{q,1}(\mathcal{E}) \cdots \Omega^{q,i}(\mathcal{E}) \xrightarrow{\bar{\nabla}} \Omega^{q,i+1}(\mathcal{E}) \cdots .$$

Suppose $\bar{\nabla}$ is a holomorphic structure. Then the $\mathcal{O}_h$-module

$$\mathcal{E}^{\bar{\nabla}} = \{e : \bar{\nabla} e = 0\}$$

is the module of holomorphic sections with respect to $\bar{\nabla}$. Notice that $D'' \in \mathcal{F}_h$.

5.3. *The gauge group action I*

The gauge group $\mathcal{G}$ acts on the almost holomorphic structures, inducing an action

$$\mathcal{G} \times \mathcal{C}_h \longrightarrow \mathcal{C}_h, \quad (g, \bar{\nabla}) \mapsto g^{-1} \bar{\nabla} g.$$

This action preserves $\mathcal{F}_h$, hence, restricts to an action $\mathcal{G} \times \mathcal{F}_h \longrightarrow \mathcal{F}_h$. The operator D'' also defines a map

$$\mathsf{D} : \mathcal{G} \longrightarrow \mathcal{C}_h, \quad \mathsf{D}(g) = g^{-1} D'' g.$$

Identifying $\mathcal{C}_h$ with $\Omega^{0,1}(\mathrm{End}(\mathcal{E}))$, there is an exact sequence of maps

$$\mathcal{G} \xrightarrow{\mathsf{D}} \Omega^{0,1}(\mathrm{End}(\mathcal{E})) \xrightarrow{\mathsf{F}} \Omega^{0,2}(\mathrm{End}(\mathcal{E})).$$

Hence the space of holomorphic structures $\mathcal{M}_h := \mathcal{F}_h / \mathcal{G}$ is a point.

Remark 5.4. A holomorphic structure $\bar{\nabla}$ produces an $\mathcal{O}_h$-module $\mathcal{E}^{\bar{\nabla}} \cong \mathcal{O}_h^{\oplus r} \subseteq \mathcal{E}$. The reverse operation is $\mathcal{E} = \mathcal{O} \otimes \mathcal{E}^{\bar{\nabla}}$.

5.4. *Holomorphic λ-connections*

At this point, we move to the holomorphic universe (or category) and make a change of notations to emphasize a certain suggestive analogy. By Remark 5.4, fix $\mathcal{E}_h = \mathcal{O}_h^{\oplus r}$ and let

$$\mathcal{T}_h = \{V \in \mathcal{T} : V(f) \in \mathcal{O}_h, \forall f \in \mathcal{O}_h\}, \quad \Omega_h^q = \{\eta \in \Omega^{q,0} : \bar{\partial}\eta = 0\},$$

and define the holomorphic de Rham complex

$$(\Omega_h^\bullet, \partial) : \Omega_h^0 \xrightarrow{\partial} \Omega_h^1 \cdots \Omega_h^i \xrightarrow{\partial} \Omega_h^{i+1} \cdots .$$

We make a slight generalization:

Definition 5.5. Let $\Omega^i_h(\mathcal{E}_h) := \Omega^i_h \otimes \mathcal{E}_h$ and $\lambda \in \mathbb{C}$. A **holomorphic λ-connection** or **λ-connection** on $\mathcal{E}_h$ is an operator

$$\nabla^\lambda : \mathcal{E}_h \longrightarrow \Omega^1_h(\mathcal{E}_h)$$

satisfying $\nabla^\lambda(fe) = \lambda\partial f \otimes e + f\nabla^\lambda(e)$ for $f \in \mathcal{O}_h$ and $e \in \mathcal{E}_h$. Denote by $\mathcal{C}^\lambda_h$ the space of all λ-connections on $\mathcal{E}_h$.

For each i, ∇^λ extends to an $\mathcal{O}_c$-linear map

$$\nabla^\lambda : \Omega^i_h(\mathcal{E}) \longrightarrow \Omega^{i+1}_h(\mathcal{E})$$

satisfying the generalized Leibniz rule

$$\nabla^\lambda(\eta \otimes e) = \lambda\partial\eta \otimes e + (-1)^i \eta \wedge \nabla^\lambda(e).$$

When $\lambda = 1$ we recover the usual definition of connection. Let

$$\mathsf{F} : \mathcal{C}^\lambda_h \longrightarrow \Omega^2_h(\mathrm{End}(\mathcal{E})), \quad \mathsf{F}(\nabla^\lambda) = (\nabla^\lambda)^2.$$

Definition 5.6. A holomorphic λ-connection $\nabla^\lambda \in \mathcal{C}^\lambda_h$ is **integrable** if $\mathsf{F}(\nabla^\lambda) = 0$. Denote by $\mathcal{F}^\lambda_h$ the space of integrable λ-connections on $\mathcal{E}_h$.

Suppose that ∇^λ and ∇^λ_0 are two λ-connections on $\mathcal{E}_h$. Then

$$(\nabla^\lambda - \nabla^\lambda_0)(fe) = \lambda\partial f \otimes e + f\nabla^\lambda e - \lambda\partial f \otimes e - f\nabla^\lambda_0 e = f(\nabla^\lambda - \nabla^\lambda_0)(e).$$

Hence $\nabla^\lambda - \nabla^\lambda_0 \in \Omega^1_h(\mathrm{End}(\mathcal{E}_h))$ and a λ-connection is of the form $\nabla^\lambda = \lambda D' + A$, where $A \in \Omega^1_h(\mathrm{End}(\mathcal{E}_h))$. Hence the space $\mathcal{C}^\lambda_h$ of λ-connections identifies with $\Omega^1_h(\mathrm{End}(\mathcal{E}_h))$.

When the integrability condition is satisfied, the λ-holomorphic de Rham complex is

$$(\Omega^\bullet_h(\mathcal{E}_h), \nabla^\lambda) : \Omega^0_h(\mathcal{E}_h) \xrightarrow{\nabla^\lambda} \Omega^1_h(\mathcal{E}_h) \cdots \Omega^i_h(\mathcal{E}_h) \xrightarrow{\nabla^\lambda} \Omega^{i+1}_h(\mathcal{E}_h) \cdots .$$

If ∇^λ is integrable, then the $\mathcal{O}_c$-module $\mathcal{E}^{\nabla^\lambda}_h = \{e : \nabla^\lambda e = 0\}$ is called the module of constant sections with respect to ∇^λ.

Remark 5.7. Vector fields in $\mathcal{T}_h$ are first order differential operators on $\mathcal{O}_h$. By operator composition, $\mathcal{T}_h$ generates $\mathcal{D}_h$, the ring of differential operators. Integrability of ∇^1 is precisely the needed condition for $\mathcal{E}_h$ to inherit a $\mathcal{D}_h$-module structure from its $\mathcal{T}_h$-module structure.

Remark 5.8. If $\lambda \in \mathbb{C}^\times$, then $\nabla^\lambda = \lambda D' + A$ is equivalent to $\nabla^1 = D' + \frac{A}{\lambda}$ in the sense that $\nabla^\lambda e = 0$ if and only if $\nabla^1 e = 0$.

5.5. *The gauge group action II*

The group G is a complex manifold; hence, it makes sense to say whether a map $g : X \longrightarrow G$ is holomorphic. Notice that each $g \in \mathcal{G}$ is an $\mathcal{O}$-module automorphism $g : \mathcal{E} \longrightarrow \mathcal{E}$. Let

$$\mathcal{G}_h = \{g \in \mathcal{G} : g : \mathcal{E}_h \to \mathcal{E}_h\},$$

i.e. elements in $\mathcal{G}_h$ restrict to $\mathcal{O}_h$-module automorphisms of $\mathcal{E}_h$. By definition, there is an action $\mathcal{G}_h \times \mathcal{E}_h \longrightarrow \mathcal{E}_h$ which induces an action

$$\mathcal{G}_h \times \mathcal{C}_h^\lambda \longrightarrow \mathcal{C}_h^\lambda, \quad (g, \nabla^\lambda) \mapsto g^{-1}\nabla^\lambda g = \lambda D' + \lambda g^{-1} D' g + g^{-1} A g,$$

where $\nabla^\lambda = \lambda D' + A$. The $\mathcal{G}_h$-action preserves $\mathcal{F}_h^\lambda$. The operator D' defines a map

$$\mathsf{D} : \mathcal{G}_h \longrightarrow \mathcal{C}_h^\lambda, \quad \mathsf{D}(g) = g^{-1}\lambda D' g$$

and there is an exact sequence of maps

$$\mathcal{G}_h \xrightarrow{\mathsf{D}} \mathcal{C}_h^\lambda \xrightarrow{\mathsf{F}} \Omega_h^2(\mathrm{End}(\mathcal{E}_h)).$$

The λ-connections in the same orbits are considered equivalent. On the unit ball X, there is only one integrable λ-connection on $\mathcal{E}_h$ up to $\mathcal{G}_h$-gauge equivalence. Hence $\mathcal{F}_h^\lambda / \mathcal{G}_h$ is a point.

Remark 5.9. The $\mathcal{O}$-module $\mathcal{E}$ together with $\bar{\nabla}$ make $\mathcal{E}^{\bar{\nabla}} \cong \mathcal{E}_h$ an $\mathcal{O}_h$-module. Furthermore, for $\lambda \in \mathbb{C}^\times$, $\mathcal{E}_h$ together with ∇^λ give an $\mathcal{O}_c$-module isomorphism $\mathcal{E}_h^\lambda \cong \mathcal{E}_c$. To summarize, for $\lambda \in \mathbb{C}^\times$,

$$\mathcal{E}_h^\lambda \subseteq \mathcal{E}_h \subseteq \mathcal{E}.$$

In the reverse direction,

$$\mathcal{E}_h \cong \mathcal{O}_h \otimes \mathcal{E}_c, \quad \mathcal{E} \cong \mathcal{O} \otimes \mathcal{E}_h, \quad \mathcal{E} \cong \mathcal{O} \otimes \mathcal{E}_c.$$

6. Local Examples in Low Dimensions

6.1. *Connections on an interval*

Let $X = (-1, 1) \subset \mathbb{R}$ and suppose that $f : X \longrightarrow \mathbb{C}$ is a smooth function. Then f is a constant function if and only if $\frac{\partial f}{\partial x} = 0$.

Consider the situation of $r = \mathrm{rank}(\mathcal{E}) = 1$. Then $G = \mathbb{C}^\times$, $\mathcal{E} \cong \mathcal{O}$, $D = d$ and $\mathcal{C}_d$ identifies with $\Omega^1(\mathrm{End}(\mathcal{E})) \cong \Omega^1$. Let $\nabla \in \mathcal{C}_d$ be a connection. Then $\nabla = D + A = d + A$ and $f \in \mathcal{E}^\nabla$ if and only if

$$\nabla f = df + Af = 0.$$

Choosing $f(0) = C$ and integrating, the solution is

$$f(x) = Ce^{-\int_0^x A}.$$

The gauge group is

$$\mathcal{G} := \Omega^0(\mathbb{C}^\times) := \mathcal{O}^\times := \{g : X \longrightarrow \mathbb{C}^\times\}$$

which acts on $\mathcal{O}$ by multiplication:

$$\mathcal{G} \times \mathcal{E} \longrightarrow \mathcal{E}, \quad (g, f) \mapsto gf.$$

Since $\dim_{\mathbb{R}}(X) = 1$, all 2-forms on X are zero. Hence $\mathcal{F}_d = \mathcal{C}_d$. Identifying $\mathcal{F}_d$ with Ω^1, there is an induced action

$$\mathcal{G} \times \mathcal{F}_d \longrightarrow \mathcal{F}_d, \quad (g, \nabla) \mapsto \nabla + d\log(g),$$

where $d\log(g)$ means $g^{-1}dg$. The set of solutions for $Df = df = 0$ is precisely $\mathcal{O}_c$. In general, the solution set for the equation $\nabla f = (d+A)f = 0$ is $\{Ce^{-\int_0^x A} : C \in \mathcal{O}_c\}$, i.e. $\mathcal{O}_c$ scaled by $e^{-\int_0^x A}$. Finally, after applying the gauge g, the solution set for the equation

$$g^{-1}\nabla g(f) = (d + A + d\log(g))f = 0$$

is $\{Cg(x)^{-1}e^{-\int_0^x A} : C \in \mathcal{O}_c\}$. Hence the $\mathcal{G}$-action simply scales the solutions.

On the unit ball $X = (-1, 1)$, every closed 1-form is exact, i.e. $A = d\log(g)$ for some $g \in \mathcal{G}$ or ∇ is equivalent to d. Hence $\mathcal{E}^\nabla \cong \mathcal{O}_c$ as an $\mathcal{O}_c$-module for all connections $\nabla \in \mathcal{C}_d$.

6.2. *Connections on the unit disk*

Next, let X be the unit disk in $\mathbb{R}^2$ and suppose $f \in \mathcal{O}$. Then $f \in \mathcal{O}_c$ (i.e. constant) if and only if

$$\frac{\partial f}{\partial x_1} = \frac{\partial f}{\partial x_2} = 0.$$

Once again, let $r = \operatorname{rank}(\mathcal{E}) = 1$ for simplicity. Then $G = \mathbb{C}^\times$ and $\mathcal{E} \cong \mathcal{O}$ as an $\mathcal{O}$-module and $D = d$. Let $\nabla \in \mathcal{C}_d$. One would like to say that $f \in \mathcal{E}$ is a constant if and only if $\nabla f = 0$. *However there is a problem:* Suppose

$$\nabla f = df + Af = 0, A \in \Omega^1.$$

Integrating, the solution is of the form

$$f(z) = Ce^{-\int_L A} = f(0)e^{-\int_L A}, \tag{6.1}$$

where L is a path connecting 0 and z in X. Hence for f to be (even locally) well-defined, the integral must depend only on z and independent of L. In other words, $\oint A = 0$ for any small loop that begins and ends at 0. By Stokes' theorem, this condition is satisfied if $dA = 0$. Since $G = \mathbb{C}^\times$ is abelian, for $\nabla = d + A$,

$$\nabla^2 = dA + A \wedge A = dA + \frac{1}{2}[A, A] = dA.$$

This is the motivation for the integrability condition: If ∇ is integrable, then one may locally solve the equation $\nabla f = 0$. Hence $\mathcal{F}_d$ identifies with

$$Z^1 = \{A \in \Omega^1 : dA = 0\},$$

the space of closed 1-forms when G is abelian. Meanwhile, the gauge $\mathcal{G}$-action is

$$\mathcal{G} \times Z^1 \longrightarrow Z^1, \quad (g, \nabla) \mapsto \nabla + d\log(g).$$

In other words, the gauge g changes the integrable connections by the log-exact form $d\log(g)$.

6.3. *The holomorphic construction*

The integrability condition $\bar{\nabla}^2 = 0$ is trivially satisfied since any $(0, 2)$-form is zero in complex dimension $n = 1$. Hence on the unit disk X, $\mathcal{F}_h = \mathcal{C}_h$. Moreover, when $r = 1$, every (almost) holomorphic structure $\bar{\nabla}$ is equivalent to $\bar{\partial}$ by a gauge transformation. Similarly, $\mathsf{F}(\nabla^\lambda) = 0$ is also trivially satisfied since all $(2, 0)$-forms are zero on X.

7. Interlude: Manifolds and Functions

We now briefly describe the concept of manifolds. Let $B \subset \mathbb{C}^n$ (or $B \subset \mathbb{R}^n$) be the open unit ball. A set X is a manifold if there exist injective maps $\{\phi_i : B \longrightarrow X\}_{i \in \mathcal{I}}$, indexed by $\mathcal{I} \subseteq \mathbb{Z}_+$, such that $X = \cup_{i \in \mathcal{I}} \phi_i(B)$. Let $U_i = \phi_i(B)$ and $\mathcal{U} = \{U_i : i \in \mathcal{I}\}$. For a subset $\mathcal{J} \subseteq \mathcal{I}$, let $U_{\mathcal{J}} = \cap_{i \in \mathcal{J}} U_i$. For $i, j \in \mathcal{I}$, define transition maps $\phi_{ij} = \phi_i^{-1} \circ \phi_j$.

The injectivity of each ϕ_i provides each U_i with the structures on B. However these structures need to be compatible on the intersections U_{ij} in order for X to inherit these structures globally. These are conditions imposed on the transition maps ϕ_{ij}.

Definition 7.1. For all $i, j \in \mathcal{I}$,

(1) X is a **topological manifold** if ϕ_{ij} is a homeomorphism.

(2) X is a **smooth manifold** if ϕ_{ij} is a diffeomorphism.
(3) X is a **complex manifold** if ϕ_{ij} is holomorphic diffeomorphism.

Let X be a smooth manifold. Then each chart U_i inherits an exterior differential operator d_i from B. Let $f : U_i \longrightarrow \mathbb{C}$ such that $f \circ \phi_i$ is smooth. Since each ϕ_{ij} is smooth, $f \circ \phi_j$ is smooth for all $j \in \mathcal{I}$. Together the local exterior differential operators $\{d_i\}_{i\in\mathcal{I}}$ define a global exterior differential operator d.

Similarly, suppose that X is a complex manifold. Then each chart U_i inherits the ∂_i and $\overline{\partial}_i$ operators with $d_i = \partial_i + \overline{\partial}_i$ from those on B. Let $f : U_i \longrightarrow \mathbb{C}$ such that $f \circ \phi_i$ is holomorphic. Since each ϕ_{ij} is a holomorphic diffeomorphism, $f \circ \phi_j$ is holomorphic for all $j \in \mathcal{I}$. Together these local operators $\{\partial_i\}_{i\in\mathcal{I}}$ and $\{\overline{\partial}_i\}_{i\in\mathcal{I}}$ define global differential operators ∂ and $\overline{\partial}$ and the exterior differential operator decomposes as $d = \partial + \overline{\partial}$.

Definition 7.2. Let X be a smooth manifold with charts $\mathcal{U} = \{U_i\}_{i\in\mathcal{I}}$. Then there is an exterior differential operator d on X. Let $U \subseteq X$ be an open set. A function $f : U \longrightarrow \mathbb{C}$ is **smooth** if $f \circ \phi_i$ is smooth for all i. Suppose that, in addition, X is a complex manifold. A function $f : U \longrightarrow \mathbb{C}$ is **holomorphic** if $f \circ \phi_i$ is holomorphic (i.e $\overline{\partial}_i(f \circ \phi_i) = 0$) for all i.

Remark 7.3. A smooth manifold X is a topological space locally homeomorphic to B with a globally defined exterior differential operator d. In addition, if X is a complex manifold, then there are globally defined operators ∂ and $\overline{\partial}$ such that $d = \partial + \overline{\partial}$.

From now on, we shall assume that X is always smooth and that X has a complex structure when discussing holomorphic objects.

8. The Čech Construction

The most direct way to define (locally) holomorphic and constant functions on X is the Čech construction. The manifold X is covered with charts. We are already familiar with functions on each chart from Sections 4 and 5 where it took only one sentence to define constant and holomorphic functions, respectively, on the unit ball. The arduous task now is to glue these local functions between charts and into systems of functions. For this, we introduce sheaves.

Let X be a smooth manifold with charts $\mathcal{U} = \{U_i\}_{i\in\mathcal{I}}$. A sheaf on a manifold X assigns to each open set $U \subseteq X$ a particular family of functions.

8.1. *Structure sheaves*

For each open set $U \subseteq X$, let

$$\mathcal{O}(U) = \{f : U \longrightarrow \mathbb{C} : f \text{ is smooth } \}.$$

In other words, the *structure sheaf* $\mathcal{O}$ on X assigns to each open set $U \subseteq X$ the ring of smooth functions on U (Recall that smoothness makes global sense since each ϕ_{ij} is smooth). Let

$$\check{C}^p(\mathcal{U}, \mathcal{O}) = \prod_{i_0 < \cdots < i_p} \mathcal{O}(U_{i_0 \cdots i_p}), \quad i_0, \cdots, i_p \in \mathcal{I}.$$

The Čech complex for $(\mathcal{U}, \mathcal{O})$ is

$$\check{C}^\bullet(\mathcal{U}, \mathcal{O}) : \ \check{C}^0(\mathcal{U}, \mathcal{O}) \xrightarrow{d} \check{C}^1(\mathcal{U}, \mathcal{O}) \xrightarrow{d} \cdots \check{C}^m(\mathcal{U}, \mathcal{O}) \cdots, \tag{8.1}$$

where

$$(d\alpha)_{i_0 \cdots i_{p+1}} = \sum_{j=0}^{p+1} (-1)^j \alpha_{i_0 \cdots \hat{i}_j \cdots i_{p+1}}.$$

8.2. *Sheaves of modules*

Local functions on charts U_i and U_j may have different descriptions on U_{ij}. This is described by Čech 1-cocycles:

Definition 8.1. The **first Čech G-cochain** with respect to $\mathcal{I}$ is

$$\check{C}^1(G) = \{\{g_{ij} : U_{ij} \longrightarrow G\}_{i,j \in \mathcal{I}}\}.$$

If the family $\{g_{ij}\}_{i,j \in \mathcal{I}}$ satisfies the additional condition $g_{ij} \circ g_{jk} \circ g_{ki} = \mathrm{id}_{U_{ijk}}$ for all $i, j, k \in \mathcal{I}$, then the family is called a 1-cocycle. The set of all such 1-cocycles is denoted by $\check{\mathcal{F}}$ (or $\check{\mathcal{Z}}^1(G)$).

A 1-cocycle $g = \{g_{ij}\}_{i,j \in \mathcal{I}}$ defines a sheaf of $\mathcal{O}$-module $\mathcal{E}$ as follows: Let $U \subseteq X$ be open. Define

$$\mathcal{E}(U) = \{\{e_i : U \cap U_i \to \mathbb{C}^{\oplus r}\}_{i \in \mathcal{I}} : e_i \text{ is smooth and } g_{ij}.e_j = e_i \ \ \forall i, j \in \mathcal{I}\}. \tag{8.2}$$

In other words, each element in $\mathcal{E}(U)$ is a collection of functions such that the function values on $U \cap U_i$ and $U \cap U_j$ differ by the transformation g_{ij}. In this way, $\mathcal{E}(U)$ is an $\mathcal{O}(U)$-module.

Remark 8.2. Sheaves over X take an open set $U \subseteq X$ and return rings and modules. Hence the usual algebraic operations such as sum, product, tensor, dualization and etc on sheaves all make sense. They just mean these operations on the rings and modules on each $U \subseteq X$.

Suppose $\mathcal{E}$ is a sheaf on X. When referring to an arbitrary open set $U \subseteq X$, we often shorten $\mathcal{E}(U)$ to $\mathcal{E}$ when no confusion arises.

For a sheaf $\mathcal{E}$, let

$$\check{C}^p(\mathcal{U}, \mathcal{E}) = \prod_{i_0 < \cdots < i_p} \mathcal{E}(U_{i_0 \cdots i_p}), \quad i_0, \cdots, i_p \in \mathcal{I}.$$

The Čech complex for $(\mathcal{U}, \mathcal{E})$ is

$$\check{C}^\bullet(\mathcal{U}, \mathcal{E}) : \ \check{C}^0(\mathcal{U}, \mathcal{E}) \overset{\nabla}{\to} \check{C}^1(\mathcal{U}, \mathcal{E}) \overset{\nabla}{\to} \cdots \check{C}^m(\mathcal{U}, \mathcal{E}) \cdots, \tag{8.3}$$

where

$$(\nabla \alpha)_{i_0 \cdots i_{p+1}} = \sum_{j=0}^{p+1} (-1)^j \alpha_{i_0 \cdots \hat{i}_j \cdots i_{p+1}}.$$

The resulting cohomologies are denoted by

$$\check{\mathrm{H}}^i(\mathcal{U}, \mathcal{E}) := \ker(\nabla)/\nabla(\check{C}^{i-1}(\mathcal{U}, \mathcal{E})).$$

When $r = 1$ and $g_{ij} = 1 \in \mathrm{GL}(1, \mathbb{C}) = \mathbb{C}^\times$, we recover the structure sheaf $\mathcal{E} = \mathcal{O}$. When $g_{ij} = 1 \in \mathrm{GL}(r, \mathbb{C})$ is the identity map, the sheaf $\mathcal{E} \cong \mathcal{O}^{\oplus r}$ is called trivial.

8.3. *Sheaf isomorphisms*

Suppose $\mathcal{E}^1$ and $\mathcal{E}^2$ are two sheaves of $\mathcal{O}$-modules over X. A homomorphism of sheaves $g : \mathcal{E}^1 \longrightarrow \mathcal{E}^2$ means that for each open $U \subseteq X$, $g_U : \mathcal{E}^1(U) \longrightarrow \mathcal{E}^2(U)$ is a homomorphism of $\mathcal{O}$-modules. In addition if $V \hookrightarrow U$ is open, then functions on U restrict to functions on V, i.e. there is a canonical map $\mathcal{E}(U) \longrightarrow \mathcal{E}(V)$ for any $\mathcal{E}$. We require the following diagram to commute

$$\begin{array}{ccc} \mathcal{E}^1(U) & \xrightarrow{g_U} & \mathcal{E}^2(U) \\ \downarrow & & \downarrow \\ \mathcal{E}^1(V) & \xrightarrow{g_V} & \mathcal{E}^2(V). \end{array}$$

The homomorphism g is an isomorphism if g_U is an isomorphism for each open $U \subseteq X$.

Denote by $\check{\mathcal{G}}$ the G-zero-cochain

$$\check{\mathcal{G}} := \check{C}^0(G) = \{\{g_i : U_i \longrightarrow G\}_{i \in \mathcal{I}}\}.$$

An element in $\check{\mathcal{G}}$ defines a set of isomorphisms $g_i : \mathcal{E}(U_i)\longrightarrow\mathcal{E}(U_i)$.

Definition 8.3. The **smooth Čech groupoid** $\check{\mathcal{F}}$ consists of

(1) Objects: Sheaves of $\mathcal{O}$-modules.
(2) Equivalence group $\check{\mathcal{G}}$-action:

$$\check{\mathcal{G}} \times \check{\mathcal{F}}\longrightarrow\check{\mathcal{F}}, \quad (\{g_i\}_{i\in\mathcal{I}}, \{h_{ij}\}_{i,j\in\mathcal{I}}) \mapsto \{g_i g_j^{-1} h_{ij}\}_{i,j\in\mathcal{I}}.$$

Definition 8.4. By replacing the word "smooth" with "constant" starting at this Section 8, we obtain the constant structure sheaf $\mathcal{O}_c$ and the definition of a constant sheaf $\mathcal{E}_c$ of $\mathcal{O}_c$-modules and the Čech complexes $\check{C}^\bullet(\mathcal{U}, \mathcal{O}_c)$ and $\check{C}^\bullet(\mathcal{U}, \mathcal{E}_c)$ and the resulting cohomologies $\check{\mathrm{H}}^i(\mathcal{U}, \mathcal{E}_c)$. If we assume that X is complex in addition, by replacing "smooth" with "holomorphic", we obtain the holomorphic structure sheaf $\mathcal{O}_h$ and the definition of a holomorphic sheaf $\mathcal{E}_h$ of $\mathcal{O}_h$-modules and the Čech complexes $\check{C}^\bullet(\mathcal{U}, \mathcal{O}_h)$ and $\check{C}^\bullet(\mathcal{U}, \mathcal{E}_h)$ and the resulting cohomologies $\check{\mathrm{H}}^i(\mathcal{U}, \mathcal{E}_h)$.

Denote the resulting locally constant and locally holomorphic Čech groupoids by $\check{\mathcal{F}}_c$ and $\check{\mathcal{F}}_h$, respectively.

A constant transition function is holomorphic and a holomorphic transition function is smooth. Hence

Remark 8.5. There are natural inclusions

$$\mathcal{O}_c \hookrightarrow \mathcal{O}_h \hookrightarrow \mathcal{O}.$$

A 1-cocycle in $\check{\mathcal{F}}_h$ defines a holomorphic sheaf $\mathcal{E}_h$ but also a smooth sheaf $\mathcal{E}$ and there is a natural inclusion $\mathcal{E}_h \hookrightarrow \mathcal{E}$. Similarly a 1-cocycle in $\check{\mathcal{F}}_c$ defines a locally constant sheaf $\mathcal{E}_c$ and also a locally holomorphic sheaf $\mathcal{E}_h$ and a smooth sheaf $\mathcal{E}$ resulting in inclusions $\mathcal{E}_c \hookrightarrow \mathcal{E}_h \hookrightarrow \mathcal{E}$. To summarize, there are inclusions

$$\check{\mathcal{F}}_c \hookrightarrow \check{\mathcal{F}}_h \hookrightarrow \check{\mathcal{F}}. \tag{8.4}$$

In the reverse direction,

$$\mathcal{E}_h = \mathcal{O}_h \otimes \mathcal{E}_c \in \check{\mathcal{F}}_h, \quad \mathcal{E} = \mathcal{O} \otimes \mathcal{E}_h \in \check{\mathcal{F}}, \quad \mathcal{E} = \mathcal{O} \otimes \mathcal{E}_c \in \check{\mathcal{F}}.$$

Important assumption: For the rest of these notes, we shall always assume that the underlying smooth sheaf, that is the final image of the maps (8.4), is isomorphic to the trivial sheaf: $\mathcal{O}^{\oplus r} \in \check{\mathcal{F}}$.

9. The Smooth de Rham Groupoid

The purpose of this section and the next is to describe the sheaves arise from the Čech constructions in terms of integrable connections, gaining other perhaps more accessible perspectives. These constructions run in parallel to those of Sections 4 and 5 with much repetition, but this is precisely the point: The conceptual leap from local construction on unit balls to global construction on manifolds is rather small. Again the running leitmotif is that a function is locally constant if and only if all its partial derivatives are zero.

9.1. *Structure sheaf*

Let X be a smooth manifold with the exterior differential operator $d : \mathcal{O} \longrightarrow \Omega^1$ (See Remark 7.3). It is then immediate that $\mathcal{O}_c = \mathcal{O}^d$ where $\mathcal{O}^d$ is the sheaf such that for $U \subseteq X$,

$$\mathcal{O}^d(U) = \{f \in \mathcal{O}(U) : df = 0\}.$$

Again this follows from the principle that f is a constant function if and only if $df = 0$.

9.2. *Tangent, cotangent sheaves and the de Rham complex*

The manifold X is covered by charts in $\mathcal{U}$ and vector fields are locally defined on each chart and together they form a sheaf over X with transition maps $g_{ij} = \mathbf{D}\phi_{ij} \circ \phi_j^{-1}$, where $\mathbf{D}$ is the standard differential operator on the unit ball. The resulting smooth sheaf $\mathcal{T}$ is called the tangent sheaf of X. Let Ω^1 be the dual sheaf of $\mathcal{T}$ (See Remark 8.2). It is the sheaf of covector fields on X. By taking exterior products of Ω^1 and extending the exterior differential operators, one obtains the de Rham complex

$$(\Omega^\bullet, d) : \Omega^0 \xrightarrow{d} \Omega^1 \cdots \Omega^i \xrightarrow{d} \Omega^{i+1} \cdots , \tag{9.1}$$

but it is now a complex of sheaves. If we take the global sections, then

$$\Omega^0(X) \xrightarrow{d} \Omega^1(X) \cdots \Omega^i(X) \xrightarrow{d} \Omega^{i+1}(X) \cdots$$

is an analogue of $\check{C}^\bullet(\mathcal{U}, \mathcal{O}_c)$.

9.3. *Connections*

By Remark 8.5, each constant sheaf $\mathcal{E}_c$ can be interpreted as a subsheaf of $\mathcal{E}$. We begin by fixing a rank-r smooth sheaf of $\mathcal{O}$-modules:

$$\mathcal{E} = \{e : X \longrightarrow \mathbb{C}^{\oplus r} : e \text{ is smooth}\}.$$

Remark 9.1. The locally constant (and later, holomorphic) sheaves we are considering embed into the trivial sheaf $\mathcal{E}$ by the assumption at the end of Chapter 8.

Definition 9.2. Let $\Omega^i(\mathcal{E}) := \Omega^i \otimes \mathcal{E}$. An $\mathcal{O}_c$-linear operator

$$\nabla : \mathcal{E} \longrightarrow \Omega^1(\mathcal{E})$$

is a **connection** if $\nabla(fe) = df \otimes e + f\nabla(e)$ for $f \in \mathcal{O}$ and $e \in \mathcal{E}$. Denote by $\mathcal{C}_d$ the set of all connections on $\mathcal{E}$.

For each i, a connection ∇ extends to an $\mathcal{O}_c$-linear map

$$\nabla : \Omega^i(\mathcal{E}) \longrightarrow \Omega^{i+1}(\mathcal{E})$$

satisfying the generalized Leibniz rule

$$\nabla(\eta \otimes e) = d\eta \otimes e + (-1)^i \eta \wedge \nabla(e).$$

Connections are local operators, operating on $\mathcal{E}(U)$ for open sets $U \subseteq X$. As before, the exterior differentiation is generalized to the rank-r trivial connection

$$D : \mathcal{E} \longrightarrow \Omega^1(\mathcal{E}) := \Omega^1 \otimes \mathcal{E}, \quad D(f) = D(\oplus_{i=1}^r f_i) = \oplus_{i=1}^r df_i.$$

Two connections on $\mathcal{E}$ differ by an element $A \in \Omega^1(\mathrm{End}(\mathcal{E}))(X)$, hence, $\mathcal{C}_d$ identifies with $\Omega^1(\mathrm{End}(\mathcal{E}))(X)$. Then ∇^2 is $\mathcal{O}$-linear, hence, $\nabla^2 \in \Omega^2(\mathrm{End}(\mathcal{E}))(X)$.

Definition 9.3. The **curvature map** is

$$\mathsf{F} : \mathcal{C}_d \longrightarrow \Omega^2(\mathrm{End}(\mathcal{E})), \quad \mathsf{F}(\nabla) = \nabla^2.$$

For $\nabla \in \mathcal{C}_d$, $\mathsf{F}(\nabla)$ is the **curvature** of ∇. A connection ∇ is **integrable** if $\mathsf{F}(\nabla) = 0$. If ∇ is integrable, then

$$\mathcal{E}^{\nabla}(U) = \{e \in \mathcal{E}(U) : \nabla e = 0\}, \quad \forall U \subseteq X$$

is the locally constant sheaf of $\mathcal{O}^d$-modules with respect to ∇. Denote by $\mathcal{F}_d$ (or $\mathcal{Z}^1(G)(X)$) the subspace of integrable connections in $\mathcal{C}_d$.

When ∇ is integrable, the de Rham complex is

$$(\Omega^\bullet(\mathcal{E}), \nabla) : \Omega^0(\mathcal{E}) \xrightarrow{\nabla} \Omega^1(\mathcal{E}) \cdots \Omega^i(\mathcal{E}) \xrightarrow{\nabla} \Omega^{i+1}(\mathcal{E}) \cdots.$$

Taking global sections

$$\Omega^0(\mathcal{E})(X) \xrightarrow{\nabla} \Omega^1(\mathcal{E})(X) \cdots \Omega^i(\mathcal{E})(X) \xrightarrow{\nabla} \Omega^{i+1}(\mathcal{E})(X) \cdots,$$

the resulting i-th cohomology is

$$\mathrm{H}^i_\nabla(\mathcal{E}) = \ker(\nabla)/\nabla(\Omega^{i-1}(\mathcal{E})(X)).$$

These are the analogues of $\check{C}^\bullet(\mathcal{U}, \mathcal{E}_c)$ and $\check{\mathrm{H}}^i(\mathcal{U}, \mathcal{E}_c)$.

The sheaf $\mathcal{T}$ generates the sheaf $\mathcal{D}$ of rings of differential operators on $\mathcal{O}$. Integrability of ∇ is precisely the condition for $\mathcal{E}$ to inherit a $\mathcal{D}$-module structure from its $\mathcal{T}$-module structure.

9.4. *The gauge group and its action*

The group G acts on $\mathbb{C}^{\oplus r}$ by matrix-vector multiplication. Define the G-gauge group

$$\mathcal{G} = \Omega^0(G)(X) := \mathcal{O}(G)(X) = \{g : X \longrightarrow G\}.$$

Then $\mathcal{G}$ acts on $\mathcal{E}$ by $\mathcal{O}$-module automorphism

$$\mathcal{G} \times \mathcal{E} \longrightarrow \mathcal{E}, \quad (g, e) \mapsto ge, \quad ge(x) = g(x)e(x),$$

where $g(x)e(x)$ means vector-matrix multiplication at $x \in X$. Suppose $g \in \mathcal{G}$. Then $g : \mathcal{E} \xrightarrow{\cong} \mathcal{E}$. If ∇ is a connection on $\mathcal{E}$, then g pulls back ∇ to a connection $g^*\nabla$ on $\mathcal{E}$ as follows:

$$g^*\nabla : \mathcal{E} \xrightarrow{g} \mathcal{E} \xrightarrow{\nabla} \Omega^1(\mathcal{E}) \xrightarrow{g^{-1}} \Omega^1(\mathcal{E}).$$

This give a $\mathcal{G}$-action on $\mathcal{C}_d$. Identifying $\mathcal{C}_d$ with $\Omega^1(\mathrm{End}(\mathcal{E}))(X)$, $\nabla = D + A$ and the $\mathcal{G}$-action is

$$\mathcal{G} \times \mathcal{C}_d \longrightarrow \mathcal{C}_d, \quad (g, \nabla) \mapsto g^{-1}\nabla g = D + g^{-1}(Dg) + g^{-1}Ag.$$

The curvature of $g^{-1}\nabla g$ is $(g^{-1}\nabla g)^2 = g^{-1}\nabla^2 g$. Hence $\nabla^2 = 0$ if and only if $g^{-1}\nabla^2 g = 0$ and the gauge action restricts to an action $\mathcal{G} \times \mathcal{F}_d \longrightarrow \mathcal{F}_d$.

Definition 9.4. The **de Rham groupoid** $\mathcal{F}_d$ is

(1) Objects: the family of integrable connections on $\mathcal{E}$.
(2) Equivalence: gauge group $\mathcal{G}$-action

$$\mathcal{G} \times \mathcal{F}_d \longrightarrow \mathcal{F}_d, \quad (g, \nabla) \mapsto g^{-1}\nabla g.$$

The connection D defines a map

$$\mathsf{D} : \mathcal{G} \longrightarrow \mathcal{C}_d, \quad \mathsf{D}(g) = g^{-1}Dg.$$

There is a sequence of maps

$$\Omega^0(G)(X) \xrightarrow{\mathsf{D}} \Omega^1(\mathrm{End}(\mathcal{E}))(X) \xrightarrow{\mathsf{F}} \Omega^2(\mathrm{End}(\mathcal{E}))(X).$$

We have already seen in Section 4 that over a small enough open ball $U \subseteq X$, this sequence is exact. However on large open sets and especially over X, this may not be the case and the resulting orbit space is our main focus.

Remark 9.5. For an integrable connection $\nabla \in \mathcal{F}_d$ and a small enough $U \subseteq X$, $\mathcal{E}^{\nabla}(U)$ is isomorphic to $\mathcal{O}_c^{\oplus r}(U)$ as an $\mathcal{O}_c(U)$-module as stated in Remark 4.8. Hence $\mathcal{E}^{\nabla}$ is isomorphic to a locally constant sheaf $\mathcal{E}_c \in \check{\mathcal{F}}_c$ on X.

10. The Holomorphic de Rham and Dolbeault Groupoid

Let X be a complex manifold with operators $d = \partial + \overline{\partial}$ (see Remark 7.3). There is a decomposition of the sheaf of exterior m-forms according to types and projections:

$$\Omega^m = \bigoplus_{q+p=m} \Omega^{q,p}, \qquad P_{q,p} : \Omega^m \longrightarrow \Omega^{q,p}. \tag{10.1}$$

For $d : \Omega^m \longrightarrow \Omega^{m+1}$,

$$d = \partial + \overline{\partial}, \quad \partial = P_{q+1,p} \circ d \quad \text{and} \quad \overline{\partial} = P_{q,p+1} \circ d.$$

Moreover, for p and q, the de Rham complex decomposes as

$$(\Omega^{q,\bullet}, \overline{\partial}) : \Omega^{q,0} \xrightarrow{\overline{\partial}} \Omega^{q,1} \cdots \Omega^{q,i} \xrightarrow{\overline{\partial}} \Omega^{q,i+1} \cdots, \tag{10.2}$$

$$(\Omega^{\bullet,p}, \partial) : \Omega^{0,p} \xrightarrow{\partial} \Omega^{1,p} \cdots \Omega^{i,p} \xrightarrow{\partial} \Omega^{i+1,p} \cdots. \tag{10.3}$$

The complex structure on X may be equivalently described by an $\mathcal{O}$-linear operator $J : \mathcal{T} \longrightarrow \mathcal{T}$ with $J^2 = -1$ such that the Nijenhuis tensor of J vanishes, i.e.

$$N_J(X,Y) := [X,Y] + J([JX,Y] + [X,JY]) - [JX,JY] = 0 \tag{10.4}$$

for all $X, Y \in \mathcal{T}$. This is the content of the Newlander-Nirenberg theorem [14]. The smooth tangent sheaf decomposes as $\mathcal{T} = \mathcal{T}^{1,0} \oplus \mathcal{T}^{0,1}$ by the eigenspaces of J with eigenvalues $\mathbf{i}$ and $-\mathbf{i}$, respectively. This induces a decomposition $\Omega^1 = \Omega^{1,0} \oplus \Omega^{0,1}$ which gives rise to the decomposition (10.1) for each m. This again gives a decomposition $d = \partial + \overline{\partial}$ which is equivalent to $\overline{\partial}^2 = 0$ and to Equation (10.4).

The sheaf $\mathcal{O}_h$ is equivalent to $\mathcal{O}^{\overline{\partial}}$, where

$$\mathcal{O}^{\overline{\partial}}(U) = \{f \in \mathcal{O}(U) : \overline{\partial} f = 0\}. \tag{10.5}$$

The holomorphic tangent sheaf is

$$\mathcal{T}_h = \{V \in \mathcal{T} : V(f) \in \mathcal{O}_h, \forall f \in \mathcal{O}_h\}.$$

The holomorphic cotangent sheaf Ω_h^1 is the $\mathcal{O}_h$-dual of $\mathcal{T}_h$ and $\Omega_h^q = \wedge_{i=1}^q \Omega_h^1$. The holomorphic de Rham complex is

$$(\Omega_h^\bullet, \partial) : \Omega_h^0 \xrightarrow{\partial} \Omega_h^1 \cdots \Omega_h^i \xrightarrow{\partial} \Omega_h^{i+1} \cdots . \tag{10.6}$$

10.1. *Holomorphic structures*

Definition 10.1. Let $\Omega^{q,p}(\mathcal{E}) := \Omega^{q,p} \otimes \mathcal{E}$. An almost holomorphic structure on $\mathcal{E}$ is an operator

$$\bar{\nabla} : \mathcal{E} \longrightarrow \Omega^{0,1}(\mathcal{E}), \quad \text{with } \bar{\nabla}(fe) = \overline{\partial} f \otimes e + f\bar{\nabla}(e)$$

for $f \in \mathcal{O}$ and $e \in \mathcal{E}$. Denote by $\mathcal{C}_h$ the space of almost holomorphic structures on $\mathcal{E}$.

For $p+q=i$, $\bar{\nabla}$ extends to an $\mathcal{O}_c$-linear map

$$\bar{\nabla} : \Omega^{p,q}(\mathcal{E}) \longrightarrow \Omega^{p,q+1}(\mathcal{E})$$

satisfying the generalized Leibniz rule

$$\bar{\nabla}(\eta \otimes e) = \overline{\partial}\eta \otimes e + (-1)^i \eta \wedge \bar{\nabla}(e).$$

The operator D also decomposes according to types $D = D' + D''$:

$$D' : \mathcal{E} \longrightarrow \Omega^{1,0}(\mathcal{E}) := \Omega^{1,0} \otimes \mathcal{E}, \quad D'(f) = D'(\oplus_{i=1}^r f_i) = \oplus_{i=1}^r \partial f_i.$$

$$D'' : \mathcal{E} \longrightarrow \Omega^{0,1}(\mathcal{E}) := \Omega^{0,1} \otimes \mathcal{E}, \quad D''(f) = D(\oplus_{i=1}^r f_i) = \oplus_{i=1}^r \overline{\partial} f_i.$$

Define map

$$\mathsf{F} : \mathcal{C}_h \longrightarrow \Omega^{0,2}(\mathrm{End}(\mathcal{E})), \quad \mathsf{F}(\bar{\nabla}) = \bar{\nabla}^2.$$

Definition 10.2. A **holomorphic structure** $\bar{\nabla}$ on $\mathcal{E}$ is an almost holomorphic structure satisfying the integrability condition $\mathsf{F}(\bar{\nabla}) = 0$. Denote by $\mathcal{F}_h$ (or $\mathcal{Z}^{0,1}(G)(X)$) the holomorphic structures on $\mathcal{E}$.

Then $D'' \in \mathcal{F}_h$ and

$$\mathcal{E}^{D''} = \{e \in \mathcal{E} : D''e = 0\} \cong \mathcal{O}_h^{\oplus r}$$

is a sheaf of $\mathcal{O}^{\overline{\partial}}$-module.

The difference of two almost holomorphic structure is $\mathcal{O}$-linear; hence, the space of almost holomorphic structures $\mathcal{C}_h$ identifies with

$\Omega^{0,1}(\mathrm{End}(\mathcal{E}))(X)$. If $\bar{\nabla} \in \mathcal{F}_h$, then $\mathcal{E}^{\bar{\nabla}} = \{f : \bar{\nabla} f = 0\}$ is the sheaf of $\mathcal{O}_h$-modules of $\bar{\nabla}$-holomorphic sections and the $\bar{\nabla}$-Dolbeault complex is

$$(\Omega^{q,\bullet}(\mathcal{E}), \bar{\nabla}) : \Omega^{q,0}(\mathcal{E}) \xrightarrow{\bar{\nabla}} \Omega^{q,1}(\mathcal{E}) \cdots \Omega^{q,i}(\mathcal{E}) \xrightarrow{\bar{\nabla}} \Omega^{q,i+1}(\mathcal{E}) \cdots .$$

Take global sections

$$\Omega^{q,0}(\mathcal{E})(X) \xrightarrow{\bar{\nabla}} \Omega^{q,1}(\mathcal{E})(X) \cdots \Omega^{q,i}(\mathcal{E})(X) \xrightarrow{\bar{\nabla}} \Omega^{q,i+1}(\mathcal{E})(X) \cdots .$$

The resulting i-th cohomology is

$$\mathrm{H}^{q,i}_{\bar{\nabla}}(\mathcal{E}) = \ker(\bar{\nabla})/\bar{\nabla}(\Omega^{q,i-1}(\mathcal{E})(X)).$$

These are the analogues of $\check{C}^{\bullet}(\mathcal{U}, \Omega^q_h(\mathcal{E}_h))$ and $\check{\mathrm{H}}^i(\mathcal{U}, \Omega^q_h(\mathcal{E}_h))$.

10.2. *The gauge group action I*

The gauge group $\mathcal{G}$ acts on the almost holomorphic structures

$$\mathcal{G} \times \mathcal{C}_h \longrightarrow \mathcal{C}_h, \quad (g, \bar{\nabla}) \mapsto g^{-1}\bar{\nabla} g.$$

This $\mathcal{G}$-action preserves $\mathcal{F}_h$.

Definition 10.3. The **groupoid of holomorphic structures** $\mathcal{F}_h$ is

(1) Objects: set of holomorphic structures on $\mathcal{E}$.
(2) Equivalence: gauge group $\mathcal{G}$-action:

$$\mathcal{G} \times \mathcal{F}_h \longrightarrow \mathcal{F}_h, \quad (g, \bar{\nabla}) \mapsto g^{-1}\bar{\nabla} g.$$

The operator D'' defines a map

$$\mathsf{D} : \mathcal{G} \longrightarrow \mathcal{C}_h, \quad \mathsf{D}(g) = g^{-1} D'' g\,.$$

Identifying $\mathcal{C}_h$ with $\Omega^{0,1}(\mathrm{End}(\mathcal{E}))(X)$, there is a sequence of maps

$$\mathcal{G} \xrightarrow{\mathsf{D}} \Omega^{0,1}(\mathrm{End}(\mathcal{E}))(X) \xrightarrow{\mathsf{F}} \Omega^{0,2}(\mathrm{End}(\mathcal{E}))(X).$$

Remark 10.4. For a holomorphic structure $\bar{\nabla} \in \mathcal{F}_h$ and a small enough $U \subseteq X$, $\mathcal{E}^{\bar{\nabla}}(U)$ is isomorphic to $\mathcal{O}_h^{\oplus r}(U)$ as an $\mathcal{O}_h(U)$-module as stated in Remark 5.4. Hence $\mathcal{E}^{\bar{\nabla}}$ is isomorphic to a holomorphic sheaf $\mathcal{E}_h \in \check{\mathcal{F}}_h$ on X.

10.3. *Holomorphic λ-connections*

A local function $f \in \mathcal{O}(U)$ is constant if and only if $\partial f = \overline{\partial} f = 0$. Hence,

$$\mathcal{O}_c(U) = \mathcal{O}_h^{\partial}(U) = (\mathcal{O}^{\overline{\partial}})^{\partial}(U) = \{f \in \mathcal{O}(U) : \overline{\partial} f = \partial f = 0\}. \qquad (10.7)$$

Fix a holomorphic structure $\bar{\nabla}$. By Remark 10.4, $\mathcal{E}^{\bar{\nabla}} \cong \mathcal{E}_h \in \check{\mathcal{F}}_h$ and we will use $\mathcal{E}_h$ to simplify notations. We emphasize here that $\mathcal{E}_h$ may no longer be isomorphic to $\mathcal{O}_h^{\oplus r}$. Let $\lambda \in \mathbb{C}$.

Definition 10.5. Let $\Omega_h^i(\mathcal{E}_h) := \Omega_h^i \otimes \mathcal{E}_h$ and $\lambda \in \mathbb{C}$. A **holomorphic λ-connection** or **λ-connection** on $\mathcal{E}_h$ is an operator

$$\nabla^\lambda : \mathcal{E}_h \longrightarrow \Omega_h^1(\mathcal{E}_h)$$

satisfying $\nabla^\lambda(fe) = \lambda \partial f \otimes e + f\nabla^\lambda(e)$ for $f \in \mathcal{O}_h$ and $e \in \mathcal{E}_h$. Denote by $\mathcal{C}_h^\lambda$ the space of all λ-connections on $\mathcal{E}_h$.

For each i, ∇^λ extends to an $\mathcal{O}_c$-linear map

$$\nabla^\lambda : \Omega_h^i(\mathcal{E}) \longrightarrow \Omega_h^{i+1}(\mathcal{E})$$

satisfying the generalized Leibniz rule

$$\nabla^\lambda(\eta \otimes e) = \lambda \partial \eta \otimes e + (-1)^i \eta \wedge \nabla^\lambda(e).$$

The curvature map is

$$\mathsf{F} : \mathcal{C}_h^\lambda \longrightarrow \Omega_h^2(\mathrm{End}(\mathcal{E})), \quad \mathsf{F}(\nabla^\lambda) = (\nabla^\lambda)^2.$$

Definition 10.6. A λ-connection $\nabla^\lambda \in \mathcal{C}_h^\lambda$ is **integrable** if $\mathsf{F}(\nabla^\lambda) = 0$. Denote by $\mathcal{F}_h^\lambda$ the space of integrable λ-connections on $\mathcal{E}_h$.

Notice that $\lambda D'$ is an integrable λ-connection on $\mathcal{O}_h^{\oplus r}$. In general, there is no such canonical λ-connection on $\mathcal{E}_h$. If $\nabla^\lambda, \nabla_0^\lambda \in \mathcal{C}_h^\lambda$, then $\nabla^\lambda - \nabla_0^\lambda \in \Omega_h^1(\mathrm{End}(\mathcal{E}_h))(X)$. Hence the space $\mathcal{C}_h^\lambda$ of integrable λ-connections identifies with $\Omega_h^1(\mathrm{End}(\mathcal{E}_h))(X)$. If ∇^λ is integrable, then $\mathcal{E}_h^{\nabla^\lambda} = \{e : \nabla^\lambda e = 0\}$ is the sheaf of $\mathcal{O}_c$-modules of constant sections with respect to ∇^λ and the ∇^λ-de Rham complex is

$$(\Omega_h^\bullet(\mathcal{E}_h), \nabla^\lambda) : \Omega_h^0(\mathcal{E}_h) \xrightarrow{\nabla^\lambda} \Omega_h^1(\mathcal{E}_h) \cdots \Omega_h^i(\mathcal{E}_h) \xrightarrow{\nabla^\lambda} \Omega_h^{i+1}(\mathcal{E}_h) \cdots .$$

Remark 10.7. Vector fields in $\mathcal{T}_h$ are first order differential operators on $\mathcal{O}_h$. By composing these operators, $\mathcal{T}_h$ generates $\mathcal{D}_h$, the sheaf of rings of holomorphic differential operators. Integrability of ∇^1 is precisely the needed condition for $\mathcal{E}_h$ to inherit a $\mathcal{D}_h$-module structure from its $\mathcal{T}_h$-module structure [1, 3, 17].

Definition 10.8. The groupoid $\mathcal{F}_d^\lambda$ on $\mathcal{E}$ consists of

(1) Objects: $\mathcal{F}_d^\lambda = \{(\bar{\nabla}, \nabla^\lambda) : \bar{\nabla} \in \mathcal{F}_h, \nabla^\lambda \in \mathcal{F}_h^\lambda, \text{ where } \mathcal{E}_h = \mathcal{E}^{\bar{\nabla}}\}$.
(2) Equivalence: gauge group $\mathcal{G}$:

$$\mathcal{G} \times \mathcal{F}_d^\lambda \longrightarrow \mathcal{F}_d^\lambda, \quad (g, (\bar{\nabla}, \nabla^\lambda)) = (g^{-1}\bar{\nabla}g, g^{-1}\nabla^\lambda g).$$

$\mathcal{F}_d^\lambda$ is called the **holomorphic de Rham groupoid** when $\lambda = 1$ and the **Dolbeault groupoid** when $\lambda = 0$.

Remark 10.9. Let $\lambda \in \mathbb{C}^\times$. For a small enough open set $U \subseteq X$,

$$\mathcal{E}^\nabla(U) \cong (\mathcal{O}^{\oplus r})^D(U) \cong (\mathcal{O}_h^{\oplus r})^{\lambda D''}(U) \cong (\mathcal{E}^{\bar{\nabla}})^{\nabla^\lambda}(U)$$

by Remarks 5.4 and 5.9. Hence, $\mathcal{F}_d$ and $\mathcal{F}_d^\lambda$ are equivalent constructions.

Remark 10.10. There is a natural map (functor)

$$\mathcal{F}_d^\lambda \longrightarrow \mathcal{F}_h, \quad (\bar{\nabla}, \nabla) \mapsto \bar{\nabla}.$$

Remark 10.11. When $\lambda = 0$, ∇^0 is $\mathcal{O}_h$-linear and $(\mathcal{E}_h, \nabla^0)$ is called a Higgs bundle. Moreover there is a natural embedding

$$\mathcal{F}_h \longrightarrow \mathcal{F}_d^0, \quad \mathcal{E}_h \mapsto (\mathcal{E}_h, 0).$$

This is not true when $\lambda \neq 0$ since there is no canonical integrable λ-connection on a general $\mathcal{E}_h$.

The holomorphic gauge group with respect to $\mathcal{E}_h$ is

$$\mathcal{G}_h = \{g \in \mathcal{G} : g : \mathcal{E}_h \to \mathcal{E}_h\}.$$

There is an action $\mathcal{G}_h \times \mathcal{E}_h \longrightarrow \mathcal{E}_h$. For fixed $\bar{\nabla} \in \mathcal{F}_h$ and $\lambda \in \mathbb{C}^\times$, define

Definition 10.12. The groupoid $\mathcal{F}_h^\lambda(\mathcal{E}_h)$ consists of

(1) Objects: the set of integrable λ-connections on $\mathcal{E}_h$.
(2) Equivalence: $\mathcal{E}_h$-holomorphic gauge group $\mathcal{G}_h$-action:

$$\mathcal{G}_h \times \mathcal{F}_h^\lambda(\mathcal{E}_h) \longrightarrow \mathcal{F}_h^\lambda(\mathcal{E}_h), \quad (g, \nabla^\lambda) \mapsto g^{-1}\nabla^\lambda g.$$

Remark 10.13. Fix $\mathcal{E}_h \in \mathcal{F}_h$ and let $\lambda \in \mathbb{C}^\times$. If $\nabla^1 \in \mathcal{F}_h^1(\mathcal{E}_h)$, then $\lambda\nabla^1 \in \mathcal{F}_h^\lambda(\mathcal{E}_h)$. This is an equivalence of $\mathcal{F}_h^1(\mathcal{E}_h)$ and $\mathcal{F}_h^\lambda(\mathcal{E}_h)$.

11. The Equivalence of Groupoids

A functor $\mathbf{F} : \mathcal{C}_1 \longrightarrow \mathcal{C}_2$ between groupoids takes objects to objects and equivalence relations to equivalence relations, satisfying the general axioms concerning categories and functors. A functor is an equivalence of groupoid if it is fully faithful and surjective onto the isomorphism classes [7,8]. In the previous sections, we described how objects in $\mathcal{F}_d, \mathcal{F}_d^\lambda$ ($\lambda \in \mathbb{C}^\times$) give rise to objects in $\check{\mathcal{F}}_c$ (Remark 9.5 and 10.4). In other words, we have described functors $\mathcal{F}_d \longrightarrow \check{\mathcal{F}}_c$ and $\mathcal{F}_d^\lambda \longrightarrow \check{\mathcal{F}}_c$. In this section, we bring in the groupoid $\mathrm{Hom}(\pi, G)$.

Let X be a manifold and $\pi = \pi_1(X, x)$ the fundamental group of X as in Section 2 and $p : \tilde{X} \longrightarrow X$ the universal cover. Let $\tilde{\mathcal{O}}$ be the structure sheaf of $\tilde{X}$, $\tilde{\mathcal{E}} = \tilde{\mathcal{O}}^{\oplus r}$ and $\tilde{d}, \tilde{D}$ be the exterior differentiation and the trivial connection on $\tilde{\mathcal{E}}$. As before, $\tilde{\mathcal{E}}^{\tilde{D}} \cong \tilde{\mathcal{O}}_c^{\oplus r}$. The projection p induces natural maps $p^* : \mathcal{O} \longrightarrow \tilde{\mathcal{O}}$, $p^* : \mathcal{O}_c \longrightarrow \tilde{\mathcal{O}}_c$. The images are π-invariant local functions.

Suppose $\mathcal{E}_c \in \check{\mathcal{F}}_c$ is a locally constant sheaf. Let γ be a loop representing an element $[\gamma] \in \pi$. Starting at x, there is a sequence $\{i_1, \cdots, i_k\} \in \mathcal{I}$ with $U_{i_1}, \cdots U_{i_k}$ following and covering γ. Since the transition maps $g_{i_j, i_{j+1}}$ are constant maps, we obtain an element $g = \prod g_{i_j, i_{j+1}}$ for $[\gamma]$. The 1-cocycle condition (See Definition 8.1) guarantees that g depends only on $[\gamma]$. This gives a representation of the fundamental group $\rho : \pi \longrightarrow G$.

Let $\rho \in \mathrm{Hom}(\pi, G)$. Then π acts on $\tilde{X}$ freely by deck transformation and on $\mathbb{C}^r$ by the linear G-action via ρ. Let $\{e_i : X \longrightarrow \mathbb{C}^{\oplus r}\}_{i=1}^r$ be a basis and $\{\tilde{e}_i : \tilde{X} \longrightarrow \mathbb{C}^{\oplus r}\}_{i=1}^r$ be an equivariant basis with respect to ρ, i.e.

$$\rho(\sigma).\tilde{e}_i(x) = \tilde{e}_i(\sigma.x).$$

Let $U \subseteq X$, $\tilde{U} = p^{-1}(U) \subseteq \tilde{X}$ and $f \in \mathcal{E}(U)$. Then $f = \sum_{i=1}^r f_i e_i$ where $f_i : U \longrightarrow \mathbb{C}$ for $1 \leq i \leq r$. Let

$$\tilde{f} : \tilde{U} \longrightarrow \mathbb{C}^{\oplus r}, \quad \tilde{f}(\tilde{y}) = \sum_{i=1}^r p^*(f_i)(\tilde{y})\tilde{e}_i(\tilde{y}).$$

Then $\tilde{D}\tilde{f} \in \Omega^1(\tilde{\mathcal{E}})(\tilde{U})$ is equivariant with respect to the π-action, hence, descends to an element $\eta \in \Omega^1(\mathcal{E})(U)$. This defines a connection $\nabla : \mathcal{E} \longrightarrow \Omega^1(\mathcal{E})$. It is integrable since $\tilde{D}$ is integrable. Notice that the construction of ∇ involves a choice of basis and is unique only up to gauge transformations. To be specific, let

$$\tilde{D}\tilde{e}_i = \sum_{j=1}^r a_{ij}\tilde{e}_j, \quad a_{ij} \in \Omega^1(\tilde{\mathcal{O}}), \text{ i.e. } a_{ij} \text{ is a 1-form on } \tilde{X}.$$

Let $\tilde{A}$ be the $r \times r$ matrix with entries a_{ij}. Then $\tilde{A} \in \Omega^1(\tilde{\mathcal{E}})$ is π-invariant. Hence $\tilde{A} = p^*(A)$ for some $A \in \Omega^1(\mathcal{E})$. Then $\nabla = D + A$ is the desired integrable connection whose monodromy representation is in $[\rho]$.

By Remark 9.5, each $\mathcal{E}^\nabla \in \mathcal{F}_d$ is isomorphic to a locally constant sheaf $\mathcal{E}_c \in \check{\mathcal{F}}_c$. By Remark 10.9, $\mathcal{F}_d$, $\mathcal{F}_d^\lambda$ are equivalent constructions for $\lambda \in \mathbb{C}^\times$.

To summarize, let $\lambda \in \mathbb{C}^\times$. Then there are the following correspondences of groupoids:

$$\begin{array}{ccccc} & & \mathrm{Hom}(\pi, G) & & \\ & \nearrow & & \searrow & \\ \check{\mathcal{F}}_c & & \longleftarrow & & \mathcal{F}_d \longleftrightarrow \mathcal{F}_d^\lambda \end{array} \tag{11.1}$$

In fact, these groupoids are all equivalent. We refer readers to [7, 8] for a proof. What we have done in effect is to show that these functors are surjective on isomorphism classes (one can go around the triangle and see that objects that one begins and ends with are equivalent).

11.1. *Kähler manifolds and projective varieties*

Let $\mathbb{P}^n$ be the projective n-space over $\mathbb{C}$. Finite sets of homogeneous polynomials in $(n+1)$ variables define algebraic varieties (loci) in $\mathbb{P}^n$.

Definition 11.1. A **smooth projective variety** X of dimension n over $\mathbb{C}$ is a projective variety that is also a manifold.

Smooth projective varieties over $\mathbb{C}$ are compact Kähler manifolds. However compact Kähler manifolds are not necessarily projective varieties. If X is a compact Kähler manifold, then Equivalence (11.1) extends to the case of $\lambda = 0$. We do not provide a proof here and refer readers to [2, 4, 7, 8, 11, 18], but emphasize that this last equivalence is complicated. For example, for the correspondence $(\mathcal{E}_h^1, \nabla^1) \mapsto (\mathcal{E}_h^2, \nabla^0)$, it is not the case that $\mathcal{E}_h^1 \cong \mathcal{E}_h^2$ in general.

12. Moduli Spaces by Simple Examples

From the groupoids $\mathrm{Hom}(\pi, G), \check{\mathcal{F}}_c, \check{\mathcal{F}}_h, \mathcal{F}_d, \mathcal{F}_h$ and $\mathcal{F}_d^\lambda$ arise the moduli spaces $\mathcal{M}_B, \check{\mathcal{M}}_c, \check{\mathcal{M}}_h, \mathcal{M}_d, \mathcal{M}_h$ and $\mathcal{M}_d^\lambda$, universal geometric objects parameterizing their respective isomorphism classes as outlined in Section 3. The word universal here means that the resulting moduli spaces represent (or corepresent) certain functors. This is firmly in the realm of Geometric Invariant Theory or GIT for short. We refer readers to [13] for a full dosage and [15] for a gentler introduction. The subject has been continuously exploding for years and we point to the reference section of [13] as evidence. The process of forming moduli spaces is a subtle and delicate one as demonstrated in Section 2 and the resulting moduli space may not be able to encompass the entire family of isomorphism classes.

The Betti moduli space $\mathcal{M}_B$ is relatively simple to construct and we have already done that in Section 2; however, recall the subtle problems when forming quotients. This space has an algebraic, analytic and smooth structure from that of G.

Here we give a few simple examples of these moduli spaces when G is the abelian $\mathrm{GL}(1,\mathbb{C}) = \mathbb{C}^\times$. This puts us back in the more familiar situation of the usual first cohomology, where $\mathcal{E}$, $\mathcal{E}_h$ and $\mathcal{E}_c$ are of rank-1. Then $\mathrm{End}(\mathcal{E}) = \mathcal{O}$, $\mathrm{End}(\mathcal{E}_h) = \mathcal{O}_h$ and $\mathrm{End}(\mathcal{E}_c) = \mathcal{O}_c$.

12.1. *Classical and abelian cohomology*

Our theory is developed for reductive groups G and is an extension of the theory of the Picard variety where $G = \mathbb{C}^\times$. This in turn arises from an understanding of the classical cohomology theory where G is replaced by the additive group $\mathbb{C}$ which is not reductive.

12.1.1. *The case of the additive group* $\mathbb{C}$

This is the setting of classical cohomologies. The constant and holomorphic complexes compute the cohomologies $\check{\mathrm{H}}^i(\mathcal{U},\mathcal{O}_c)$ and $\check{\mathrm{H}}^p(\mathcal{U},\Omega^q_h)$:

$$\check{C}^\bullet(\mathcal{U},\mathcal{O}_c) : \check{C}^0(\mathcal{U},\mathcal{O}_c) \xrightarrow{d} \check{C}^1(\mathcal{U},\mathcal{O}_c) \xrightarrow{d} \cdots \check{C}^j(\mathcal{U},\mathcal{O}_c)\cdots, \quad (12.1)$$

$$\check{C}^\bullet(\mathcal{U},\Omega^q_h) : \check{C}^0(\mathcal{U},\Omega^q_h) \xrightarrow{d} \check{C}^1(\mathcal{U},\Omega^q_h) \xrightarrow{d} \cdots \check{C}^j(\mathcal{U},\Omega^q_h)\cdots. \quad (12.2)$$

There is the notion of *acyclic cover* (§3, Chapter 0, [9]). These acyclic covers exist under mild assumptions. When $\mathcal{U}$ is an acyclic cover with respect to Ω^q_h, $\check{C}^\bullet(\mathcal{U},\Omega^q_h)$ computes the Čech cohomologies $\check{\mathrm{H}}^p(\Omega^q_h)$ by definition. When $\mathcal{U}$ is an acyclic cover with respect to $\mathcal{O}_c$, $\check{C}^\bullet(\mathcal{U},\mathcal{O}_c)$ computes the Čech cohomologies $\check{\mathrm{H}}^i(\mathcal{O}_c)$ by definition. The fact that Čech cohomologies are independent of acyclic covers is known as the Leray theorem.

Given a sheaf or a complex of sheaves $F^\bullet$ on X, there is a notion of acyclic resolution (§1, Chapter 3, [10]). Let $F^\bullet \longrightarrow I^\bullet$ be such an acyclic resolution. Then the sheaf cohomologies $F^\bullet$ are computed via $I^\bullet$ after taking global sections. As an example, when $\mathcal{U}$ is an acyclic cover, $\check{C}^\bullet(\mathcal{U},\mathcal{O}_c)$ and $\check{C}^\bullet(\mathcal{U},\Omega^q_h)$ should be thought of as the results of applying the global section functor to the sheafified acyclic Čech complex resolutions (§4, Chapter

3, [10]). A fundamental result of homological algebra is that the resulting cohomological groups are independent of the acyclic resolutions.

The de Rham and Dolbeault complexes are respectively:

$$(\Omega^\bullet, d) : \Omega^0 \xrightarrow{d} \Omega^1 \cdots \Omega^j \xrightarrow{d} \Omega^{j+1} \cdots, \tag{12.3}$$

$$(\Omega^{q,\bullet}, \overline{\partial}) : \Omega^{q,0} \xrightarrow{\overline{\partial}} \Omega^{q,1} \cdots \Omega^{q,j} \xrightarrow{\overline{\partial}} \Omega^{q,j+1} \cdots. \tag{12.4}$$

The Dolbeault complex is an acyclic resolution of the holomorphic sheaf Ω^q_h, hence, computes the holomorphic Čech cohomology:

$$\mathrm{H}^{q,p}_{\overline{\partial}}(\mathcal{O}) \cong \check{\mathrm{H}}^p(\Omega^q_h).$$

This is the content of Dolbeault theorem.

The de Rham complex is an acyclic resolution of the constant sheaf $\mathcal{O}_c = \mathbb{C}$, hence, computes the constant Čech cohomology:

$$\mathrm{H}^i_d(\mathcal{O}) \cong \check{\mathrm{H}}^i(\mathcal{O}_c).$$

This is the content of de Rham theorem.

The de Rham complex also resolves the holomorphic de Rham complex

$$\begin{array}{ccccccccc} (\Omega^\bullet, d) : & \Omega^0 & \xrightarrow{d} & \Omega^1 & \cdots & \Omega^j & \xrightarrow{d} & \Omega^{j+1} & \cdots \\ & \uparrow & & \uparrow & & \uparrow & & \uparrow & \\ (\Omega^\bullet_h, \partial) : & \Omega^0_h & \xrightarrow{\partial} & \Omega^1_h & \cdots & \Omega^j_h & \xrightarrow{\partial} & \Omega^{j+1}_h & \cdots. \end{array} \tag{12.5}$$

The decomposition (10.1) gives a filtration of the de Rham complex

$$F^k\Omega^i = \bigoplus_{q+p=i, q\geq k} \Omega^{q,p}.$$

This then induces a filtration on the de Rham cohomology which gives rise to a double complex and a spectral sequence with $E^{0,1}_1 = \mathrm{H}^{0,1}_{\overline{\partial}}(\mathcal{O})$. The projection to the graded piece gives a map

$$\mathrm{H}^1_d(\mathcal{O}) \longrightarrow E^{0,1}_1 = \mathrm{H}^{0,1}_{\overline{\partial}}(\mathcal{O}). \tag{12.6}$$

If X is smooth projective, then the spectral sequence degenerates at E_1 and the Hodge theorem states that the de Rham cohomologies decompose as

$$\mathrm{H}^i_d(\mathcal{O}) \cong \bigoplus_{q+p=i} \mathrm{H}^{q,p}_{\overline{\partial}}(\mathcal{O}) \cong \bigoplus_{q+p=i} \check{\mathrm{H}}^p(\Omega^q_h).$$

By covering based loops with open charts as in Section 11, there is an isomorphism $\check{\mathrm{H}}^1(\mathcal{O}_c) \cong \mathrm{Hom}(\pi, \mathbb{C})$. To summarize, we have the following equivalences:

$$\mathrm{Hom}(\pi, \mathbb{C}) \cong \check{\mathrm{H}}^1(\mathcal{O}_c) \cong \mathrm{H}^1_d(\mathcal{O}) \overset{X \text{ projective}}{\cong} \mathrm{H}^1(\Omega^0_h) \oplus \mathrm{H}^0(\Omega^1_h). \tag{12.7}$$

12.1.2. $G = \mathbb{C}^\times$

The moduli spaces are generalizations of classical cohomologies. Let $\mathcal{O}_c^\times$ (resp. $\mathcal{O}_h^\times$) be the sheaf of (multiplicative) locally constant (resp. holomorphic) functions that are never zero. These sheaves are related by exact sequences

$$0 \longrightarrow \mathbb{Z} \longrightarrow \mathcal{O}_c \stackrel{exp}{\longrightarrow} \mathcal{O}_c^\times \longrightarrow 0, \tag{12.8}$$

$$0 \longrightarrow \mathbb{Z} \longrightarrow \mathcal{O}_h \stackrel{exp}{\longrightarrow} \mathcal{O}_h^\times \longrightarrow 0. \tag{12.9}$$

Here $\mathbb{Z}$ denotes the sheaf of constant functions with integer values. The long cohomological sequence associated with the short exact sequence (12.8) is

$$\cdots \stackrel{0}{\longrightarrow} \check{\mathrm{H}}^1(\mathbb{Z}) \longrightarrow \check{\mathrm{H}}^1(\mathcal{O}_c) \longrightarrow \check{\mathrm{H}}^1(\mathcal{O}_c^\times) \stackrel{\delta}{\longrightarrow} \check{\mathrm{H}}^2(\mathbb{Z}) \cdots . \tag{12.10}$$

Then $\check{\mathcal{M}}_c := \ker(\delta) \cong \check{\mathrm{H}}^1(\mathcal{O}_c)/\check{\mathrm{H}}^1(\mathbb{Z})$. Here δ is the Chern class. Replacing the subscript c with h in the above construction, the long exact sequence associated with sequence (12.9) is

$$\cdots \check{\mathrm{H}}^1(\mathbb{Z}) \longrightarrow \check{\mathrm{H}}^1(\mathcal{O}_h) \longrightarrow \check{\mathrm{H}}^1(\mathcal{O}_h^\times) \stackrel{\delta}{\longrightarrow} \check{\mathrm{H}}^2(\mathbb{Z}) \cdots . \tag{12.11}$$

The cohomology $\check{\mathrm{H}}^1(\mathcal{O}_h^\times)$ is the moduli space of all rank-1 holomorphic sheaves on X and $\check{\mathcal{M}}_h := \ker(\delta) \cong \check{\mathrm{H}}^1(\mathcal{O}_h)/\check{\mathrm{H}}^1(\mathbb{Z})$ which is the moduli space of rank-1 sheaves that embed into the trivial smooth sheaf (see Remark 8.5).

12.2. *The punctured disk*

Let X be the punctured unit disk in $\mathbb{C}$ and let $z = x + iy$.

12.2.1. *The Betti moduli space*

The fundamental group π is isomorphic to $\mathbb{Z}$ and $\mathrm{Hom}(\pi_1(X), \mathbb{C}^\times) = \mathbb{C}^\times$.

12.2.2. *The Čech moduli*

From the long exact sequence (12.10), the moduli space of rank-1 locally constant sheaves is

$$\check{\mathcal{M}}_c = \ker(\delta) \cong \check{\mathrm{H}}^1(\mathcal{O}_c)/\check{\mathrm{H}}^1(\mathbb{Z}) \cong \mathbb{C}/\mathbb{Z}$$

which is a 1-dimension complex torus as is $\mathcal{M}_B$.

From the long exact sequence (12.11), the moduli space of all rank-1 holomorphic sheaves on X is $\check{\mathcal{M}}_h = \ker(\delta)$ which is the moduli space of rank-1 sheaves that embed into the trivial smooth sheaf (see Remark 8.5)

$$\mathcal{E} = \{e : X \longrightarrow \mathbb{C} : e \text{ is smooth}\}.$$

It is well-known that $\check{\mathrm{H}}^1(\mathcal{O}_h) = \{0\}$ (§3, Chapter 0, [9]). Hence $\check{\mathcal{M}}_h$ consists of a single point.

12.2.3. *The smooth de Rham moduli space*

A smooth integrable connection ∇ on $\mathcal{E}$ is of the form $d + \eta$ where η is a closed 1-form in $\Omega^1(\mathrm{End}(\mathcal{E})) \cong \Omega^1$. The de Rham moduli space $\mathcal{M}_d = \ker(d)/\operatorname{im}(d\log)$ is the sheaf cohomology of

$$\mathcal{G} \xrightarrow{d\log} \Omega^1 \xrightarrow{d} \Omega^2,$$

where $\mathcal{G}$ is the gauge group consisting of maps $g : X \longrightarrow \mathbb{C}^\times$. Hence it induces a map on the fundamental group $g_* : \pi_1(X) \longrightarrow \pi_1(\mathbb{C}^\times)$ which is an endomorphism of $\mathbb{Z}$. In other words $g(z) = z^k g'(z)$, where $g'_* : \pi_1(X) \longrightarrow \pi_1(\mathbb{C}^\times)$ is the trivial map. By homotopy lifting, we can consider $g' \in \Omega^0 = \mathcal{O}$. In other words, $\mathcal{G} \cong \mathbb{Z} \times \mathcal{O}$. There is a decomposition

$$\Omega^1 = \{\frac{adz}{z}\} \oplus d(\mathcal{O}), \quad a \in \mathbb{C}.$$

The de Rham cohomology $\mathrm{H}^1_d(\mathcal{O}) \cong \mathbb{C}$ is generated by the cocycle $[\frac{dz}{z}]$ and $\mathcal{M}_d \cong \mathbb{C}/\mathbb{Z}$ is a torus as expected. One can also obtain this from the de Rham theorem which states that $\mathrm{H}^1_d(\mathcal{O}) \cong \check{\mathrm{H}}^1(\mathcal{O}_c)$.

12.2.4. *The holomorphic structures*

A holomorphic structure on $\mathcal{E}$ is also defined by $\bar{\partial} + \eta$ where $\eta \in \Omega^{0,1}(\mathrm{End}(\mathcal{E}))$. The space $\mathcal{M}_h$ is the sheaf cohomology of

$$\mathcal{G} \xrightarrow{\bar{\partial}\log} \Omega^{0,1} \xrightarrow{\bar{\partial}} \Omega^{0,2}.$$

Since X has dimension 1, $\Omega^{0,2} = \{0\}$. One can compute this directly or use the Dolbeault theorem which states that $\mathrm{H}^{0,1}_{\bar{\partial}} \cong \check{\mathrm{H}}^1(\mathcal{O}_h)$. Thus $\mathcal{M}_h \cong \check{\mathrm{H}}^1(\mathcal{O}_h^\times)$ and consists of a single point.

12.2.5. *Holomorphic integrable connections ($\lambda = 1$)*

Since there is only one holomorphic structure up to isomorphism, we assume that it is $\overline{\partial}$. The moduli space of integrable connections is the sheaf cohomology of

$$\mathcal{G}_h \xrightarrow{\partial \log} \Omega_h^1 \xrightarrow{\partial} \Omega_h^2.$$

Since X is of complex dimension 1, $\Omega_h^2 = \{0\}$. The holomorphic gauge group $\mathcal{G}_h$ consists of holomorphic maps $g : X \longrightarrow \mathbb{C}^\times$. Again, the induced map $g_* : \pi_1(X) \longrightarrow \pi_1(\mathbb{C}^\times)$ is an endomorphism on $\mathbb{Z}$. Hence $g(z) = z^k g'(z)$, where $g'_* : \pi_1(X) \longrightarrow \pi_1(\mathbb{C}^\times)$ is the trivial map. By homotopy lifting, one may assume that $g' \in \Omega_h^0 \cong \mathcal{O}_h$. In other words, $\mathcal{G}_h \cong \mathbb{Z} \times \mathcal{O}_h$.

To further elaborate, the $\mathcal{O}_h$-module of closed holomorphic 1-forms is of the form $f(z)dz$ where f is a holomorphic function on X. Let $A = \frac{adz}{z} \in \Omega_h^1$. Since G is abelian, $\mathsf{F}(\nabla) = \partial A = 0$ and $\nabla = \partial + A$ is an integrable connection. However A is not exact because the anti-derivative of $\frac{1}{z}$ is $\log(z)$ which is not defined on the entire X. The equation $\nabla f = (\partial + A) f = 0$ has solutions of the form $f = Cz^{-a}$ which is multi-valued on X. Hence the equation has no global solution on X. This is reminiscent of the path dependence of formula (6.1), except the dependence is only on isotopy classes. The multi-valued function f on X is then by definition a function $f : \tilde{X} \longrightarrow \mathbb{C}$. The π-action on $\tilde{X}$ induces a π-action on $p^{-1}(x)$, hence, an action on $\{f(y) : y \in p^{-1}(x)\}$, hence, a (monodromy) representation

$$\pi \longrightarrow \mathrm{Aut}(\mathbb{C}) \cong \mathbb{C}^\times.$$

Since X is a punctured disk, $\pi \cong \mathbb{Z}$. Therefore the representation variety $\mathrm{Hom}(\pi_1(X), \mathbb{C}^\times)$ is also $\mathbb{C}^\times$. In the end, $\mathcal{M}_d^1 \cong \mathbb{C}/\mathbb{Z}$ just as before.

12.2.6. *The Dolbeault moduli space ($\lambda = 0$)*

The Dolbeault moduli space $\mathcal{M}_d^0$ is parameterized by the trivial holomorphic structure $\overline{\partial}$ and $\check{\mathrm{H}}^0(\Omega_h^1)$. The gauge group acts by the adjoint action, hence, trivially. Thus $\mathcal{M}_d^0 \cong \check{\mathrm{H}}^0(\Omega_h^1) \cong \mathbb{C}$ which is not diffeomorphic to $\mathcal{M}_B \cong \mathbb{C}^\times$.

12.3. *Compact Riemann surfaces*

Now we assume that X is a compact Riemann surface of genus g.

12.3.1. *The Betti moduli space*

This is already done in Section 2 with $G = \mathbb{C}^\times$. The commutator relation is trivial; hence, $\mathcal{M}_B = (\mathbb{C}^\times)^{2g}$.

12.3.2. *The Čech moduli space*

By the long exact sequence (12.10), the moduli space of rank-1 locally constant sheaves is

$$\check{\mathcal{M}}_c = \ker\delta \cong \check{\mathrm{H}}^1(\mathcal{O}_c)/\check{\mathrm{H}}^1(\mathbb{Z}) \cong \mathbb{C}^{2g}/\mathbb{Z}^{2g} \cong (\mathbb{C}^\times)^{2g}.$$

Similarly,

$$\check{\mathcal{M}}_h = \ker\delta \cong \mathbb{C}^g/\mathbb{Z}^{2g}$$

is the moduli space of rank-1 holomorphic sheaves that embed into the trivial smooth sheaf (see Remark 8.5). The space $\check{\mathcal{M}}_h$ is the Jacobi variety of X. Observe that the complex structure on $\check{\mathcal{M}}_h$ depends on how $\mathbb{Z}^{2g}$ injects into $\mathbb{C}^g$. This in turn depends on the complex structure on X. This is in direct contrast to $\check{\mathcal{M}}_c$ where the embedding $\mathbb{Z}^{2g} \hookrightarrow \mathbb{C}^{2g}$ is canonical and independent of the complex structure on X.

12.3.3. *The smooth de Rham moduli space*

A smooth integrable connection ∇ on $\mathcal{E}$ is of the form $d + \eta$ where $\eta \in \Omega^1(\mathrm{End}(\mathcal{E}))(X) \cong \Omega^1(X)$. The de Rham moduli space $\mathcal{M}_d = \ker(d)/\operatorname{im}(d\log)$ is the sheaf cohomology of

$$\mathcal{G} \xrightarrow{d\log} \Omega^1 \xrightarrow{d} \Omega^2,$$

where $\mathcal{G}$ is the gauge group consisting of maps $g : X \longrightarrow \mathbb{C}^\times$. The de Rham theorem states that $\mathrm{H}^1_d(\mathcal{O}) \cong \check{\mathrm{H}}^1(\mathcal{O}_c)$ and from this, we conclude that

$$\mathcal{M}_d \cong \mathbb{C}^{2g}/\mathbb{Z}^{2g} \cong (\mathbb{C}^\times)^{2g}.$$

12.3.4. *The holomorphic structures*

A holomorphic structure on $\mathcal{E}$ is also defined by $\overline{\partial} + \eta$ where $\eta \in \Omega^{0,1}(\mathrm{End}(\mathcal{E}))$. The space $\mathcal{M}_h$ is the sheaf cohomology of

$$\mathcal{G} \xrightarrow{\overline{\partial}\log} \Omega^{0,1} \xrightarrow{\overline{\partial}} \Omega^{0,2}.$$

Since X has complex dimension 1, $\Omega^{0,2} = \{0\}$. The Dolbeault theorem states that $\mathrm{H}^{0,1}_{\bar\partial}(\mathcal{O}) \cong \check{\mathrm{H}}^1(\mathcal{O}_h)$. The moduli space of holomorphic structure is

$$\mathcal{M}_h \cong \check{\mathcal{M}}_h \cong \ker(\delta) \cong \mathbb{C}^g/\mathbb{Z}^{2g}.$$

Again this is the Jacobi variety of X.

12.3.5. *Holomorphic integrable connections ($\lambda = 1$)*

An object in this groupoid consists of a pair $(\bar\nabla, \nabla^1)$ where $\bar\nabla \in \mathcal{F}_h$ and ∇^1 is a holomorphic integrable connection on $\mathcal{E}_h \cong \mathcal{E}^{\bar\nabla}$.

However, with this construction, there is a canonical rational map $\mathcal{M}^1_d \dashrightarrow \mathcal{M}_h$ arising from (see (12.6))

$$\mathcal{F}^1_d \longrightarrow \mathcal{F}_h, \quad (\bar\nabla, \nabla^1) \mapsto \bar\nabla.$$

In this case of $G = \mathbb{C}^\times$, this rational map is actually defined every where, hence, holomorphic. To be more explicit, there is a canonical holomorphic map

$$\mathcal{M}^1_d \cong (\mathbb{C}^\times)^{2g} \longrightarrow \mathbb{C}^g/\mathbb{Z}^{2g} \cong \mathcal{M}_h \cong \check{\mathcal{M}}_h.$$

Recall again that $\mathcal{M}_h$ is actually the Jacobi variety whose complex structure depends on the complex structure on X while the complex structure on $\mathcal{M}^1_d$ does not depend on the complex structure on X.

12.3.6. *The Dolbeault moduli space ($\lambda = 0$)*

The Dolbeault moduli space $\mathcal{M}^0_d$ parameterizes pairs $(\bar\nabla, \nabla^0)$ where $\bar\nabla \in \mathcal{F}_h$ and $\nabla^0 \in \Omega^1_h(X)$. Hence there are respectively canonical holomorphic and rational maps $\mathcal{M}^0_d \overset{\dashrightarrow}{\leftarrow} \mathcal{M}_h$ arising from

$$\mathcal{F}_h \longrightarrow \mathcal{F}^0_d, \quad \bar\nabla \mapsto (\bar\nabla, 0) \quad \text{and} \quad \mathcal{F}^0_d \longrightarrow \mathcal{F}_h, \quad (\bar\nabla, \nabla^0) \mapsto \bar\nabla.$$

The general theory states that when X is projective, there are the following topological (diffeomorphic) equivalences

$$\mathcal{M}^0_d \cong \mathcal{M}^1_d \cong \mathcal{M}_d \cong \check{\mathcal{M}}_c.$$

In this case of $G = \mathbb{C}^\times$, the general theory effectively reduces to the classical Hodge theory. Recall that the classical Hodge theorem states that

$$\mathrm{H}^1_d(\mathcal{O}) \cong \mathrm{H}^{0,1}_{\bar\partial}(\mathcal{O}) \oplus \mathrm{H}^{1,0}_{\bar\partial}(\mathcal{O}) \cong \check{\mathrm{H}}^1(\mathcal{O}_h) \oplus \check{\mathrm{H}}^0(\Omega^1_h).$$

From this, we obtain a direct sum decomposition

$$\mathcal{M}^0_d \cong \mathcal{M}_h \oplus \mathrm{H}^{1,0}_{\bar\partial}(\mathcal{O}) \cong \check{\mathcal{M}}_h \oplus \check{\mathrm{H}}^0(\Omega^1_h).$$

We want to emphasize here that the identification of $\mathcal{M}_d^0$ and $\mathcal{M}_d$ is homeomorphic (diffeomorphic), but neither algebraic nor holomorphic – $\mathcal{M}_d^0$ contains a compact complex torus while $\mathcal{M}_d$ does not.

13. Some Final Remarks

The moduli spaces $\mathcal{M}_B$ and $\mathcal{M}_d$ have the same underlying topology. When X is smooth projective, so does the Dolbeault moduli space $\mathcal{M}_d^0$. Denote by $\mathcal{M}$ the underlying topological space. When G is non-abelian, the topology of $\mathcal{M}$ is complicated. One can see this already from the Betti moduli space with its inherited variety structure from G. When X is smooth projective, the Dolbeault moduli space $\mathcal{M}_d^0$ is homeomorphic (diffeomorphic) to $\mathcal{M}_d$, hence, to $\mathcal{M}_B$ [2, 4, 7, 8, 11, 18]. This gives an extra complex structure on $\mathcal{M}$. The smooth part of the Betti moduli space $\mathcal{M}_B$ also has a natural underlying symplectic structure [6] which, together with the two complex structures on $\mathcal{M}$, give rise to a hyperkähler structure on $\mathcal{M}$. This hyperkähler structure yields rich topological information of $\mathcal{M}$ [11] and the most pleasant surprise occurs when the structure group is real:

Let $K = \mathrm{U}(r) \subseteq \mathrm{GL}(r, \mathbb{C}) = G$, the maximum compact subgroup of G. Readers can immediately check that the constructions for the Betti, the smooth and constant Čech and the smooth de Rham groupoids are still valid. Denote them as $\mathrm{Hom}(\pi, K), \check{\mathcal{F}}(K), \check{\mathcal{F}}_c(K), \mathcal{F}_d(K)$, respectively. The inclusion $P : \check{\mathcal{F}}_c \longrightarrow \check{\mathcal{F}}_h$ of Remark 10.10 on objects is injective, but the induced map on the isomorphism classes $P : \mathrm{Iso}(\check{\mathcal{F}}_c) \longrightarrow \mathrm{Iso}(\check{\mathcal{F}}_h)$ may neither be injective nor surjective.

Continuing from last chapter, we now assume that X is a compact Riemann surface of genus g and $r = 1$, i.e. $K = \mathrm{U}(1) \subseteq \mathbb{C}^\times = G$. Then $\check{\mathcal{M}}_h \cong \mathbb{C}^g/\mathbb{Z}^{2g}$ is a complex torus and so is $\mathcal{M}_B(K) = \mathrm{Hom}(\pi, K)/K$ and the composition map

$$\mathrm{Iso}(\check{\mathcal{F}}_c(K)) \longrightarrow \mathrm{Iso}(\check{\mathcal{F}}_c) \xrightarrow{P} \mathrm{Iso}(\check{\mathcal{F}}_h)$$

is one-to-one and onto. More specifically, this gives the following diffeomorphisms

$$\mathrm{U}(1)^{2g} \cong \mathrm{Hom}(\pi, \mathrm{U}(1))/\,\mathrm{U}(1) \cong \check{\mathcal{M}}_h.$$

This provides $\mathcal{M}_B(\mathrm{U}(1)) := \mathrm{Hom}(\pi, \mathrm{U}(1))/\,\mathrm{U}(1)$ with a complex structure.

In general, for a smooth complex projective variety X and a compact Lie group K, the representation variety $\mathcal{M}_B(K)$ acquires the complex variety structure of $\mathcal{M}_h$ [11, 21]. It so happens that the same is true for any real form $H \subset G$. That is if H is a real form of G, then $\mathcal{M}_B(H) := \mathrm{Hom}(\pi_1(X), H)^+/H$ acquires a complex variety structure [11, 21]. Notice that in the case of $G = \mathbb{C}^\times$, there are diffeomorphisms

$$\mathcal{M}_B(\mathrm{U}(1)) \oplus \mathcal{M}_B(\mathbb{R}^\times) \cong \mathcal{M}_B \cong \check{\mathcal{M}}_h \oplus \check{\mathrm{H}}^0(\Omega^1_h).$$

To summarize, if G is complex reductive, then $\mathcal{M}$ has a hyperkähler structure; moreover, if $H \subset G$ is a real form, then $\mathcal{M}(H)$ (with underlying topological space $\mathcal{M}_B(H)$) acquires a complex variety structure.

Acknowledgments

The author thanks Nan-Kuo Ho for proof-reading the manuscript.

The author gratefully acknowledges partial support by the National Science Council, Taiwan with grants 97-2115-M-006-001-MY3 and 100-2115-M-006-004-MY2 and the Institute of Mathematical Sciences of the National University of Singapore. He also thanks the latter's hospitality during his visit.

References

1. Borel, A.; Grivel, P.-P.; Kaup, B.; Haefliger, A.; Malgrange, B.; Ehlers, F. Algebraic D-modules. Perspectives in Mathematics, 2. *Academic Press, Inc., Boston, MA*, 1987.
2. Corlette, Kevin Flat G-bundles with canonical metrics. *J. Differential Geom.* **28** (1988), no. 3, 361–382.
3. Coutinho, S. C. A primer of algebraic D-modules. London Mathematical Society Student Texts, 33. *Cambridge University Press, Cambridge*, 1995. xii+207 pp.
4. Donaldson, S. K. Twisted harmonic maps and the self-duality equations. *Proc. London Math. Soc. (3)* **55** (1987), no. 1, 127–131.
5. Goldman, William M. Higgs bundles and geometric structures on surfaces. *The many facets of geometry,* 129-163, *Oxford Univ. Press, Oxford,* 2010.
6. Goldman, William M. The symplectic nature of fundamental groups of surfaces. *Adv. in Math.* **54** (1984), no. 2, 200–225.
7. Goldman, William M.; Millson, John J. The deformation theory of representations of fundamental groups of compact Kähler manifolds. *Inst. Hautes Études Sci. Publ. Math. No.67* (1988), 43–96.
8. Goldman, William M.; Xia, Eugene Z. Rank one Higgs bundles and representations of fundamental groups of Riemann surfaces. *Mem. Amer. Math. Soc.* **193** (2008), no. 904, viii+69 pp.
9. Griffiths, Phillip; Harris, Joseph Principles of algebraic geometry. Pure and Applied Mathematics. *Wiley-Interscience [John Wiley & Sons], New York,* 1978. xii+813 pp.

10. Hartshorne, Robin Algebraic geometry. Graduate Texts in Mathematics, No. 52. *Springer-Verlag, New York-Heidelberg,* 1977. xvi+496 pp.
11. Hitchin, N. J. The self-duality equations on a Riemann surface. *Proc. London Math. Soc. (3)* **55** (1987), no. 1, 59–126.
12. Katz, Nicholas M. Rigid local systems. Annals of Mathematics Studies, **139**. Princeton University Press, *Princeton, NJ,* 1996. viii+223 pp.
13. Mumford, D.; Fogarty, J.; Kirwan, F. Geometric invariant theory. Third edition. Ergebnisse der Mathematik und ihrer Grenzgebiete (2) [Results in Mathematics and Related Areas (2)], **34**. *Springer-Verlag, Berlin,* 1994. xiv+292 pp.
14. Newlander, A.; Nirenberg, L. Complex analytic coordinates in almost complex manifolds. Ann. of Math. (2) **65** (1957), 391–404.
15. Newstead, P. E. Introduction to moduli problems and orbit spaces. Tata Institute of Fundamental Research Lectures on Mathematics and Physics, **51**. *Tata Institute of Fundamental Research, Bombay; by the Narosa Publishing House, New Delhi,* 1978. vi+183 pp.
16. Riemann, Georg Friedrich Bernhard Deux theéorèmes généraux sur les équations différentielles linéares a coefficients algébriques, uvres mathmatiques de Riemann. (French) Traduites de l'allemand par L. Laugel, avec une prface de C. Hermite et un discours de Flix Klein. Nouveau tirage *Librairie Scientifique et Technique Albert Blanchard, Paris* 1968, 353-368.
17. Schneiders, Jean-Pierre An introduction to D-modules. Algebraic Analysis Meeting (Liège, 1993). *Bull. Soc. Roy. Sci. Liège* **63** (1994), no. 3-4, 223–295
18. Simpson, Carlos T. Constructing variations of Hodge structure using Yang-Mills theory and applications to uniformization. *J. Amer. Math. Soc.* **1** (1988), no. 4, 867–918.
19. Simpson, Carlos T. Higgs bundles and local systems. *Inst. Hautes Études Sci. Publ. Math. No. 75* (1992), 5–95.
20. Simpson, Carlos T. Moduli of representations of the fundamental group of a smooth projective variety. I. *Inst. Hautes Études Sci. Publ. Math. No. 79* (1994), 47–129.
21. Simpson, Carlos T. Moduli of representations of the fundamental group of a smooth projective variety. II. *Inst. Hautes Études Sci. Publ. Math. No. 80* (1995), 5–79.